EFFECTIVE TEACHING OF TECHNICAL COMMUNICATION

THEORY, PRACTICE, AND APPLICATION

Foundations and Innovations in Technical and Professional Communication

Series Editor: Lisa Meloncon

Series Associate Editors: Kristin Marie Bivens and Sherena Huntsman

The Foundations and Innovations in Technical and Professional Communication series publishes work that is necessary as a base for the field of technical and professional communication (TPC), addresses areas of central importance within the field, and engages with innovative ideas and approaches to TPC. The series focuses on presenting the intersection of theory and application/practice within TPC and is intended to include both monographs and co-authored works, edited collections, digitally enhanced work, and innovative works that may not fit traditional formats (such as works that are longer than a journal article but shorter than a book).

The WAC Clearinghouse, Colorado State University Open Press, and University Press of Colorado are collaborating so that these books will be widely available through free digital distribution and low-cost print editions. The publishers and the series editors are committed to the principle that knowledge should freely circulate. We see the opportunities that new technologies have for further democratizing knowledge. And we see that to share the power of writing is to share the means for all to articulate their needs, interests, and learning into the great experiment of literacy.

EFFECTIVE TEACHING OF TECHNICAL COMMUNICATION

THEORY, PRACTICE, AND APPLICATION

Edited by Michael J. Klein

The WAC Clearinghouse
wac.colostate.edu
Fort Collins, Colorado

University Press of Colorado
upcolorado.com
Louisville, Colorado

The WAC Clearinghouse, Fort Collins, Colorado 80523

University Press of Colorado, Louisville, Colorado 80027

© 2021 by Michael J. Klein. This work is licensed under a Creative Commons Attribution-NonCommercial-NoDerivatives 4.0 International.

ISBN 978-1-64215-112-1 (PDF) | 978-1-64215-113-8(ePub) | 978-1-64642-189-3 (pbk.)

DOI 10.37514/TPC-B.2020.1121

Library of Congress Cataloging-in-Publication Data

Names: Klein, Michael J., 1969– editor.
Title: Effective teaching of technical communication : theory, practice, and application / edited by Michael J. Klein.
Description: Fort Collins, Colorado : The WAC Clearinghouse ; Louisville, Colorado : University Press of Colorado, [2021] | Series: Foundations and innovations in technical and professional communication | Includes bibliographical references and index.
Identifiers: LCCN 2021003565 (print) | LCCN 2021003566 (ebook) | ISBN 9781646421893 (paperback) | ISBN 9781642151121 (adobe pdf) | ISBN 9781642151138 (epub)
Subjects: LCSH: Communication of technical information—Study and teaching.
Classification: LCC T10.5 .E345 2021 (print) | LCC T10.5 (ebook) | DDC 601/.4076—dc23
LC record available at https://lccn.loc.gov/2021003565
LC ebook record available at https://lccn.loc.gov/2021003566

Copyeditor: Meg Vezzu
Designer: Mike Palmquist
Series Editor: Lisa Melonçon
Series Associate Editors: Kristin Marie Bivens and Sherena Huntsman

The WAC Clearinghouse supports teachers of writing across the disciplines. Hosted by Colorado State University, and supported by the Colorado State University Open Press, it brings together scholarly journals and book series as well as resources for teachers who use writing in their courses. This book is available in digital formats for free download at wac.colostate.edu.

Founded in 1965, the University Press of Colorado is a nonprofit cooperative publishing enterprise supported, in part, by Adams State University, Colorado State University, Fort Lewis College, Metropolitan State University of Denver, University of Colorado, University of Northern Colorado, University of Wyoming, Utah State University, and Western Colorado University. For more information, visit upcolorado.com.

Contents

Introduction to Effective Teaching of Technical Communication. 3
    Michael J. Klein

Part One: Expanding Pedagogy

1. Beyond Situated Learning: Rethinking Internship Theory and Practice in the Distributed Workplace. 13
    Jennifer L. Bay

2. Interstitial Design Processes: How Design Thinking and Social Design Processes Bridge Theory and Practice in TPC Pedagogy 29
    Liz Lane

3. Engaging Plain Language in the Technical Communication Classroom 45
    Kira Dreher

4. (Teaching) Ethics and Technical Communication. 67
    Derek G. Ross

Part Two: Shaping Curriculum

5. Confronting Methodological Stasis: Re-Examining Approaches to Technical Communication Pedagogical Literacy Frameworks 91
    Halcyon M. Lawrence and Liz Hutter

6. Trial and Error: Designing an Introductory Course to Technical Communication. 111
    Chen Chen

7. Regenerating a Once Fallow Ground: Theorizing Process and Product in Twenty-First-Century Technical Communication Ecologies 131
    Adrienne Lamberti and David M. Grant

8. Creating the "Through-Line" by Engaging Industry Certification Standards in SLO Redesign for a Core Curriculum Technical Writing Course 147
    Julianne Newmark and Joseph Bartolotta

Part Three: Incorporating Technology

9. The Rhetoric, Science, and Technology of Twenty-First Century Collaboration. 169
    Ann Hill Duin, Jason Tham, and Isabel Pedersen

10. Using the Community of Inquiry Theory to Assess Online Programs and Help Students to Analyze Their Learning 193
    Julie Watts

11. Designing a Team-Based Online Technical Communication Course 209
    Luke Thominet

12. Preparing Future Professionals in and for a Global Context: A Case for Telecollaborative Educational Initiatives 231
    Elisabet Arnó-Macià and Tatjana Schell

Part Four: Engaging Communities

13. Visual Communication in Community Contexts 255
    Elise Verzosa Hurley

14. Competing Mentalities: Situating Scientific Content Literacy Within Technical Communication Pedagogy 271
    Lisa DeTora

15. Technical Communication Pedagogy and Layered Literacies in Workplace Training Courses 287
    Elizabeth L. Angeli

16. Hidden Arguments: Rhetoric and Persuasion in Diverse Forms of Technical Communication 303
    Jessica McCaughey and Brian Fitzpatrick

Contributors 319

Index 325

EFFECTIVE TEACHING OF TECHNICAL COMMUNICATION

THEORY, PRACTICE, AND APPLICATION

Introduction to Effective Teaching of Technical Communication

Michael J. Klein
JAMES MADISON UNIVERSITY

Exigency for a New Collection

Katherine Staples and Cezar M. Ornatowski's *Foundations for Teaching Technical Communication* (1997) has served and still serves as an entry point into the discipline for many students. Because of its influence, *Foundations* also serves as a conceptual starting point for this collection. However, much has changed in the past 24 years: the institutional structures that support and house technical communication programs, the social and technical contexts in which technical communicators do their work, and the discourse communities technical communicators engage with. These new dynamics have manifested in a number of ways and provided us with the impetus to develop a new collection to reflect these changes.

Over the past two decades, the field of technical and professional communication (TPC) has continued to flourish, with degree programs of all types in TPC at four-year institutions experiencing a 17 percent growth rate in the last five years (Melonçon & Schreiber, 2018). Given this growth, it is important that we reflect upon what and how we are teaching students who will become the next generation of technical communicators. This collection brings together diverse scholarly voices and perspectives in both freestanding technical communication programs and as part of larger departments. By doing so, the collection endeavors to broaden our understanding of current effective teaching and pedagogical methods by facilitating a discussion of important and innovative theories, concepts, and practices related to the teaching of technical communication.

Thus, in this new collection, we seek to address the similar overarching themes of theory, practice, and application that Staples and Ornatowski's foundational work grappled with over 20 years ago in light of the changes that have accompanied the growth of the field. Three primary changes guide the structure and makeup of this work.

First, there has been a need for more attention to issues of accessibility and inclusivity as the audiences we communicate with are more varied and are more globally and culturally diverse (Melonçon, 2013/2017; Thatcher & St.Amant, 2011). For example, one way this is evidenced in the technical communication (TC) literature is by the field's social justice turn. Writing in their 2019 monograph *Technical Communication After the Social Justice Turn,* authors Rebecca Walton, Kristen Moore, and Natasha N. Jones explain that "[a] turn comprises not only a wave of scholarship engaging with a particular concept, theory, or topic but also a more substantial shift, a transformation in thinking and meaning making" (6). While acknowledging that "[t]he precise beginning point of such a turn is debatable" (7), they trace the turn back to the work of Carolyn Rude (2009) mapping the field of technical communication. Selections in the first part of this collection—Expanding Pedagogy—address these and other issues related to expanding our curriculum to be more inclusive and comprehensive.

Second, as previously mentioned, we have seen the growth of technical communication programs, especially as stand-alone units in universities and colleges. This, in turn, has allowed us to change the way we teach. For example, we now have more freedom to design our curricula without having to accede to the constraints/directives of other disciplines such as English literature, composition, or engineering (Melonçon & Henschel, 2013). Selections in the second part of the collection—Shaping Curriculum—grapple with the advantages and added responsibilities we encounter as our field maintains its independence and grows.

Third, there has been a significant rise in the use of communication technologies, especially the use of social media by professional organizations, governmental organizations, and universities. This greater variety of technologies means designing for different platforms and writing for more varied audiences in specialized contexts (Hea, 2013; Vie, 2017). Furthermore, as online and electronic collaboration across cultures and national borders have become more commonplace, the way organizations communicate internally and with their constituents/clients has also changed, thus increasing specialization of both program foci and employment positions (Andersen, 2013; Virtaluoto et al., 2016). Selections in the third—Incorporating Technology—and the fourth—Engaging Communities—parts of the collection provide readers with examples of how these technologies can be leveraged both in the traditional academic classroom as well as when working with external communities.

These changes have altered the tools that technical communicators use and the situations and contexts in which the tools are used (Hovde & Renguette, 2017; Wilson & Wolford, 2017). We have also seen changes expand the organizational need for those proficient in the latest technical communication best practices (Brumberger & Lauer, 2015). Additionally, the workplace itself has changed: no longer relegated to a physical space and set hours, the traditional office has expanded its confines as telecommuting and virtual interaction become more commonplace (Ferro & Zachry, 2013; Frith, 2017; Wilson & Wolford, 2017). This has especially been evident in 2020 as the COVID-19 pandemic has forced us to move our teaching into a virtual environment with little to no warning and required us to confront assumptions about teaching that guide our pedagogy (Williamson et al., 2020).

About This Collection

The 16 chapters collected here coalesce around four topics: expanding pedagogy, shaping curriculum, incorporating technology, and engaging communities. Through a diverse collection of methods and perspectives, the authors provide readers with concrete suggestions and examples of how to reconceptualize, revise, and reimplement their teaching in technical communication courses that grow beyond the traditional four walls of the college classroom. Readers are invited to use any or all of these writings to aid them in teaching the next generation of technical writers, editors, and communicators.

Part One: Expanding Pedagogy

The selections in the first part of this collection address the need to follow the social justice turn in TC as we reflect and revise our pedagogical curriculum. They examine ways in which we can broaden the types of pedagogy we employ in the classroom through a focus on social justice, among other topics (Jones, 2016).

In her chapter on situated learning, Jennifer L. Bay retheorizes internships as rhetorical opportunities for students to learn soft skills. She argues that the internship course and experience typically employed by the programs allow students to learn these skills so they can be competitive and successful in the global workplace.

Liz Lane's discussion of interstitial design processes—the blending of several design theories into one for use in the TC classroom—posits that such an approach is appropriate for teaching issues of social justice, which have been recently shaping our field's research and pedagogy. She does so by offering an example of how an assignment based in interstitial design can be combined with genre-based assignments with a social justice focus.

In her examination of plain language, Kira Dreher encourages instructors to use plain language strategies in TC courses, as plain language strategies overlap with aims of technical communication. By offering specific in-class applications on the use of plain language, she demonstrates how expertise in TC lends itself well to engaging with the plain language movement.

Finally, Derek G. Ross suggests that the incorporation of instruction on ethical decision-making into our classrooms helps us better teach students how understanding how we make decisions allows us to better communicate our decisions to others. A rich version of ethics decision-making, which he proposes, would consider multiple ethical models, including feminist ethics and ethics of care, when seeking to creatively solve problems.

Part Two: Shaping Curriculum

The chapters found in this part of the collection deal with curricular development, both at the course and program level, for new approaches to TC instruction. They also demonstrate the practical application of using reimagined TC theories—like layered literacies (Cargile Cook, 2002)—in the classroom and learning environments (e.g., Bourelle et al., 2015; Sapp & Crabtree, 2002) that technical communicators use to train the next generation of practitioners.

In the first chapter in part two, Halcyon Lawrence and Liz Hutter call on instructors to critically engage with how the field develops and uses pedagogical literacy frameworks. Rather, they argue that because of our field's growth, there needs to be an inclusion of other qualities, including responsiveness, multidimensionality, and sustainability.

Next, Chen Chen examines her experiences in developing an introductory course in technical communication for a university without a technical communication degree program. Chen discusses how a class with a problem-solving perspective and social justice orientation helps students develop the core conceptual skills of TC.

Adrienne Lamberti and David Grant discuss how TC pedagogy often privileges application at the expense of theory. In participating in their program's curricular revision, the two found that a balance of theory and practice helped faculty develop a theoretical framework that embodied ecologies of practice and civility.

In the final chapter of this section, Julianne Newmark and Joseph Bartolotta recount the development of new student learning outcomes (SLOs) for a sophomore-level TC course. Their work provides insights into developing SLOs through the engagement of industry standards and practices.

Part Three: Incorporating Technology

The four chapters in part three provide guidance for effectively using technology in the TC classroom. In their chapter, Ann Hill Duin, Jason Tham, and Isabel Pedersen focus on the need for collaboration and the open-access tools instructors can use for this. They argue that this preparation for students is essential to their future success as practitioners in the field as collaboration is critical in working with clients and customers in a global context.

Julie Watts argues that community of inquiry (COI), a theoretical framework for online learning, helps instructors and learners determine the best way to achieve deep learning in a community. Additionally, she demonstrates that when paired with an outcomes learning approach, COI provides a key component to successful program assessment of both the online learning environment and what outcomes students achieve.

Luke Thominet's chapter also examines online learning, advocating for a team-based learning (TBL) approach for group projects. In his chapter, Thominet shows that TBL affords learners effective collaborative learning experiences in courses with units with repeated cycles of evaluation and analysis.

Lastly, Elisabet Arnó-Macià and Tatjana Schell present their evaluation of the educational practices of the Trans-Atlantic and Pacific Project (TAPP), a multinational collaborative network. Based upon their findings, the authors maintain that including telecollaboration can help instructors design more internationalization-focused college curricula and support students in strengthening their skills beyond oral and written communication.

Part Four: Engaging Communities

The final section of the collection provides exemplars of students and instructors engaging with communities outside the academy by outlining cases for utilizing pedagogical training in workplaces (e.g., Kohn, 2015; Pickering, 2017; Zachry & Thralls, 2007) while contending with various organizational dynamics.

Elise Verzosa Hurley uses visual communication in her course as a means of facilitating the creation of community-based projects. These critical appraisals of the visual help learners understand the larger social and cultural contexts of which communication documents are a part.

Lisa DeTora's chapter focuses on the norms of the science community and the responsibilities for technical communicators to successfully engage with that field. She advocates for those in TC to improve their literacy of the scientific materials and their modes of production as a means of improving their pedagogical practices.

In her examination of the workplace training classroom for paramedics and firefighters, Elizabeth L. Angeli explores the role layered literacies play in better understanding how workplace communicators learn outside the traditional TC classroom. Her findings in the workplace training classroom suggest that literacies are more than just layered and may sometimes be in tension with one another.

Lastly, Jessica McCaughey and Brian Fitzpatrick provide the results from a case study of three professionals who perform technical writing daily. Their findings suggest that the types of persuasion our students encounter in the workplace are complex and potentially implicit.

Acknowledgments

I would like to thank all the authors who trusted me with their individual manuscripts when assembling this collection. I would also like to thank Kristin Marie Bivens and Sherena Huntsman for their invaluable support during this project and the anonymous reviewers who provided feedback to the authors. Finally, a very big thank you to Lisa Melonçon, who entrusted this project to me. I have learned so much from the experience and from her guidance these past two years.

References

Andersen, R. (2013). Rhetorical work in the age of content management: Implications for the field of technical communication. *Journal of Business and Technical Writing*, 28(2), 115–57. https://doi.org/10.1177/1050651913513904.

Bourelle, A., Bourelle, T. & Jones, N. (2015). Multimodality in the technical communication classroom: Viewing classical rhetoric through a 21st century lens. *Technical Communication Quarterly, 24*(4), 306–27. https://doi.org/10.1080/10572252.2015.1078847.

Brumberger, E. & Lauer, C. (2015). The evolution of technical communication: An analysis of industry job postings. *Technical Communication, 62*, 224–243.

Cargile Cook, K. (2002). Layered literacies: A theoretical frame for technical communication pedagogy. *Technical Communication Quarterly, 11*(1), 5–29. https://doi.org/10.1207/s15427625tcq1101_1.

Ferro, T. & Zachry, M. (2013). Technical communication unbound: Knowledge work, social media, and emergent communicative practices. *Technical Communication Quarterly, 23*(1), 6–21. https://doi.org/10.1080/10572252.2014.850843.

Frith, J. (2017). Forum design and the changing landscape of crowd-sourced help information. *Communication Design Quarterly Review, 4*(2), 12–22. https://doi.org/10.1145/3068698.3068700.

Hea, A. C. K. (2013). Social media in technical communication. *Technical Communication Quarterly, 23*(1), 1–5. https://doi.org/10.1080/10572252.2014.850841.

Hovde, M. R. & Renguette, C. C. (2017). Technological literacy: A framework for teaching technical communication software tools. *Technical Communication Quarterly, 26*(4), 395–411. https://doi.org/10.1080/10572252.2017.1385998.

Jones, N. N. (2016). The technical communicator as advocate: Integrating a social justice approach in technical communication. *Journal of Technical Writing and Communication, 46*(3), 342–361. https://doi.org/10.1177/0047281616639472.

Kohn, L. (2015). How professional writing pedagogy and university-workplace partnerships can shape the mentoring of workplace writing. *Journal of Technical Writing and Communication, 45*(2), 166–188. https://doi.org/10.1177/0047281615569484.

Melonçon, L. (Ed.). (2017). *Rhetorical accessibility: At the intersection of technical communication and disability studies*. Routledge. (Original work published 2013).

Melonçon, L. & Henschel, S. (2013). Current state of U.S. undergraduate degree programs in technical and professional communication. *Technical Communication, 60*, 45–64.

Melonçon L. & Schreiber, J. (2018). Advocating for sustainability: A report on and critique of the undergraduate capstone course. *Technical Communication Quarterly, 27*(4), 322–335. https://doi.org/10.1080/10572252.2018.1515407.

Pickering, K. (2017). Navigating discourses of power through relationships: A professional and technical communication intern negotiates a meaningful identity within a state legislature. *Journal of Technical Writing and Communication, 48*(4), 441–470. https://doi.org/10.1177/0047281617732019.

Rude, C. D. (2009). Mapping the research questions in technical communication. *Journal of Business and Technical Communication, 23*(2), 174–215. https://doi.org/10.1177/1050651908329562.

Sapp, D. A. & Crabtree, R. D. (2002). A laboratory in citizenship: Service learning in the technical communication classroom. *Technical Communication Quarterly, 11*(4), 411–432. https://doi.org/10.1207/s15427625tcq1104_3.

Staples, K. & Ornatowski, C. M. (Eds.). (1997). *Foundations for teaching technical communication: Theory, practice, and program design*. Ablex.

Thatcher, B. & St. Amant, K. (Eds.). (2011). *Teaching intercultural rhetoric and technical communication: Theories, curriculum, pedagogies, and practices*. Routledge.

Vie, S. (2017). Training online technical communication educators to teach with social media: Best practices and professional recommendations. *Technical Communication Quarterly, 26*(3), 344–359. https://doi.org/10.1080/10572252.2017.1339487.

Virtaluoto, J., Sannino, A. & Engeström, Y. (2016). Surviving outsourcing and offshoring: Technical communication professionals in search of a future. *Journal of Business and Technical Writing, 30*(4), 495–532. https://doi.org/10.1177/1050651916651908.

Walton, R., Moore, K. & Jones, N. (2019). *Technical communication after the social justice turn*. Routledge.

Williamson, B., Eynon, R. & Potter, J. (2020). Pandemic politics, pedagogies and practices: Digital technologies and distance education during the coronavirus emergency. *Learning, Media and Technology, 45*(2), 107–114. https://doi.org/10.1080/17439884.2020.1761641.

Wilson, G. & Wolford, R. (2017). The technical communicator as (post-postmodern) discourse worker. *Journal of Business and Technical Writing, 31*(1), 3–29. https://doi.org/10.1177/1050651916667531.

Zachry, M. & Thralls, C. (Eds.). (2007). *Communicative practices in workplaces and the professions: Cultural perspectives on the regulation of discourse and organizations*. Baywood.

Part One: Expanding Pedagogy

1. Beyond Situated Learning: Rethinking Internship Theory and Practice in the Distributed Workplace

Jennifer L. Bay
PURDUE UNIVERSITY

Abstract: This chapter retheorizes internships as pedagogical moments for students to learn what we have commonly called "soft skills." I argue that soft skills, which consist of communication, collaboration, ethics, work ethic, critical thinking skills, and the like, are fundamentally rhetorical skills that require individuals to learn how to read and respond effectively to different workplace situations, people, technologies, and problems. Workplace supervisors and human resource professionals across all disciplines agree that soft skills are highly desired by employers, but there is little agreement on if and how they can be taught to students. New internship configurations such as virtual internships, global internships, and online education make soft skills all the more necessary for success. If the most desirable quality in job seekers is soft skills, then we should actively teach and cultivate them. Today's internship practicum should include more about how to function effectively in a job using soft skills; as such we need to develop different pedagogical interventions that cultivate these skills.

Keywords: internships, soft skills, experiential learning

Key Takeaways:

- Internship practica can cultivate soft skills via opportunities to read, respond, and critically reflect on a variety of different workplace rhetorical contexts.
- Soft skills instruction is more relevant and "teachable" in the context of the kinds of distributed technical communication work and internship arrangements that interns face.
- Case study pedagogy, attention to diversity, and modular course structures can foster soft skills development by asking students to critically reflect on how they can deploy soft skills to address workplace issues.

The 1990s and early 2000s saw a surge in scholarship theorizing how technical and professional communication (TPC) students learn through internships (e.g., Anson & Forsberg, 1990; Beaufort, 1999; Freedman & Adam, 1996; Gaitens, 2000; Little, 1993; Savage, 1997; Smart & Brown, 2002; St.Amant, 2003; Tovey, 2001). Long integrated into TPC undergraduate majors, internships have been seen as a critical component for learning technical communication principles and workplace practices. Melonçon & Henschel (2013) note that over half of all TPC undergraduate programs require an internship course. Internships, then, serve as a critical bridge between the academy and industry, allowing students, supervisors, and academics to forge connections. Situated learning and legitimate peripheral participation (Lave & Wenger, 1991) became key concepts for theorizing how TPC students learn via internship opportunities, but these theories often accounted for learning in traditional office environments, in which interns were mentored by veteran technical communicators, and internships were coordinated by faculty.

Times have changed for TPC interns. In our global, distributed workplace environments, internship work is vastly different. What was once a one-time internship in a traditional office environment has been replaced by new arrangements such as virtual internships, internships with start-ups, global and study-abroad internships, micro-internships, contract work, unpaid internships, and the expectation of multiple internships over a student's college career (Durack, 2013; Gates, 2014; Leath, 2009; Perlin, 2012; Ruggiero & Boehm, 2016; Suzuki et al., 2016; Yarbrough, 2016). In these new arrangements, technical communication interns may not have seasoned TPC professionals as mentors or may not even have any academic or industry mentor. Moreover, a student today will be expected to have at least 12–15 jobs during their lifetime (U.S. Department of Labor, 2019). While excellent research on internships has continued in the field (Baird & Dilger, 2017; Bourelle, 2015; Katz, 2015; Kramer-Simpson, 2018), how interns learn and how we might teach them differently in these new TPC contexts and work arrangements is less understood.

In light of these shifts, this chapter retheorizes internships as pedagogical moments for students to learn what we have commonly called "soft skills." I argue that soft skills, which consist of communication, collaboration, ethics, work ethic, critical thinking skills, and the like, are fundamentally rhetorical skills that require individuals to learn how to read and respond effectively to different workplace situations, people, technologies, and problems. Workplace supervisors and human resource professionals across all disciplines agree that soft skills are one of the most desirable job qualifications (Robles, 2012), but there is little agreement on if and how they can be taught (Shuman et al., 2005). If the most desirable quality in job seekers is soft skills, then why aren't we actively teaching, or at least cultivating, them? In short, today's internship practicum should include more about how to function effectively in a job using soft skills; as such, we need to develop different pedagogical interventions that cultivate these skills.

I start with a brief overview of internship theory in TPC, then move to how TPC work and internships have shifted with a global, distributed landscape. I outline emerging issues that TPC interns face, including working remotely, working without TPC mentors or direct supervision, self-directed learning, taking initiative, using distributed technologies, gendered and racial conflict, cultural differences, global communication, entrepreneurship, and collaboration. Using evidence from internship courses taught over the past 15 years, I then offer three specific pedagogical approaches to soft skills training that can help students to negotiate these emerging issues in productive and ethical ways. The key takeaway here is that the TPC internship and its corresponding practicum should be less about introducing students to the field and more about teaching students the essential workplace soft skills that they will need to succeed in today's global workplace.

Internship Theory and Pedagogy in Professional and Technical Communication

The field of professional and technical communication has produced a rich body of research on internships, how developing writers benefit from internships, and how they make the transition from school to work. Early scholarship in the field was invested in identifying and describing the internship experience and what it can do for the student (Gloe, 1983; Hull, 1977; Little, 1993; Southard, 1988; Tessier, 1975; Wyld, 1978). Much like internships are theorized in engineering and other STEM fields, this early internship scholarship in TPC created an opening for integrating experiential learning in TPC. Scholarly topics included how to set up internship programs, what a successful internship looks like, and different examples of internships (Bosley, 1988; Coggin, 1989; Hager, 1990). This early work operated in a kind of epideictic fashion to argue that internships are important learning opportunities but did not necessarily theorize what was being learned and how interns were learning.

Several strands of research on internships soon emerged. Brenton Faber (2002), Aviva Freedman and Christine Adam (1996), and Tiffany Bourelle (2014), for instance, each focus on how the internship is an opportunity for students to see themselves as "professionals" on the job. The professional internship is "an aspect of professional certification that prevails in one form or another across most professions" (Savage & Seible, 2010, p. 53). A second and related dimension of internship research is the transition from school to work. Research by Chris M. Anson and L. Lee Forsberg (1990), Neil Baird and Bradley Dilger (2017), Anne Beaufort (1999), Doug Brent (2012), and Susan M. Katz (1998) each present different ways in which interns move from seeing themselves as students to seeing themselves as professionals. Anson and Forsberg (1990), for instance, describe a specific learning process through which interns move from academically oriented writing to workplace-specific discourses and processes. Beaufort (1999) similarly provides a shift from one identity (student) to another (intern/worker).

Most of the scholarship that theorizes how interns make these shifts has been based on Jean Lave and Etienne Wenger's concept of situated learning. Lave and Wenger (1991) argue that learning is situated in a community of practice. An apprentice enters into a community of practice as a newcomer and is initiated, so to speak, into that professional community. In this model, "legitimate peripheral participation" introduces newcomers through immersion in the community and absorbing its modes of action and meaning. In many instances, this immersion occurs through in-person apprenticeships. Aviva Freedman and Christine Adam (1996) posit that in legitimate peripheral participation, learners do not fully participate and that the learning that occurs is incidental, drawn forth more by the community of practice (p. 399). While Freedman and Adam theorize the learning as "authentic," the emphasis on peripheral still holds. In this model, interns are working with or alongside a professional who models processes and practices; thus, the model is predicated on professionals in the workplace doing the modeling or "teaching" of the intern.

The "teaching" in an internship, then, comes largely from the internship supervisor, not the faculty member. Faculty tend to be in the role of coordinating internships and making sure the academic parameters are in place. This focus on internship coordination over teaching, mentoring, or coaching leaves the faculty member as an observer of what is happening in the internship, which he or she then uses to apply to future interns (see Anson & Forsberg [1990] for instance). In TPC, we assume supervisors will train interns because they are in the writing and communication professions; as such, what shines forth in our scholarship is often what the student is doing/learning in the internship rather than how we are teaching interns. As educators and scholars, we need to develop innovative ways to support students in their internships, coach them to develop the professional skills they need on the job, and help them to make explicit connections between learning on the job and their academic coursework.

While there has been a robust discussion of experiential learning in internships, little explicit connection has been made to soft skills. Ironically, the description of situated learning and much internship literature related to legitimate peripheral participation and communities of practice imply soft skill development. Unlike academic learning, in which students attend classes, listen to lectures, read textbooks, complete assignments, and take tests, internships are a form of applied learning, or learning in which students apply theories and concepts to a real-world environment. Another way of understanding internships is as a situated learning activity; in this case, learning is situated not in a classroom environment where objective knowledge is conveyed but is situated within the activity in which someone participates. Lave's early work on situated learning is foundational for this understanding. Lave (1991) posits that learning is always situated in a place, among people, technologies, and cultural factors: "learning, thinking, and knowing are relations among people engaged in activity in, with, and arising from the socially and culturally structured world" (p. 67). Epistemologically, situated learning is contextual, emplaced, and active. Lave (2009) further claims that "theories of situated activity do not separate action, thought, feeling, and value and their collective, cultural-historical forms of located, interested, conflictual, meaningful activity" (p. 202). I highlight this sentence specifically because of the inclusion of terms normally associated with soft skills—value and feeling. Feeling and value also influence one's ability to learn in and from a situation. And while we might want to see academic learning as objective learning, all learning is in fact situated.

For instance, college students might do very well in class academically but have difficulty in applying what they know to real-world situations. Or they struggle with traditional academic work and do extremely well applying what they know in real-world situations. It is the rare student who can move between both kinds of epistemological frameworks seamlessly. This may be partly because there are differences in the skills needed to succeed in the university and the soft skills needed in the workplace. Academic skills are honed by college students over sixteen years of schooling and are different from workplace soft skills where there is no safety net and employees must communicate and negotiate with multiple stakeholders at once. As instructors, we can create the conditions of possibility for students to understand these differences and invent out of them. As Lave points out, "Doing and knowing are inventive in another sense: They are open-ended processes of improvisation with the social, material, and experiential resources at hand" (Lave, 2009, p. 204). That is, part of what we are cultivating in internship experiences is gathering the resources necessary to invent meaning in a new workplace context.

Since Lave and Wenger's theories were introduced into the field, new internship formations have challenged and increased the pedagogical need for faculty to facilitate internship learning. These new formations highlight a gap in the TPC scholarship on internships: a lack of attention to instructor pedagogy, or at least the pedagogy of the internship course or arrangement. In the new landscape described below, this lack of attention to specific pedagogical approaches stands out as a crucial need in current theories on internships.

■ Shifting Landscapes for Internships

Thus far, our pedagogies for internships have not changed over the past 30 years, even though our students, their workplaces, and their internships have changed. In *Internships: Theory and Practice* (2017), Charles Sides and Ann Mrvica predict that trends such as managed education, the student as consumer, and internships outside of an academic support system may affect the future of internships as a way of learning by doing. Today, some of these trends have manifested, but other trends have also emerged. Internships outside of an academic environment have definitely blossomed as more and more students take on multiple internships while in college. By the time my students have taken the internship practicum their senior year, they have already been in at least two to three internships. Popular websites and career advice suggest that students undertake multiple internships while in college. We've also witnessed an expanded discourse about the exploitation of interns, the problems with unpaid internships, and sexual harassment and discrimination against interns in multiple industries.

Other new internship formations include virtual internships, micro-internships, study abroad internships, entrepreneurships, and service-learning internships. In these situations, it's no longer realistic to think we're teaching students to write in one particular industry or type of workplace. What we're teaching is how to navigate and succeed in an ever-changing workplace in which students may undertake more than 15 jobs in their lifetime, often in different careers and industries that have yet to be imagined. In such formations, new pedagogical challenges emerge: remote work with little to no supervision, lack of TPC mentors or experts on the job, self-directed learning, use of distributed technologies, gendered or racial conflicts, cultural differences, global communication practices, and collaboration. How do we cultivate learning in emerging internship formations? With the rise of TPC programs comes the rise of technical knowledge in TPC theory and practice. That is, TPC students are getting the "hard skills" they need in their major courses; what's needed is more instruction in the soft skills they need to succeed in the workplace. Internship practica are the ideal place for practicing and reflecting on the soft skills that they need to become successful in any professional environment.

Rhetoric and Soft Skills in Emerging Internship Configurations

Soft skills has become a buzzword to describe a specific skill set desired by companies. As opposed to the "hard skills" of engineering, coding, building, accounting, and other specific technical skills or expertise, soft skills refers largely to the communication and human interactive skills needed to succeed in a workplace. Marcel Robles (2012) describes soft skills as "interpersonal qualities, also known as people skills, and personal attributes that one possesses" (p. 453). Robles surveyed business executives to come up with the ten soft skills most desired in today's workplace: "integrity, communication, courtesy, responsibility, social skills, positive attitude, professionalism, flexibility, teamwork, and work ethic" (p. 453). As many scholars note, these skills are not easily "taught" (Robles, 2012; Shuman et al., 2005) because they are ways of being toward others, toward work, and toward environments, which develop over time through situations, self-reflection, and applications. While students can read about soft skills, they must have opportunities to examine their own abilities, see where they need work, gain feedback from others, and test out these new skills as they are in development. As employees experience different and new workplace situations, solve different problems, and learn to work with different populations and colleagues from different cultural backgrounds, they (hopefully) develop, refine, and hone these interpersonal and soft skills.

Students with the best soft skills already have some meta-awareness of the behaviors needed to succeed because they can "read" people well and/or interact in large groups to produce consensus while also being able to advocate for themselves and others. This kind of rhetorical sensibility can be cultivated. I avoid using the term "taught" because it implies that soft skills can be obtained and held like rote learning, when they are only truly understood in practice, much like rhetorical ability. Rhetorical ability, because it involves assessing the rhetorical situation, learning about or understanding your audience, and adjusting your writing/language/behavior to meet that audience, underlies all of the soft skills listed above; there is a strong performative and improvisational dimension that requires cultivation.

▌ Pedagogical Cultivation of Soft Skills

If the TPC internship and its corresponding practicum are now less about introducing students to the field and more about teaching students the essential workplace soft skills that they will need to succeed in today's global workplace, what assignments or approaches should we use? Programs have a variety of approaches to supervising interns. Three common configurations that have emerged are one-on-one counseling (independent study) with an internship coordinator; a traditional course where students must have an internship during the class—this is often called a practicum; or an online course (distance education) with students either in one major or across majors (Bay, 2017). As part of these configurations, different conceptions of pedagogy are implicit. When internship supervisors offer independent study credit, there is often the assumption that the internship supervisor is teaching and mentoring the intern; as such, the internship coordinator serves as almost a gatekeeper to ensure the student has learned or undertaken the work. In a classroom situation, the assumption is that students have much to teach each other in a more formal institutional structure. The online class also provides that support but recognizes that interns can learn from students in other majors and backgrounds.

Most of these pedagogical configurations still require weekly writing assignments where students reflect on what they are doing/learning. The end of the semester internship report is a common genre where the student reports on what they did in the internship in order to earn credit. They may provide written updates to internship coordinators throughout the internship, either in person or online. The final report may contain samples of the internship work and may often be a report already composed as a requirement of their internship supervisor. While TPC programs may have one or more of these configurations, students are undertaking internships outside of those configurations as well, which means there is no pedagogical infrastructure for internships.

At Purdue, our TPC internship course meets once a week for a two-hour block. The first hour consists of engagement and reflection on the internship experience. Students are required to complete an internship work agreement form, on which they collaborate with their supervisors on goals and supporting activities for the internship. Worklogs, or guided reflections on the internship experience, are shared with the class on a bi-weekly basis. Students bring samples of their internship work to share and receive feedback from classmates. Short readings and articles about internship experiences are also discussed. The second hour focuses on professional identity; students work on developing an online portfolio, LinkedIn and networking profiles, and presenting themselves as emerging professionals. While students do receive mid-semester and final evaluations from their supervisors, I have downplayed these components in their final grades. Very rarely do internship supervisors provide critical comments or evaluations of work. As many interns have reported, their supervisors do not want to provide negative comments in case they might impact student grades. For students with virtual internships, it is sometimes difficult to speak directly to supervisors or even maintain sustained contact. For these reasons, plus the shifting dynamics of internships, I have slowly shifted the course more toward a critical reflection on the internship experience.

Following Kristen Lucas and Jacob D. Rawlins (2015), I present a curricular overview of how soft skills can be integrated into internship pedagogy. I focus specifically on the use of case studies, approaches to diversity and difference through reflection, and self-directed online modules that students can complete inside or outside of an official course or credit. These components respond to contemporary exigencies such as COVID-19, the #metoo movement, and racial conflict, all of which impact the development of an intern's professional identity.

Case Study Approach

One approach that has been useful for cultivating soft skill development is the case study method. In this situation, interns do not read about artificial or even real-world based experiences with other professionals, as in Gerald J. Savage and Dale L. Sullivan's textbook (2001); rather, they write their own case studies to share with their classmates and use as reflective development tools. Using cases in TPC is a well-documented approach to teaching problem solving, critical thinking, and decision making (Balzotti & Hansen, 2019; Mara, 2006; Pennell & Miles, 2009), but little attention has been paid to having students write their own cases. Having students write their own cases asks them to reflect on their own soft skill development and to see how workplace problems are often embedded in a web of communication, collaboration, and "people" issues. That is, a procedural problem in the workplace—say an ineffective workplace process—is just as much a structural issue, as in using the wrong process for the job, as it is a soft skills problem—being able to collaboratively develop a different process and communicate that change to people who may not be amenable to a change.

An example I have referenced before (Bay, 2017) might help. A summer engineering intern has trouble getting things done because she doesn't know how to communicate with the hourly workers on the shop floor. She problem solves with her instructor and peers on ways to collaborate more effectively with this group of co-workers. One of the suggestions provided is to bring something like donuts to share with these workers in order to create goodwill and buy-in for her ideas. She ultimately develops more camaraderie with these workers based on this small, thoughtful act.

This instance is a perfect example of a case study at work. In order to be effective, case studies must describe a situation in which there is no clear or obvious solution to the issue or problem that is being discussed. In such a situation, soft or interpersonal skills are often part of the solution since much of the execution of any solution in the workplace involves how it's deployed, the personalities involved, and the values being advocated. This intern had to figure out what the shop floor workers needed in order to see her as someone they could trust and respect. By collaborating on the problem in class, she was presented with several different options but figured out ultimately that there were particular approaches she could take to gain trust—all of which involved the successful deployment of soft skills. Having students write their own cases on real-world workplace problems and issues can help them become aware of how they need to cultivate soft skills as much as writing skills on the job.

David H. Jonassen (2006) provides a typology of case types that can be used as case studies for different rhetorical purposes: exemplars, analogies, problems to solve, and even student constructed cases. In his taxonomy, exemplar cases are often used as examples to be followed; in such a situation, students study the case and learn by reading through the thought process and solution to the problem. In problem-based cases, students read through an authentic case with no clear solution and then work to try to solve the problem on their own; such an approach is often used in legal and medical education. Student-generated cases, in Jonassen's approach, are often done in hypertext and allow others to move through the problem as they see it.

In an internship environment, case studies can be based on a student's personal internship experience or in combination with other workplace experiences. They should be written as issues emerge on the job. In this approach, interns are assigned to write and present a case study of a problem or issue on the job based on their own experiences. Students must produce a detail-rich narrative of a problem without judgment or bias and present it to the class. Everyone in class writes a response or solution to the problem and shares it with the other interns. Because each student writes a response based on their own internship formation or orientation, they consider the issue from different angles. The intern chooses an approach that best fits their situation and with which they feel most comfortable. If possible, the intern actually enacts the approach in their internship to see what happens. The intern is thus exposed to different rhetorical approaches and learns to adapt rhetoric to the circumstance. The intern reports back to class and reflects on the outcome. Throughout this process, the intern learns what works and what does not in their particular situation, while also reflecting on their rhetorical skill development.

Ideally, interns will write these cases as they emerge, so they could occur throughout the semester without a defined due date. Often, in the writing of the cases, interns may be able to solve their own issue, which teaches them how to use writing and reflection to solve issues on their own without advice from others or the teacher. More often when I have taught this assignment, students mull over a situation at work for several weeks before they recognize it's an issue to be addressed. A common example are students who does not feel challenged by the work they have been assigned in their internships and don't understand why others are getting more sophisticated work projects. In this situation, we often brainstorm that it's not that the intern is not competent or able but that they are not actively and successfully communicating that competency with workplace cues or other soft skills. Once the student hears how others might solve the issue on the job, they have an arsenal of approaches that they can either try out or adapt to fit their personality and situation. In this case, it might be providing written progress reports to their supervisor on a daily basis, or it could be participating in breaks with co-workers rather than continuing to work through them. In both of these possible solutions, an intern must demonstrate competency through more than just the final products, which is an example of rhetorical expertise in action.

Attention to Diversity and Difference

The social justice turn in TPC has highlighted the ways that we need to do better at making students aware of structural and institutional inequities. The internship course is an ideal location for asking students to reflect on racial and gendered dynamics in the workplace. One approach that has proven useful is to reconsider the experiences of interns as explored in some older scholarship in TPC. For instance, I have students read Sherry G. Southard's "Protocols and Human Relations in the Corporate World" (1989), a 30-year-old essay that outlines expectations for interns in the workplace. While we might say this is a dated essay—and it is—many of the points still hold since they often focus on soft skills cultivation: observing behaviors, paying attention to interactions, and noticing how professionals interact. Translating those behaviors to actions, though, is different than reading about them. We look at that essay and others to examine the gendered, racial, and economic assumptions that are happening there. We then talk about how workplaces are different today, or not. Similarly, students read classic essays that detail intern experiences, such as Anson and Forsberg (1990) and D. Kathleen Stitts (2006), to look at unexamined assumptions about interns and co-workers. For example, a series of recent articles by Kristin Pickering (2018a, 2018b, 2019) provides fascinating analysis of how interns manage emotions in the workplace. However, reframing those experiences from the perspective of gender and race provides starkly different conclusions about emotions in the workplace. I reinforce these points through focused worklog prompts about how diversity and difference appear in students' workplaces and on readings about microaggressions and conflict in the workplace; when students try to deny or elide discussions of difference, we have a foundation to discuss why they are unable to see that difference.

After reading through and re-analyzing these narratives, students are much more aware of how different behaviors and expectations emerge in the workplace. On multiple occasions, I have had students bring up examples of sexual harassment and microaggressions they have experienced in their internships. Before they have the opportunity to read about and discuss these issues, they often do not see their experiences as problematic and worthy of intervention. Reading and reflection opportunities provide them with a vocabulary and a way of thinking about these experiences. It also makes other students aware of these experiences and how they would address them in their own internships. While not a "soft skill" in Robles' list, social awareness of microaggressions and harassment is part of the interpersonal relations that are at the heart of soft skills training.

Modular Approaches to Internship Pedagogy

With the rise of virtual internships and recent shifts in higher education because of COVID-19, we need to take a modular approach to internship pedagogy. Viewing internships as capstone courses taken in a major's final semester or as the sole experience in a program does not work when students are undertaking internships throughout their entire time in the university. Positioning internships throughout the curriculum might be one approach to cultivating soft skills throughout a major or program. In order to support such an approach, we might need to take a modular approach to internship education. We have enacted this approach partially at Purdue with our summer internship course for any majors throughout the university. As I have written (Bay, 2017), internship courses that focus on soft skills can be offered to students from different majors and programs across the university. We've done this through an online summer internship course that functions like a business writing "on the job" course. By focusing on soft skills, students are able to cross the programmatic and disciplinary boundaries that might limit learning. In fact, the approach works quite well because students constantly have to attend to different audiences and expectations.

A modular approach would allow for pedagogical interventions in situations where there is no mechanism or ability to obtain course credit. Some students, for instance, decline to obtain credit for their internship experiences because they cannot afford the tuition expense. Micro-internships, like those offered by Parker-Dewey, are short-term projects that don't fit a traditional internship credit mechanism. Likewise, study abroad internships are often encapsulated in a study abroad program and don't always provide a mechanism for learning on the job. Providing modules online that would allow students to gain proficiency in different areas of soft skills might be a way for students to reflect on their learning and feel supported. Much like a badge system, modules could ask students to reflect on different aspects of the internship experience or solve a problem, which could provide some evidence of their ability to deploy different soft skills. Another model would be for programs to create discussion boards or learning management systems open to any TPC student undertaking an internship. While someone might need to monitor these sites, such a configuration would provide students with support for their learning outside of an academic structure. These kinds of institutional structures can allow for support for students, while also allowing faculty and staff to better understand what is happening for interns at various points in their academic careers.

Future Directions

New developments such as virtual internships, global internships, and multiple internships over time have shifted the opportunities for faculty to facilitate internship learning. We must move internship theory beyond situated learning and legitimate peripheral participation to an understanding of internships as rhetorical phenomena in which students learn "soft skills," regardless of the discipline or field of study. Soft skills, which consist of communication, collaboration, ethics, work ethic, critical thinking skills, and the like, are fundamentally rhetorical skills that require individuals to learn how to read and respond effectively to diverse situations, people, and problems. Workplace supervisors and human resource professionals across all disciplines agree that soft skills are one of the most sought job qualifications, but there is little literature about how and if they can be taught. Internships provide the perfect opportunity to cultivate soft skills because they are opportunities to read, respond, and critically reflect on a variety of different workplace situations. Such cultivation can occur in a variety of internship configurations, whether they be course based or independent study. But reorienting our internship pedagogy toward soft skills as rhetorical work requires us to incorporate attention to diversity and difference, bring in more examples and case studies, and implement a modular approach to internship education. Coupled together, these additions can help move the internship practicum forward to address the new realities of student interns.

References

Anson, C. M. & Forsberg, L. L. (1990). Moving beyond the academic community: Transitional stages in professional writing. *Written Communication, 7*(2), 200–231.

Baird, N. & Dilger, B. (2017). How students perceive transitions: Dispositions and transfer in internships. *College Composition and Communication, 68*(4), 684–712.

Balzotti, J. & Hansen, D. (2019). Playable case studies: A new educational genre for technical writing instruction. *Technical Communication Quarterly, 28*(4), 407–421.

Bay, J. (2017). Training technical and professional communication educators for online internship courses. *Technical Communication Quarterly, 26*(3), 329–343.

Beaufort, A. (1999). *Writing in the real world: Making the transition from school to work*. Teachers College Press.

Bosley, D. (1988). Writing internships: Building bridges between academia and business. *Iowa State Journal of Business and Technical Communication, 2*(1), 103–113.

Bourelle, T. (2014). New perspectives on the technical communication internship: Professionalism in the workplace. *Journal of Technical Writing and Communication, 44*(2), 171–189.

Bourelle, T. (2015). Writing in the professions: An internship for interdisciplinary students. *Business and Professional Communication Quarterly, 78*(4), 407–427.

Brent, D. (2012). Crossing boundaries: Co-op students relearning to write. *College Composition and Communication, 63*(4), 558–592.

Coggin, W. O. (Ed.). (1989). *Establishing and supervising internships*. Association of Teachers of Technical Writing.

Durack, K. T. (2013). Sweating employment: Ethical and legal issues with unpaid student internships. *College Composition and Communication, 65*(2), 245–272.

Faber, B. (2002). Professional identities: What is professional about professional communication? *Journal of Business and Technical Communication, 16*(3), 306–337.

Freedman, A. & Adam, C. (1996). Learning to write professionally: "Situated learning" and the transition from university to professional discourse. *Journal of Business and Technical Communication, 10*(4), 395–427.

Gaitens, J. (2000). Lessons from the field: Socialization issues in writing and editing internships. *Business Communication Quarterly, 63*(1), 64–76.

Gates, L. (2014). The impact of international internships and short-term immersion programs. *New Directions for Student Services, 2014*(146), 33–40.

Gloe, E. M. (1983). Setting up internships in technical writing. *Journal of Technical Writing and Communication, 13*(1), 7–27.

Hager, P. J. (1990). Mini-internships: Work-world technical writing experiences without leaving campus. *Technical Writing Teacher, 17*(2), 104–113.

Hull, L. C. (1977). Internships in technical writing: A sponsor's view. *Technical Communication, 24*(1), 7–9.

Jonassen, D. H. (2006). Typology of case-based learning: The content, form, and function of cases. *Educational Technology, 46*(4), 11–15.

Katz, S. M. (1998). A newcomer gains power: An analysis of the role of rhetorical expertise. *The Journal of Business Communication, 35*(4), 419–442.

Katz, S. M. (2015). Making the most of your internship program. *Programmatic Perspectives, 7*(1), 43–66.

Kramer-Simpson, E. (2018). Moving from student to professional industry mentors and academic internship coordinators supporting intern learning in the workplace. *Journal of Technical Writing and Communication, 48*(1), 81–103.

Lave, J. (1991). Situating learning in communities of practice. *Perspectives on socially shared cognition, 2*, 63–82.

Lave, J. (2009). The practice of learning. In K. Illeris (Ed.), *Contemporary theories of learning* (pp. 200–208). Routledge.

Lave, J. & Wenger, E. (1991). *Situated learning: Legitimate peripheral participation*. Cambridge University Press.

Leath, B. (2009). Solving the challenges of the virtual workplace for interns. *Young Scholars in Writing, 6*, 3–11.

Little, S. B. (1993). The technical communication internship: An application of experiential learning theory. *Journal of Business and Technical Communication*, *7*(4), 423–451.

Lucas, K. & Rawlins, J. D. (2015). The Competency pivot: Introducing a revised approach to the business communication curriculum. *Business and Professional Communication Quarterly*, *78*(2), 167–193.

Mara, A. (2006). Pedagogical approaches: Using charettes to perform civic engagement in technical communication classrooms and workplaces. *Technical Communication Quarterly*, *15*(2), 215–236.

Melonçon, L. & Henschel, S. (2013). Current state of US undergraduate degree programs in technical and professional communication. *Technical Communication*, *60*(1), 45–64.

Pennell, M. & Miles, L. (2009). "It actually made me think": Problem-based learning in the business communications classroom. *Business Communication Quarterly*, *72*(4), 377–394.

Perlin, R. (2012). *Intern nation: How to earn nothing and learn little in the brave new economy*. Verso Books.

Pickering, K. (2018a). Learning the emotion rules of communicating within a law office: An intern constructs a professional identity through emotion management. *Business and Professional Communication Quarterly*, *81*(2), 199–221.

Pickering, K. (2018b). Navigating discourses of power through relationships: A professional and technical communication intern negotiates a meaningful identity within a state legislature. *Journal of Technical Writing and Communication*, *48*(4), 441–470.

Pickering, K. (2019). Emotion, social action, and agency: A case study of an intercultural, technical communication intern. *Technical Communication Quarterly*, *28*(3), 238–253.

Robles, M. M. (2012). Executive perceptions of the top 10 soft skills needed in today's workplace. *Business Communication Quarterly*, *75*(4), 453–465.

Ruggiero, D. & Boehm, J. (2016). Design and development of a learning design virtual internship program. *The International Review of Research in Open and Distributed Learning*, *17*(4), 105–120.

Savage, G. J. (1997). Doing unto others through technical communication internship programs. *Journal of Technical Writing and Communication*, *27*(4), 401–415.

Savage, G. J. & Seible, M. K. (2010). Technical communication internship requirements in the academic economy: How we compare among ourselves and across other applied fields. *Journal of Technical Writing and Communication*, *40*(1), 51–75.

Savage, G. J. & Sullivan, D. L. (2001). *Writing a professional life: Stories of technical communicators on and off the job*. Longman.

Shuman, L. J., Besterfield-Sacre, M. & McGourty, J. (2005). The ABET "professional skills"—Can they be taught? Can they be assessed? *Journal of Engineering Education*, *94*(1), 41–55.

Sides, C. & Mrvica, A. (2017). *Internships: Theory and practice*. Routledge.

Smart, G. & Brown, N. (2002). Learning transfer or transforming learning? Student interns reinventing expert writing practices in the workplace. *Technostyle*, *18*(1), 117.

Southard, S. G. (1988). Experiential learning prepares students to assume professional roles. *IEEE Transactions on Professional Communication*, *31*(4), 157–159.

Southard, S. G. (1989). Protocol and human relations in the corporate world: What interns should know. In W. O. Coggins (Ed.), *Establishing and supervising internships* (pp. 65–77). Association of Teachers of Technical Writing.

St. Amant, K. (2003). Expanding internships to enhance academic-industry relations: A perspective in stakeholder education. *Journal of Technical Writing and Communication, 33*(3), 231–241.

Stitts, D. K. (2006). Learning to work with emotions during an internship. *Business Communication Quarterly, 69*(4), 446–449.

Suzuki, R., Salehi, N., Lam, M. S., Marroquin, J. C. & Bernstein, M. S. (2016, May). Atelier: Repurposing expert crowdsourcing tasks as micro-internships. In *Proceedings of the 2016 CHI Conference on Human Factors in Computing Systems* (pp. 2645–2656). ACM.

Tessier, D. (1975). The value of a summer internship in technical writing. *The Technical Writing Teacher, 2*(2), 16–18.

Tovey, J. (2001). Building connections between industry and university: Implementing an internship program at a regional university. *Technical Communication Quarterly, 10*(2), 225–239.

U.S. Department of Labor, Bureau of Labor Statistics. (2019). *Number of Jobs, Labor Market Experience, and Earnings Growth: Results from a National Longitudinal Survey* (USDL-19-1520). https://www.bls.gov/news.release/pdf/nlsoy.pdf.

Wyld, L. D. (1978). Practical professional training through internships. *Journal of Technical Writing and Communication, 8*(2), 89–95.

Yarbrough, C. (2016). STEM students go abroad for research and internships. *International Educator, 25*(2), 44.

2. Interstitial Design Processes: How Design Thinking and Social Design Processes Bridge Theory and Practice in TPC Pedagogy

Liz Lane
University of Memphis

Abstract: This chapter explores interdisciplinary concepts of design theory: design thinking and social design. The author presents interstitial design as a combined, process-based approach for exploring social justice issues through design within technical and professional communication, offering insight into how interstitial design can work within our classrooms and genre-based assignments.

Keywords: design thinking, social design, theory, praxis, pedagogy

Key Takeaways

- Design thinking provides an interdisciplinary approach to confronting communication design problems within and beyond the classroom.
- Social design offers a process for encouraging exploration of broader issues and audiences through an explicitly social lens.
- Both design thinking and social design processes can enhance our pedagogies to engage with social justice issues in classroom settings.

At the core of technical and professional communication (TPC) pedagogy, design literacy contributes to foundational knowledge of TPC best practices, including best practices regarding document design (Sánchez, 2017; Williams, 2015), user-centered design (Redish & Barnum, 2011; Salvo, 2001), and accessible design (Melonçon, 2014; Walters, 2010)—what Kelli Cargile Cook (2002) terms "layered literacies,"—multifaceted skill sets that address increasingly complex industries and competencies. Frequently, our students go on to produce deliverables for specific audiences and design communication for a myriad of workplace or community contexts, yet as our field's social justice turn continues to reshape our pedagogy, research, and engagement (Haas & Eble, 2018; Jones et al., 2016), our understanding of design for what and for whom is continually evolving, responding to and designing for users in an increasingly globalized workplace and community. With the rise in popularity of design thinking as a framework for approaching problem solving, design terminology and theory has further proliferated disciplines like TPC that explicitly inhabit the aforementioned sectors of different types of design and the social impact of those pursuits. In this chapter, I explore a layered approach to synthesizing these interdisciplinary and adjacent approaches to "design," presenting interstitial design as a bridge to connect and extend our field's strengths in theory and praxis of design and critical thinking. I explore the complex uses of "design" within TPC and outline the benefits of both design thinking and social design, as these theories enhance our extant work within TPC and chart flexible paths for continued engagement with action-oriented, socially just foci.

Design, when considered as a core concept of making process-based decisions toward crafting a solution or deliverable, involves recursive critical thinking toward a goal. Within writing studies classrooms that range from rhetoric and composition to TPC, recursive processes shape how we frame concepts to students and influence how students demonstrate learning. Sally Henschel and Lisa Melonçon (2014) investigated common practical and conceptual skills valued by industry and academia, mapping the overlaps to propose "critical system thinking," which they define as the ability to "understand the processes by which parts are linked together; the ethical responsibility to consider ideological/power stances of those structures and critique when necessary" (p.13). Though as the authors note, the model of critical system thinking they put forth is difficult, if not impossible, to implement into one TPC course, as they outline an inventory of TPC courses to dispatch a layered approach to the process. This chapter explores the broad and flexible nature of *interstitial design* as an adaptable practice to implement in the classroom and frames this discussion through this main question: In what ways might our pedagogies better promote a practice focused on multifaceted design processes to reach these ends?

■ Contemplating the Work of Design in TPC

In TPC, it is no longer enough for us to use the term "design" as shorthand. We must interrogate and articulate what we mean when we say design, as we know from our roots in rhetoric, user experience, and effective communication writ large that there is never a neutral approach to "design." Fernando Sánchez, in his 2017 study of how writing studies scholars approach and write about design in the field's major journals, writes that ". . . design has become an understood facet of technical communication, it continues to be a subject of study within our field that gains importance and complexity—a complexity that can generate multiple (and sometimes contradictory) terms stemming from our own and borrowed from other fields" (p. 360). Further, he writes, "Essentially, the proliferation of design in technical communication has led to different terminology and varying starting points in the rich literature of design" (p. 361). Similarly, Charles Kostelnick identified that "the field of creative problem-solving—design—can shed light on the evolution and future direction of the writing paradigm," again pointing to interdisciplinary offerings of design (1989, p. 267). Our pursuit of design is always inherently interdisciplinary, but it is time for TPC to mark its unique take on design that echoes the humanistic roots, adaptable potentials, and creative processes that proliferate our pedagogy and scholarship.

Though terms such as "design thinking" and "creative thinking" are enjoying a popular (and at times buzzworthy or trivial) moment in our field and many others, this chapter examines an interstitial approach to teaching design processes by way of the design thinking process and social design. An interstitial approach to teaching design, defined below, will allow our students to analyze cultural and social justice issues in rapidly evolving "real [workplace] settings" that demand flexibility and responsiveness to increasingly connected and global audience needs (Henry, 2000). As this collection emphasizes the changing nature of the technical communication field and classroom, I argue a renewed approach to our pedagogical processes can only benefit our students and ourselves in the dynamic technical communication field today. I focus specifically on social design (Resnick, 2016; Shea, 2012) and elements of design thinking (Brown & Wyatt, 2010) as interstitial tools to support our theories, practices, and approaches within TPC. These design theories are each process-driven and recursive and, when joined, allow for a flexible cognitive process that can benefit the many types of problems TPC curricula are sculpted to solve, or offer students the experience of researching and exploring to solve. By exploring both the recent influx of design thinking and social design-focused curricula and considering the field's evolving pedagogy toward socially engaged design, this chapter specifically ponders questions in the following sections such as: Where does design thinking and social design fit in TPC pedagogy? How might the design thinking, and social design processes, encourage our pedagogy to be more accountable to local and global user needs? What benefits does an interstitial approach to teaching design offer our students?

Defining Interstitial Design Processes

The concept of *interstitiality* considers the forming or occupying of interstices, a space between boundaries or merely "in-betweenness," the "borders of genre," and articulates an idea of not fitting perfectly into one exact category (Schanoes, 2004). It is a concept that sees usage in biology, the arts, computing, and architecture that centers on connections, a term I employ here to examine the junctures that emerge between designer, situations, and the problems that we seek to solve through design. An interstitial framing of design reflects the term's frequent state of flux across disciplines (Kaufer & Butler, 2013). I term the practice of a dynamic approach to teaching design in TPC as *interstitial design* because of the multifaceted, interdisciplinary benefits of blending several theories of design into one broad, flexible recursive process for TPC pedagogy that is itself evolving rapidly. Interstitiality strengthens and illuminates the connections between two junctures, buttressing an adaptable approach to unique situations.

As TPC instructors and scholars frequently turn to interdisciplinary sources to reinforce our field's work, taking an interstitial approach to teaching design through and for social justice issues is a natural progression. I explore interstitial design as a conduit for these interdisciplinary habits we already employ as TPC teachers and scholars, how our pedagogies often sit at the interstices of, for example, visual rhetoric, universal design, and usability studies (Greenwood et al., 2019; Holsinger, 2012; Redish & Barnum, 2011; Shea, 2012). In 2019, a special issue of the *Journal of Business and Technical Writing* addressed this emerging approach to design thinking in TPC, including an article by April Greenwood, Benjamin Lauren, Jessica Knott, and Dànielle N. DeVoss that explored the different ways instructors apply design thinking principles as a rhetorical methodology in their various classrooms. An interstitial design process also bridges our theories and practices of applying design in TPC more apparently to contexts within and beyond our classrooms, where we challenge students to apply TPC best practices through applications and software as they learn. The familiar genres of TPC curricula—that is, assignments such as employment documents; technical instructions and descriptions; white papers; usability studies, surveys, and reports; and collaboratively edited and written documentation, to name a few—are grounded in similar recursive foundational concepts of defining, drafting, iterating, revising, and delivering.

▋ Interstitial Design in the Classroom

In her research on teaching justice issues in the university classroom, criminal justice scholar Kristi Holsinger discusses traditional teaching practices' singular approach, such that values "conformity, an individualistic approach, and competition," a method that oftentimes encourages "passive learning and even creates passive learners" (Cameron, 2002; Campbell & Smith, 1997, as cited in Holsinger, 2012, p. 14). Yet, effective principles for undergraduate education, such as those presented by Arthur W. Chickering and Zelda F. Gamson (1987) center "cooperation and collaboration among students in the classroom and encourage active participation in order to maximize student learning"—a socially collaborative approach to creative problem solving by doing and exploring (Holsinger, p. 15). An interstitial design approach flourishes when collaboration rests at the center of its application, echoing James Purdy's claim that "design projects require multiple hands and minds, and a design thinking approach to writing makes such collaboration standard, accepted, and unquestioned" in his study of design thinking in writing studies (2014, p. 633). Indeed, the TPC classroom is often wildly collaborative, in that we prepare our students to collaborate and test project management skills for eventual implementation in their professional lives beyond the classroom. We value recursive, iterative composing processes, borrowed from our roots in composition and writing studies, to apply process-driven composing to the dynamic genres of TPC, such as white papers, websites, extended usability studies and reports (Bay, 2010; Purdy, 2014). Thus, it is a natural progression to reconsider these collaborative values and pedagogical practices as a process of designing toward effective and dynamic student needs and learning outcomes, lest we "eclipse the possible connections" that enable us to adapt and evolve with the changing field (Bay, 2010, p. 33).

In her research on cognitive psychology's bearings on technical communication, Ginny Redish examines constructivism, arguing that "each reader, writer, and student has his or her own schemas and mental models that affect how he or she perceives and remembers what happens in a document or writing assignment," ultimately claiming that "listening to lectures is seldom as useful a learning experience as actually doing relevant work" (1997, p. 70). Instead, Redish presents schemas as flexible, unstructured processes through which users can "link information" by working with their best "mental model . . . a changeable collection of associations in people's minds" (1997, p. 71). As thinkers, composers, writers, and practitioners, we are process-driven at our cores, applying our unique preferences for approaching and solving a problem in order to make the most sense of the association in our mind. Likewise, the social and collaborative component of applying such schemas within classrooms allows students to build upon and explore schemas in new contexts and challenges (Redish, 1997, p. 72).

A 2002 study on collaboration and creativity (Madjar et al.) found that it is possible to boost and increase "employee's creativity if supervisors and coworkers are trained and encouraged to provide explicit support" (765), reiterating the value of collaborative problem solving and idea generation that TPC instructors are familiar with in the structure of their courses. In the sections below, I discuss how two different design theories can coalesce into an interstitial approach to foster the benefits of collaborative cognitive schemas. Though their foundations include dividing the design process into meaningful and manageable compartments or a process-based approach, interstitial design processes can also flow together and create a strong, interdisciplinary practice that can enhance our pedagogies in TPC.

∎ Overview of Design Thinking and Benefits to TPC

A term first thought to have been used in 1987 by a professor of architecture at the Harvard School of Design, Peter Rowe used the term *design thinking* to "account for the underlying structure and focus on inquiry directly associated with those rather private moments of 'seeking out' on the part of designers" (Rowe, 1987, as cited in Nixon, 2016). Interstitiality appears to permeate Rowe's early definition of the term, with special emphasis on structures of support and the affective process of inquiring about design problems and audience needs. Since that time, the term has only grown in usage through workshops in academic fields and mainstream business publications, particularly focused toward corporate or economic success (Nixon, 2016). Popularized by the design consultancy firm IDEO, the design thinking process most widely adopted in current scholarship and application references Stanford University's design school (or "d school") and encourages innovation and human-centric perspectives (Brown & Wyatt, 2010). The process encourages optimism and suppleness, prioritizing "constructive experimentation [that] allows high-impact solutions to bubble up from below rather than being imposed from the top, a process more about doing than thinking," and embracing messy or wild ideas (Brown & Wyatt, 2010, p. 3).

IDEO CEO and president Tim Brown (2015) articulates the firm's design thinking methodology as a cognitive process that "allows people who aren't trained designers to use creative tools to address a vast range of challenges." Brown describes design thinking as an alternative to "conventional problem-solving practices," valuing the process's emphasis on intuition, analysis and pattern recognition, and affective idea generation that allows designers to "construct ideas that are emotionally meaningful as well as functional." Brown recognizes that "nobody wants to run an organization on feeling, intuition, and inspiration, but an overreliance on the rational and the analytical can be just as risky," presenting design-thinking as a process that explores "multiple possible solutions" and combines useful elements of traditional problem-solving schemas. When brought into TPC classrooms, design thinking holds the potential to bridge technical communication best practices and usability scholarship with creative needs analyses for myriad contexts and audiences.

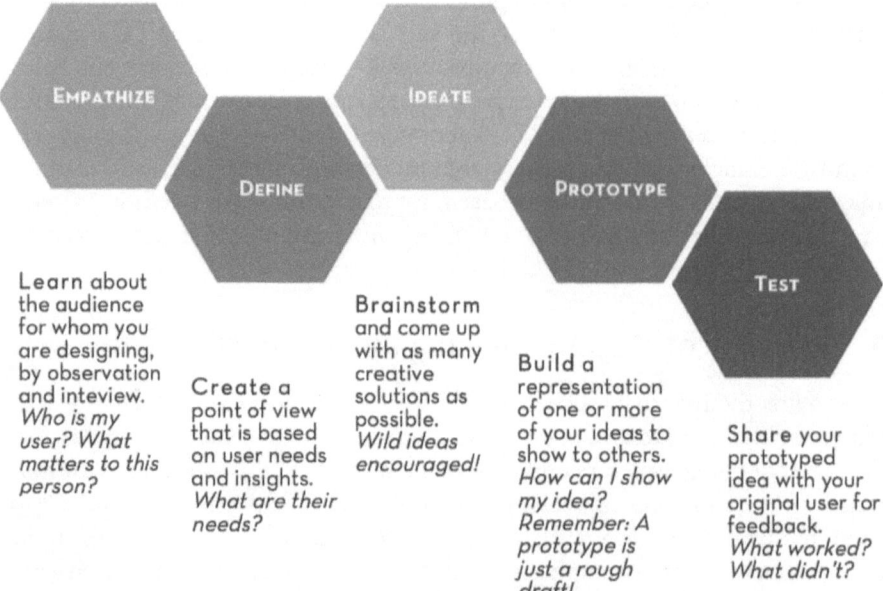

Figure 2.1. The design thinking process as articulated by Stanford's d.school via the Stanford d.school bootcamp bootleg document (Institute of Design at Stanford, n.d.).

Indeed, many aspects of the design thinking process are already familiar territory within TPC, as Tom Lockwood defines the approach as "a human-centered innovation process that emphasizes observation, collaboration, fast learning, visualization of ideas, rapid concept prototyping, and concurrent business analysis" (Lockwood, 2009, as cited in Nixon, 2016, p. 13). Design thinking is an inherently recursive process, encouraging its users to "build in order to think . . . designers learn by doing," which is a pedagogical value many TPC instructors echo in their classrooms and their emphasis on practical, skills-based learning (Nixon, 2016, p. 15). Additionally, design thinking is frequently used in industry contexts, and preparing our students to work with this schema in our TPC classrooms allows them to practice applying its stages and process and preparing to articulate their experience working with design thinking concepts, before they move beyond the classroom boundaries. Tim Brown and Jocelyn Wyatt caution users to not think of the design thinking process as a rote, intractable process but rather as "a system of overlapping spaces" that, as shown in Figure 2.1, can flow into one another, work cyclically, and complement many recursive outcomes within TPC classrooms (2010, p. 33).

The first stage, "empathize," asks designers to identify their user and ponder what matters most to them, echoing the foundations of usability and user experience scholarship (Redish & Barnum, 2011; Rose et al., 2018; Salvo, 2001). Likewise, the "define" stage of the process advises designers to adopt a persona by creating a point of view based on their users' needs and insights, echoing core elements of usability studies and UX scholarship (Melonçon, 2014; Redish & Barnum, 2011). The "ideate" and "prototype" stages are incredibly elastic, as they encourage messiness and wild ideas to find the best design solution for the user, emphasizing rapid prototyping to learn and fail quickly while adjusting prototypes with that new knowledge. With social issues in mind, design thinking offers instructors and students the freedom to err, stumble, and revise within comfortable boundaries of an iterative, recursive cycle. All ideas should be valued and placed into conversation with other prototypes, drafts, and versions. Finally, the "test" stage of the process requires sharing prototyped ideas with the original user for constructive feedback, a usability test hallmark.

As a tool for communication deliverables common to TPC classrooms, the design thinking process emphasizes local expertise and embedded knowledge to best design for specific user needs, placing high value on the user while considering inclusivity and marginalized audiences as a part of the empathizing, defining, and ideation stages of the research process. Many TPC and writing studies courses echo the design thinking process by emphasizing audience analysis, synthesis and critical thinking, and iterative progress toward a final product. This same approach shapes our composing and thinking habits, both inside and outside institutional and classroom environments. Yet the design thinking process urges its adopters to more visibly grapple with and combine these often-invisible cognitive phases at each step. In short, design thinking provides an interdisciplinary approach to confronting communication design problems within and beyond the classroom, making the process well-suited for an interstitial combination with TPC design best practices, as well as other interdisciplinary design theories, as described below.

▍ Exploring Social Design After TPC's Social Justice Turn

Our field is at a crucial and exciting moment of reshaping how we equip students with the skills to respond to a range of dynamic workplace and community needs, blurring the intersections between theory and praxis, workplace and educational institution—a perennial issue our field has brought up frequently (Bay, 2010; Haas & Eble, 2018; Staples & Ornatowski, 1997). In the changing workplace, our students are entering industries where collaborative and project-based work, remote or asynchronous work, and the option of flexible hours are growing common, reflecting a "work environment that values solutions . . . and allows for user- and experience-based solutions rather than those based around budgets and unit-based targets" (Nixon, 2016, p. 10). Employers are increasingly searching for creative innovation in their employees regardless of field, as some researchers suggest that creativity and innovation factor into hiring and advancement decisions in many industries (Florida, 2002; Pink, 2005). And in our globalized world, industries are seeking solutions for addressing social issues through their work (Bay, 2010; Purdy, 2014). A secondary theory to bring into interstitial design approach is that of *social design*, "the practice of design where the primary motivation is to promote positive social change within society" (Resnick, 2016, p. 12). Social design, also known as public interest design, social impact design, and humanitarian design, is widely used in the field of graphic and universal design. I use social design here as an entry point to exploring social issues in TPC classrooms, an area that Natasha N. Jones identifies as especially pressing as TPC moves beyond its social justice turn. Jones argues that as we integrate issues of diversity and social justice[1] into our pedagogy and scholarship, "we must examine the design and dissemination of communication critically with a focus on understanding how oppressive conditions can be rearticulated and reinforced" (2016, p. 346). Further, Jones encourages TPC scholars to boldly question the social structures of power behind communication in their pedagogy and scholarship, arguing,

> social justice in technical communication investigates how communication broadly defined can amplify the agency of oppressed people—those who are materially, socially, politically, and/or economically under-resourced. Key to this definition is a collabora-

---

1. Jones defines diversity as "a focus on the inclusion of various perspectives and viewpoints" and social justice as "critical reflection and action that promotes agency for the marginalized and the disempowered," respectively (2016, p. 343).

tive, respectful approach that moves past description and exploration of social justice issues to taking action to redress inequities. (2016, p. 347)

Broaching such subjects in a TPC classroom can be challenging, yet social design offers a process for encouraging instructors and students to explore broader issues and audiences through an explicitly social lens, highlighting cultural, economic, and racial disparities as a part of the design process. Graphic designer Andrew Shea describes the altruistic approach to social design as using design skills to "support civic and cultural causes," an approach that values "designing *with*, not for" communities (2012, p. 9). His *Designing for Social Change* (2012) offers design case study tactics for designers and scholars seeking actionable methods for bridging design theory with social issues that best benefit the target audiences. Shea's work is especially illuminating for TPC classrooms that might partner with community organizations and practice interstitial design through community engagement-based projects.

Recent scholarship on the field's social justice turn charges TPC scholars to "actively integrate" social justice perspectives into our pedagogy and research so that we can demonstrate how the values of TPC can "promote social change on a broader level," equipping our students with the analytical and creative design skills that can address complex social issues through communicative means (Jones, 2016, p. 343). Social design is an ideal theory to apply in our field's post-social justice turn era, encouraging students to critically consider complex social justice problems and to design empathetic, engaging technical communication materials toward various needs. Graphic designer Elizabeth Resnick presents the broader purpose of social design as urging students to study how their "research, analysis, discourse, and creation [of designed components] at local, national, and even international levels" may impact these distributed audiences (2016, p. 13). Inherent in the social design process is the flexibility of its users to "redefine what it means to be a designer," focusing intently on "improving the way [humans] interact and communicate with each other and within their communities as citizen designers," never losing sight of the communal component of design (Resnick, 2016, p. 13). Therefore, calling upon interdisciplinary approaches to design in order to examine social justice issues in the classroom is a natural entry point, extending TPC's focus on the humanistic impact of communication.

■ Potentials of Interstitial Design in TPC Pedagogies

Given the flexible and adaptable nature of both design thinking and social design, how might an approach to interstitial design, which combines both theories, manifest in TPC pedagogies? Within TPC, we frequently build our curricula around process-driven composing practices, requiring that students propose topics, outline and brainstorm, submit rough drafts and conduct peer review, and finalize their writing for final submission—common patterns of composing that directly echo TPC's roots in rhetoric and writing studies (Bay, 2010). Therefore, it is a natural move to more explicitly name this process-driven approach as *interstitial design*, showcasing how TPC instructors are creatively bridging theory and practice through design and adapting traditional genres taught in our classrooms with design tactics.

Consider an instructor that seeks to bring more social justice-focused issues into their TPC classroom in an effort to engage students with the actionable concepts of technical writing and to demonstrate how technical writing can create calls to action. For example, consider a group of Introduction to Technical Writing students researching a local civic issue in order to compose a white paper report on the topic, an assignment that initially tasks students to explore a social justice issue in their local community. Each student will collaboratively define an issue in small teams: lack of access to healthcare resources in their city or underfunded public education sites and redlining in school districts, for example. An assignment would frame this broad task as a social justice communication problem and break down components of the assignment using interstitial design, from empathizing with audiences, defining topic proposals, research activities and reports, drafting and peer review recursive processes, and prototyping and revising toward a final product (see Figure 2.2). Over the course of the collaborative project, students will share their own thought processes and knowledge of the issue as they draft ideas and research together, following steps of the design thinking process and using social design as a launchpad, eventually crafting an agreed-upon association that aims to inform and persuade their audience toward their communicative goals (see Figure 2.2 for a sample assignment sequence).

Note that in Figure 2.2, the ideate/iterate, prototype, and test stages can work as a cycle (as can the entire process), but these steps in particular allow for interstitial considerations from extant TPC scholarship and social design work. It is in teaching contexts such as the one described in this example that interstitial design processes allow for the flexibility of cognitive and critical thinking processes that best address the dynamic communication scenarios and genres we confront in TPC. The interstitial design process can be explicitly shared with students at the outset of an assignment to demonstrate the emphasis on a recursive thinking/drafting/creating process or can implicitly underpin one's pedagogical approach through classroom engagement and instruction. Yet, I assert one must openly discuss the phases of interstitial design and listen to student feedback about the process, integrating ideas and input into the process along the way. Resistance or uncertainty from students may occur and should be embraced as part of the complex interstitial process, asking students and instructors to explore why resistance emerged and to note the uncertainties as part of the recursive process. It is through such transparent metaprocess conversations that interstitial design strengthens the conceptual phases that encourage one to deliberately interrogate social justice issues closely and carefully.

Empathize: Research communities impacted by lack of healthcare access *(interview, observe, define, and question who is unable to easily access care)*.		Define: Consider social design and consider the specific needs of the audience impacted by this issue *(i.e., will our process and design benefit the community at the center of this issue?)*.		Ideate/Iterate: Generate multiple ideas or solutions to address this issue *(i.e., how can we raise awareness about healthcare inaccess in our city?)*.

Test: Share design with intended user, gather feedback on what works or needs improvement *(i.e., map shows disparity in public transit lines, work to integrate info on public transit into healthcare access map and white paper/map at next iteration)*.		Prototype: How can we represent our findings and ideas? *(i.e., will visual maps of neighborhood demographics help communicate our issue to a broad readership? Will these representations benefit the audience at the center of this issue?)*

Figure 2.2. A sample interstitial design process, using design thinking and social design, as applied to a traditional technical writing white paper assignment with a social justice issue focus.

■ Looking Ahead: Our Charge as Educators and Designers

Over 25 years ago, Jennifer D. Slack, David J. Miller, and Jeffrey Doak identified a pressing need for technical communicators to become more aware of how the manner in which they articulate meaning bears ethical weight upon the audiences consuming the communication they produce, writing:

> Most educators acknowledge that it would be a good idea for students to understand politics, power, and ethics, but there is very little explanation offered to suggest what they might do with that knowledge on the job. But one thing is certain: a technical communicator cannot be just a technical writer anymore. (1993, p. 25).

Going back to Resnick, she posits, "How can design educators help students engage in a world that is considerably interconnected and immediate, yet disturbingly more fractured, unstable, and totally disconnected from what really matters?" (2016, p. 12). Resnick raises many of the questions that perplex instructors teaching TPC, wondering how to help students see the applicability and transferability of their classroom work to their communities and daily lives, or challenging students to design communicative materials toward pressing and affective issues. I argue employing interstitial design enables TPC educators to both explore design as an actionable practice within a flexible recursive process in our classrooms and to interrogate how design can illuminate social justice issues and carry the lesson gained from interstitial design in praxis into TPC's evolving future.

As our field's social justice turn charged each of us with the task of bringing social justice issues more apparently into our pedagogy and research, the time is ideal for our field to begin showing how we can apply our unique design theories, practices, and knowledge to pursuing this call. Technical communicators, in praxis, are by their nature interstitial, occupying multiple intersections of expertise while adapting to the changing needs and expectations of global audiences.

As TPC instructors, we can make interstitial design a part of our curricula and pedagogical practices. I suggest the following practices, a list that is certainly not exhaustive, to bring these practices more apparently into the work we already take on:

- **Question core definitions of design along with technical communication:** Early in our courses, TPC instructors often ask students to discuss their definitions or approaches to technical communication. I argue we should include "design" with that discussion and query students about their prior knowledge of it, experiences with it, and questions for its application. We must challenge students to explore the public work of design as connected to technical communication and urge students to think about the material impact of design (Holsinger, 2012; Purdy, 2014; Rose et al., 2018; Sánchez, 2017).

- **Integrate interdisciplinary design processes into our pedagogies:** Turn to industry design sources, case studies, and knowledge articles to complement our academic texts and TPC scholarship. An interdisciplinary, interstitial approach to building design materials will better equip our students to apply interstitial design to their broad career endeavors and enable a dynamic, creative thinking background that enhances TPC's already highly adaptable goals, means, and outcomes (Brown & Wyatt, 2010; Resnick, 2016; Shea, 2012).

- **Diversify the texts students read:** Include resources from design disciplines in courses from introductory TPC courses to upper-level or major-specific courses. Offer students a range of perspectives from graphic design, industrial design, and design consultancies to offer varying perspectives on how design work is actively applied in dynamic scenarios (Shea, 2012; Williams, 2015).

- **Deploy interstitial design processes in core assignments and curricula:** Challenge students to use the design thinking process and social design in their brainstorming, conceptual sketching, pre-writing, and drafting stages, and ask them to reflect on how the processes complement their writing processes. For curricula and assignments that are built upon recursive or creative goals, ask students to map their defining moments, empa-

thizing, and prototyping ideas and sketches, or ask students to define how their work can benefit a particular social group.
- **Emphasize collaboration more frequently:** As a cognitive process, the collaborative elements of interstitial design urge students to encounter experiences and information previously unfamiliar to them by nature of the recursive process. In our classrooms, structure assignments to more apparently value collaboration in the understanding and defining stages and especially at the prototyping and testing stages where students can learn the most from others' feedback, insights, and perceptions of their design solutions (Purdy, 2014; Redish, 1997)

The TPC classroom is a place for radical design, where we say just as much with plain and simple language as we do with adhering to and bending design principles. Our genre-based assignments that explore documents common to technical communication industry and practice are ideal spaces to engage with social justice issues of audience lived experience. As TPC educators, we owe it to our students and the future of our field to tackle social justice issues in our genre-focused assignments in order to equip our students with the process-based skills to address pressing issues in their lives beyond the classroom. Learning from and continuing to be attuned to the intersections of various design theories can only enhance our pedagogies in the evolving technical communication field and industry.

■ References

Bay, J. (2010). Writing beyond borders: Rethinking the relationship between composition studies and professional writing. *Composition Studies, 38*(2), 29–46.
Brown, T. (2009). *Change by design: How design thinking creates new alternatives for business and society.* Collins Business.
Brown, T. (2015). *IDEO Design Thinking.* https://designthinking.ideo.com.
Brown, T. & Wyatt, J. (2010). Design thinking for social innovation. *Development Outreach, 12*(1), 29–43.
Cargile Cook, K. (2002). Layered literacies: A theoretical frame for technical communication pedagogy. *Technical Communication Quarterly, 11*(1), 5–29.
Chickering, A. W. & Gamson, Z. W. (1987). Seven principles for good practice in undergraduate education. *American Association for Higher Education Bulletin, 39*(7), 3–7.
Florida, R. L. (2002). *The rise of the creative class: And how it's transforming work, leisure, community and everyday life.* Basic Books.
Greenwood, A., Lauren, B., Knott, J. & DeVoss, D. N. (2019). Dissensus, resistance, and ideology: Design thinking as a rhetorical methodology. *Journal of Business and Technical Communication, 33*(4), 400–424.
Haas, A. M. & Eble, M. F. (2018). *Key theoretical frameworks: Teaching technical communication in the twenty-first century.* Utah State University Press.
Henry, J. (2000). *Writing workplace cultures: An archaeology of professional writing.* Southern Illinois University Press.

Henschel, S. & Melonçon, L. (2014). Of horsemen and layered literacies: Assessment instruments for aligning technical and professional communication undergraduate curricula with professional expectations. *Programmatic Perspectives, 6*(1), 3–26.

Holsinger, K. (2012). *Teaching justice: Solving social justice problems through university education.* Routledge.

Institute of Design at Stanford. (n.d.). *The d.school bootcamp bootleg.* https://static1.squarespace.com/static/57c6b79629687fde090a0fdd/t/58890239db29d6cc6c3338f7/1485374014340/METHODCARDS-v3-slim.pdf.

Jones, N. N. (2016). The technical communicator as advocate: Integrating a social justice approach in technical communication. *Journal of Technical Writing and Communication, 46*(3), 342–361.

Jones, N. N., Moore, K. R. & Walton, R. (2016). Disrupting the past to disrupt the future: An antenarrative of technical communication. *Technical Communication Quarterly, 25*(4), 211–229.

Kaufer, D. S. & Butler, B. S. (2013). *Rhetoric and the arts of design.* Routledge.

Kostelnick, C. (1989). Process paradigms in design and composition: Affinities and directions. *College Composition and Communication, 40*(3), 267–281.

Madjar, N., Oldham, G. R. & Pratt, M. G. (2002). There's no place like home? The contributions of work and nonwork creativity support to employees' creative performance. *The Academy of Management Journal, 45*(4), 757–67.

Melonçon, L. (2014). *Rhetorical accessibility: At the intersection of technical communication and disability studies.* Routledge.

Nixon, N. W. (Ed.). (2016). *Strategic design thinking: Innovation in products, services, experiences, and beyond.* Fairchild Books.

Pink, D. (2005). *A whole new mind: Why right brainers will rule the future.* Penguin.

Purdy, J. P. (2014). What can design thinking offer writing studies? *College Composition and Communication, 65*(4), 612–641.

Redish, J. G. (1997). Understanding people: The relevance of cognitive psychology to technical communication. In K. Staples & C. M. Ornatowski (Eds.), *Foundations for teaching technical communication: Theory, practice, and program design* (pp. 67–84). Greenwood.

Redish, J. G. & Barnum, C. (2011). Overlap, influence, intertwining: The interplay of UX and technical communication. *Journal of Usability Studies, 6*(3), 90–101.

Resnick, E. (Ed.). (2016). *Developing citizen designers.* Bloomsbury Academic.

Rose, E. J., Edenfield, A., Walton, R., Gonzales, L., McNair, A. S., Zhvotovska, T., Jones, N., Garcia de Mueller, G. I. & Moore, K. (2018, August). Social Justice in UX: Centering marginalized users. In *Proceedings of the 36th ACM International Conference on the Design of Communication* (p. 1–2). ACM.

Salvo, M. J. (2001). Ethics of engagement: User-centered design and rhetorical methodology. *Technical Communication Quarterly, 10*(3), 273–290.

Sánchez, F. (2017). The roles of technical communication researchers in design scholarship. *Journal of Technical Writing and Communication, 47*(3), 359–391.

Schanoes, V. (2004). Critical theory, academia, and interstitiality. *Journal of the Fantastic in the Arts, 15*(3), 243–247.

Shea, A. (2012). *Designing for social change: Strategies for community-based graphic design.* Princeton Architectural Press.

Slack, J. D., Miller, D. J. & Doak, J. (1993). The technical communicator as author: Meaning, power, authority. *Journal of Business and Technical Communication, 7*(1), 12–36.
Staples, K. & Ornatowski, C. M. (Eds.). (1997). *Foundations for teaching technical communication: Theory, practice, and program design.* Ablex.
Walters, S. (2010). Toward an accessible pedagogy: Dis/ability, multimodality, and universal design in the technical communication classroom. *Technical Communication Quarterly, 19*(4), 427–454.
Williams, R. (2015). *The non-designer's design book: Design and typographic principles for the visual novice.* Pearson Education.

3. Engaging Plain Language in the Technical Communication Classroom

Kira Dreher
CARNEGIE MELLON UNIVERSITY, QATAR

Abstract: This chapter encourages instructors to engage with plain-language strategies in technical communication courses. Robust plain-language strategies overlap substantively with core aims of technical communication. They prioritize users' needs through effective content, style, and design, and by involving users themselves. By exploring plain language in a course context, instructors can also pursue with students a theoretical inquiry into the fraught concepts of "plain" and "clear," the capacity and limitations of the movement to promote social justice, and the intersections of technical communication and rhetoric. In practical terms, instructors can show that technical communication expertise is central to the plain-language movement, which is well-anchored and recognizable across fields (business, law, health, the sciences) and across the globe. This chapter provides introductory information to understand, frame, and further explore plain language through a technical communication lens, as well as five in-class applications that engage plain language in theory and practice.

Keywords: plain language, plain-writing movement, plain writing

Key Takeaways

- Effective plain-language guidelines overlap many of the goals of technical communication, such as prioritizing users' needs and interests, involving users in producing texts, and using effective organization and design.
- Instructors can treat plain language as a practical application in technical communication courses, as well as an object of critical inquiry for students to explore its contextualized history, its potential support of social justice, and its rhetorical assumptions.
- Plain-language experience offers students a marketable, recognizable skill that they can strategically use to contextualize other specialized technical communication knowledge in their future careers.

In recent decades, the use of plain-language guidelines has dramatically increased in government communications, law, business, healthcare, and elsewhere (Schriver, 2017; Willerton, 2015). Plain-language guidelines have become a strategy in these fields to solve communication problems, effectively prioritize audiences, and save resources. Effective plain-language strategies, such as those showcased by the U.S. federal government and organizations like the Center for Plain Language, coincide with the goals and best practices of technical communication. I argue that plain language is an important framework for technical communication teachers and students because it is an opportunity to see our field anchored to an established public movement, to use our disciplinary knowledge to critique and address the movement's limitations, as well as to interrogate assumptions about social justice, "plainness," and access with our students. Given the global popularity of plain language, I also suggest that it offers technical communication courses a new way to engage with international communication practices and policies. In short, this chapter is intended to persuade instructors to engage theoretically and practically with plain language in their courses and to provide the introductory content, questions, and resources for doing so.

What exactly are plain-language guidelines? Many approaches have developed over time, and while some approaches use readability formulas or decontextualized rules, this chapter will focus on current, popular guidelines that are robust, rhetorical frameworks. These guidelines address audience and context, written style, information design, and user-testing. For example, the Center for Plain Language (CPL) defines plain language in the following way: "A communication is in plain language if its wording, structure, and design are so clear that the intended readers can easily find what they need, understand what they find, and use that information" (2019).

The CPL then offers the following five steps, each with nuanced sub-steps, to communicate in plain language:

Step 1: Identify and describe the target audience

Step 2: Structure the content to guide the reader through it

Step 3: Write the content in plain language

Step 4: Use information design to help readers see and understand

Step 5: Work with the target user groups to test the design and content

In Figures 3.1 and 3.2, I include before-and-after examples to briefly illustrate these revision steps in action. The revision reflects attention to direct language, organization, document design, and key actions for the reader.

Figure 3.1 gives a preliminary look at the way the CPL's guidelines emphasize an audience's ability to locate, understand, and use texts to complete tasks, effectively paralleling some of the key goals of technical communication. Other popular strategies, such as those found at plainlanguage.gov, support a similar approach. These strategies are much more than readability formulas or decontextualized rules aiming at a shallow concept of "the public." Indeed, plain language has become a critical and highly relevant site where students can grapple with technical communication's theories, practices, and effects in organizations, governments, and various other field contexts.

> Dear _____,
>
> City Water of XYZ is implementing a water meter replacement plan in partnership with our city's Green Utility Projects. It is intended that this plan, as well as several others, will minimize our city's use of unnecessary utilities, including water, waste, and power. A mandatory part of this initiative is to replace old water meters with more efficient "smart" meters that can be read automatically and can better detect water usage that is likely due to leaking faucets and infrastructure. This letter provides options for city residents regarding their meter replacement.
>
> **Water Meter Replacement Plan Options**
>
> Standard Replacement Appointments:
> - City Water of XYZ has hired ABC Utility Suppliers to install the meter upgrades. Each city district will be addressed in sequence over the next 12 months. Each district has been assigned a 2-month period during which their meters will be replaced. Each household will sign up for their appointment in the month preceding their replacement period. Please see our website for the sign-up and replacement periods and for appointment sign-ups: www.citywaterxyz.com. You will also receive a reminder.
> - City Water Resident Help Line can answer any questions or concerns.
>   - (xxx) xxx-xxxx from 8am-5pm Monday through Friday.
>
> Special Circumstances appointments:
> - If your household or meter has special circumstances (such as having a recently updated meter, having a meter in a location that requires special accommodations, or being unable to be present during your allotted Ward installation period), you may coordinate with City Water of XYZ by contacting us via phone or email.
>
> There are no fees or charges associated with this water meter upgrade. We do not anticipate any changes to the monthly billing process for residents. Any problems or concerns can be directed to City Water Resident Help Line [(xxx) xxx-xxx] or info@citywaterxyz.com.
>
> Sincerely,
>
> City Water of XYZ

Figure 3.1. "Before" water meter upgrade letter.

Over the past decade, technical and professional communication (TPC) researchers have considered empirically whether plain-language communication benefits readers and their attitudes in the areas of health literacies (Grene & Marcus-Quinn, 2017), environmental texts (Derthick et al., 2009; Jones et al., 2012), and in city governments (Dreher, 2017). Other scholars have worked to theorize and interrogate the ethical (Ross, 2015; Willerton, 2015) and social justice (Cheung, 2017; Jones & Williams, 2017; Williams, 2010) work that plain language does—or does not—do in TPC. Further, Karen Schriver (2017) has documented recent trends in plain language that move beyond comprehension and task completion to building user trust. Plain-language research also necessarily intersects with design (Mazur, 2000). A great deal of research connected to the Center for Civic Design has well established the design/plain language relationship through effective voting ballots. In some of this work, such as Summers et al. (2014) and Ramchandani et al. (2017), the authors couple "plain language" with "plain interaction" (Summers et al, 2014, p. 22), which helps to capture the breadth and scope of what "plain language" has come to mean.

City Water of XYZ

<DATE>

<Resident's Name>
<Address>

Dear <Resident's Name>,

> Contact City Water of XYZ:
> info@citywaterxyz.com
>
> Resident Help Line: (xxx)xxx-xxxx
> Mon-Fri | 8am-5pm
>
> **www.citywaterxyz.com**

City Water of XYZ is upgrading all water meters to meet the goals of XYZ's new Green Utility Project. In this upgrade, ABC Utility Supplier will install "smart" water meters in all city households over the next 12 months. Below we describe what the upgrade and installation process means for you.

Does the water meter upgrade apply to me?
Yes, this city-wide upgrade is mandatory for all households.

What do I need to do?
Schedule an appointment for the meter installation through our website or Resident Help Line. A household member will need to be present during the 60-minute appointment.

When do I need to schedule my appointment?

If you live in district…	Visit our website in…	To schedule an appointment in…
1	January	February or March
2	March	April or May
3	May	June or July
4	July	August or September

*Watch for a reminder in the mail before your sign-up period.

If you have special conditions, such as a recently replaced meter or limited availability, please contact us.

Will this cost me anything?
No, you will not pay for the upgrade or have any billing interruptions.

Why are we upgrading our water meters?
The city's Green Utility Project aims to minimize wasted resources in XYZ City through better management, maintenance, and awareness of our city utilities. The new "smart" water meters better identify leaks and problems and they automatically report water consumption levels to residents and City Water of XYZ. More information about the Green Utility Project initiatives and the meter upgrades is available on our website.

Thank you for your cooperation in this exciting initiative! Please contact us with any questions.

Sincerely,

_____
Vice President of City Water of XYZ

Figure 3.2. "After" water meter upgrade letter.

Plain-language experience can help students characterize their expertise in recognizable, marketable ways outside the classroom. It can serve as a launching point to introduce more specialized technical communication expertise in areas like healthcare, government, and the sciences, since these fields and others have developed specific forms of plain language for their unique audiences and tasks.

This chapter introduces teachers to the plain-language movement and its current trends and goals, as well as five in-class applications. I begin with a brief but necessary discussion of readability formulas in the mid-twentieth century—formulas that we resist in technical communication for good reason, but that do persist and can affect conceptions of what is "plain." Following this brief address, I explore plain language as a rhetorical strategy to 1) prioritize users across different fields, 2) support social justice, and 3) save resources. In each of these sections, I consider the successes and limitations in these efforts from a technical communication perspective. I then propose intersections between plain language and the rhetorical tradition for instructors who seek, as I do, to contextualize the movement in this way for students, and I address some objections to plain language. Lastly, I detail five specific activities for instructors to use in the classroom, as well as a list of further resources.

A Precursor to the Plain-Language Movement: Readability Formulas

In the 1940s, readability formulas emerged as a strategy to quantify and evaluate text in relation to audience comprehension. Readability formulas like the Flesch Reading Ease Formula and the Gunning-Fog Formula used surface features like syllables, word-length, and sentence counts to rate texts at different education levels. Longo (2004) describes these formulas as "cultural artifacts" emerging in a post-World War II moment that championed the idea that "[a]n educated citizenry would be better prepared to understand and act on rapidly changing social, technological, and political situations" (p. 166). In short, these formulas were thought to better equip the US to disseminate complex information for wide audiences.

Readability formulas have had a significant influence on the concept of clear or plain language since the 1940s. They link plainness to surface textual features and length—metrics they believe measure an audience's presumable comprehension. Researchers and practitioners have shown the severe limitations of readability formulas, citing their lack of attention toward real audiences, material contexts, document organization, and design, as well as a host of other rhetorical considerations (Redish, 2000; Schriver, 2000; Selzer, 1983). Readability formulas are often presumed to be part of the plain-language movement, but they can be inimical to what plain language has come to represent.

Despite the evidence against them, readability formulas and their metrics of plainness continue to thrive, so students should be aware of them. Organizations seeking to write plain documents continue to use these formulas because they are quick, inexpensive, and offer seemingly concrete evidence of improved writing. They are also easy to access. For instance, many versions of Microsoft Word come with the Flesch-Kincaid readability formula already built in for users. For these reasons, readability formulas and their limitations are important to address in technical communication courses, even briefly. Readability formula limitations also help to show students that the concepts of "clear" and "plain" are not neutral, but rather dynamic concepts that reflect the paradigms, histories, and constraints of different communities.

∎ Plain Language as a Strategy to Prioritize Users

In a 1991 edited collection entitled *Plain Language: Principles and Practice*, Irwin Steinberg defines plain language as "language that reflects the interests and needs of the reader and consumer rather than the legal, bureaucratic, or technological interests of the writer or the organization the writer represents" (p. 7). Steinberg's definition captures the prevailing motive of the plain-language movement. Researchers and practitioners then and now are working to parse through what it means to prioritize the audience effectively. What began with sentence-level readability metrics now routinely includes user testing, information design, and organization, as well as an ever-deepening understanding of the relationship between readers, writers, and contexts.

One way to explore how plain language prioritizes users is through an ethical lens. In *Plain Language and Ethical Action*, Russell Willerton (2015) theorizes the extent to which plain language can constitute ethical communication in technical writing. Drawing on Martin Buber, Willerton proposes that in plain language, the relationship between writer and reader can reflect an "I-You" relationship rather than "I-it," allowing the writer to partner or dialogue with the reader to prioritize their goals and enable their important actions (p. 53). As such, he proposes that plain language can be used to promote ethical communication, especially in contexts that are bureaucratic, unknown, rights-oriented, and critical for users. That said, ethical communication is not inherent to plain language (Ross, 2015), and in teaching, we must explicitly recognize that using plain language doesn't guarantee ethical communication. Yet, a plain-language framework like the one supported by the Center for Plain Language remains a useful tactic for writers and organizations who seek to build an ethical, dialogic relationship with users.

Various fields prioritize users by using plain language for a range of reasons. Sometimes plain language is tied to government mandates and regulation, to building users' trust of a brand or company, to disseminating knowledge effectively for the greater good, or often simply to being cost-effective (see the section "Saving Resources"). In the remainder of this section, I address the motivations for using plain language in four loosely clustered, and sometimes intersecting, fields of practice, all of which produce a great deal of technical communication: government and law, business and finance, health-related fields, and the sciences. The barriers to users that exist in these fields are not all issues plain language can address; many issues are deeply systemic and social. But the plain-language movement supports the assumption that effective communication can still drastically improve access, action, or trust for users in these and other contexts.

Within Government and Law

Government policies and regulations have prompted much of the development of plain language in the US. Clear, plain communication has long been championed as a way for the government to prioritize the needs of the population, a way to serve the public. Often cited as an initiator of plain language in U.S. government, Congressman Maury Maverick called for an end to "gobbledygook language" in 1944, claiming it "fouls people up" (cited in Greer, 2012). Several presidents have issued executive orders regarding clear and direct prose, including Jimmy Carter's 1978 Executive Order 12174 and Bill Clinton's Executive Orders 12988 and 12866 in the 1990s. These calls all support the notion that unclear writing impedes citizens' and the government's abilities to function effectively. Former Vice President Al Gore furthered the obligation of the government to be clear in his often-cited 1998 statement that "Clear writing from your government is a civil right" (https://www.plainlanguage.gov/resources/quotes/government-quotes/).

In a more recent and crucial step of the plain-language movement, Barack Obama signed into law the Plain Writing Act of 2010, the purpose of which was "to improve the effectiveness and accountability of Federal agencies to the public by promoting clear Government communication that the public can understand and use" (US House, 111th Congress). Note the explicit link made here between serving the public and clear communication. This act requires all government agencies and departments to adhere to federal plain-language guidelines. While the act has been critiqued for not having the "teeth" it needs to enforce these requirements, it has prompted a significant increase in funding for quality guidelines and resources. Plainlanguage.gov, developed and curated by The Plain Language Action and Information Network (PLAIN), houses public guidelines for federal agencies to compose and revise documents, guidelines supporting the same kind of principles emphasized by The Center for Plain Language in the example earlier in this chapter. Through the Plain Writing Act, federal government communication became linked to the plain-language movement and, arguably, to technical communicators' expertise.

Dozens of local and state governments enforce various plain-language requirements as well. The state of New York led the way in 1977 with the first state-based plain-language mandate for certain legal contracts, and many states have followed since, including California, New Jersey, Washington, Florida, Oregon, Minnesota, and many others (Kimble, 1992, p. 33). Numerous city and local governments have used or required plain language in multiple ways as well. A large city government in the Midwest, for instance, recently revised its city charter with plain language, and city government insiders reported significant improvements in the use of the document for internal processes (Dreher, 2017).

In parallel, plain legal writing has made great strides through the efforts of folks like Joseph Kimble and Bryan Garner, as well as long-running publications like *The Clarity Journal*. Lawyers and law scholars have raised many questions about plain language and its effects on laws and legal documents (see, for example, Assy, 2010), but the plain-language camp has claimed its ground, prompting significant changes in the field. Willerton's (2015) chapter on restyling the Federal Rules of Evidence offers a look into the stakeholders, process, and negotiations involved in revising these sorts of texts. Willerton's chapter helps to show the robust possibilities the plain-language movement offers for deepening the relationship between technical communication and law.

Within Business and Finance

Businesses use plain language to serve many different goals. One goal has been to meet the requirements of government mandates that enforce plain-language standards. These mandates, like those discussed above, link plain language to prioritizing and *protecting* users. Recent laws linked to finance, like the 2010 Dodd-Frank Wall Street Reform and Consumer Protection Act, make plain and easily accessible documents a requirement in many financial contexts. Schriver (2017) describes this Act as an "important step forward in assessing the quality of financial disclosures" because the Act requires user testing (p. 361). Other requirements for plain language are included in laws regulating business communication, such as the 2009 Credit Card Act, which calls for plain language in credit agreements with consumers.

Another goal in business for using plain language is consistent with Schriver's (2017) claim that plain language has become a way to build user trust. Plain language helps to advance the broader trend in business to build audience trust and personal loyalty through transparency. For example, important documents such as end-user license agreements for applications (Kunze, 2008; Willerton, 2015) or company privacy policies (Center for Plain Language, 2015) are now scrutinized for their easy access, clarity, and design. In a privacy policy analysis, the Center for Plain Language suggests that the access and language of these texts denotes whether or not a company wants users to read them, showing the trustworthiness (or not) of that company.

Within Health Fields

Across health fields, communicating complex medical information to patients and other non-experts is a perennial challenge. Unclear information can severely inhibit a patient's or caretaker's ability to understand and make decisions about health, as well as their ability to logistically navigate medical care and insurance. Research shows that using plain language can make health communication more effective, especially for those with low levels of health literacies (Grene et al.,

2017), and plain language has been framed as a tactic for patient advocacy and empowerment (Bonk, 2015).

The Federal Plain Writing Act of 2010 has far-reaching implications for the health fields through the National Institute of Health, the Department of Health and Human Services, and other health-related government offices under the purview of the Act. These offices are obligated to meet plain-language standards and have developed resources for revising and composing health communication accordingly, such as the publicly available NIH Plain Language Training and the Office of Disease Prevention and Health Promotion's Health Literacy Improvement materials (available in the chapter resources), all of which prioritize audience needs and tasks and encourage writers to work with users. Further, many documents that fall under the regulation of HIPPA law are required to be in plain language. In his research exploring ethics in technical communication, Willerton (2015) traces the uses of plain language in several medical communication-related nonprofits, stating that many of the high-stake situations in which plain language can support ethical communication are related to health. The meteoric rise of e-health sites and applications has further amplified the use of plain, accessible language in health fields.

Within the Sciences

The sciences have seen a dramatic increase in the need to make complex, specialized scientific findings more available to new and wider audiences. Scientific fields have taken up plain language and other similar strategies as a framework for making science more usable and accessible. For instance, many scientific journals now require authors to compose plain-language summaries or abstracts of their research. Scientists are, in these cases, responsible for framing and plainly communicating their work to a much bigger pool of readers, who bring with them different goals, histories, and expertise. This kind of access helps non-experts—ranging from any interested individual, experts from other fields, journalists, public officials, and others—use scientific findings to inform voting and policy-making (American Geophysical Union, n.d.). The Alan Alda Center for Communicating Science takes an interesting approach to science communication that is founded on empathy, dialogue, and personal connection—an approach that I think can help to deepen the potential scope of plain language in courses. As science plays an increasingly visible role in public life and government deliberation and legislation, effective and responsible communication becomes equally as critical in scientific disciplines.

■ Plain Language as a Strategy to Advance Social Justice

Any deep, substantive consideration of audience, especially in light of the recent plain-language movement focus on user trust (Schriver, 2017), demands that technical communication scholars confront the complex system of language and power that surrounds access to information. How is the notion of "plain" or "clear" language bound to race, class, and linguistic privilege? Does the plain-language movement support social justice? The recent turn of the technical communication field toward social justice and accountability further emboldens us to ask these questions. Natasha N. Jones (2016) defines social justice in technical communication as "critical reflection and action that promotes agency for the marginalized and disempowered" (p. 343). Insofar as plain language is intended to prioritize users and not institutions or writers, the plain-language movement may contribute to this kind of social justice work. However, few studies directly inquire into what extent plain language can promote the agency of marginalized and disempowered audiences. Plain language may offer an important strategy for advancing social justice, but if used shallowly, it may deflect attention from vulnerable audience groups or other issues of access, and it may re-inscribe existing marginalization.

A few technical communication scholars have begun to conduct this important research. Miriam Williams (2010), for example, found in a study that African American business owners felt increased trust toward a city government due to regulations written in plain language. Williams' project extended the conversation about plain language—especially in the context of government regulatory writing—into more specific histories of institutionalized discrimination and disadvantage. She revealed the work plain language could accomplish in generating trust and familiarity within historically marginalized groups. In another study, Jones and Williams (2017) consider the history of marginalization of African American homebuyers to help explore the ways plain-language ARM mortgage disclosure statements can affect vulnerable homebuyers. They found that fine-grained issues in plain-language revisions can subtly re-inscribe systematic biases, as well as reinforce the mistrust marginalized audiences already feel toward institutions. The authors call for a wider consideration of contextual and historical factors that may link to textual features and that may inhibit or promote user agency. Iva W. Cheung (2017), using cognitive load theory, argues that social justice ends may be pursued through plain language, calling it an ethical imperative. In short, the plain-language movement offers strong, publicly anchored strategies of communication that scholars and practitioners can explore as a potential way to advance social justice work, but they must constantly interrogate their practices and assumptions, always remaining alert to the way plain language may deflect attention from systemic and social issues.

■ Plain Language as a Strategy to Save Resources

An argument for plain language at play throughout the preceding overview is that it is cost- and resource-efficient. The promise of efficiency helped fuel the readability formulas, as well as the later plain-language movement. For example, in the 1940s, the U.S. federal government worried about the expanding paperwork and documentation following World War II and the New Deal, and plain writing spoke to that need (Longo, 2004, p. 167). This concern has only grown over time, and ultimately the Paperwork Reduction Act, which was passed in 1980, was heavily tied to calls for plain language. More recently, the Federal Plain Writing Act of 2010 and other regulations regarding plain writing also have been linked to saving practical resources.

The plain-language movement finds an anchor in the fact that effective communication prevents problems and saves time and money for both users and organizations. Joseph Kimble (2012), a leading plain-language expert in law and policy, devotes a large portion of his book *Writing for Dollars, Writing to Please* to fifty examples where revising in plain language saved immense resources. For instance, Kimble details the $4.4 million saved by the Veterans Administration Bureau in 1999 after they revised a single letter instructing veterans to update their life insurance beneficiaries. He points out that small, targeted plain-language revisions to documents that serve thousands, or millions, of users can save incredible resources. Kimble tracks multiple areas of saved resources: employee time necessary to complete tasks, reduced materials, and retained comprehension by users. In other examples of medical instructions, manuals, and tax forms, Kimble mentions the increased positive attitudes and decreased frustrations that coincide with plain language—changes that also connect to better bottom lines for companies and organizations. Apart from the fact that these examples serve as compelling arguments to use plain language, they provide insight into the way technical communication intertwines with the material, financial, and personnel resources of different contexts.

▍ Plain Language, Rhetoric, and Technical Communication

For those instructors inclined to intersect technical communication with rhetoric in their courses, as I am, plain-language guidelines serve as a useful, practical opportunity to do so. In this section, I offer three ways that instructors can invoke the rhetorical tradition to help deepen students' approach to plain language and technical communication. First, I offer a very brief look at the plain style in the rhetorical tradition. Second, I discuss audience as it links to clarity and plainness. Third, I suggest that instructors and upper-level students consider the turn toward user trust in the plain-language movement in terms of persuasion.

The concept of plainness has a long history in the rhetorical tradition; the plain style has been among the most durable categories over the past two millennia. Deployed for different ends across periods, the plain style was initially linked to the teaching or instructive portions of orations by Cicero and Quintilian. In English traditions, there are various accounts of the roots of plain style, including the well-known narrative of Francis Bacon, Thomas Sprat, and the Royal

Society of London during the early modern period to use plain style in science. They sought to remove ornamentation and ambiguity in order to foreground uninhibited scientific truth (see Halloran & Whitburn [1982] for further discussion). This narrative, which I only briefly touch on here, is often positioned as a precursor to technical and professional writing practices. But a counternarrative by Elizabeth Tebeaux (2004) roots technical communication in older utilitarian writing, including instructional, administrative, and record-keeping documents, among others. Unlike the scientific writing under Bacon's purview, much of this utilitarian writing was intended to be accessible and comprehensible by wide audiences and even spoken aloud. "Plainness" in this case reflected everyday speech and everyday needs. These two accounts provide only the briefest glimpse into the myriad of ways "plainness" has been deployed to meet different goals over time. Introducing even brief histories of plainness can reveal for students the idea that through plain language, we are promoting a conception of the term that embeds and conceals contemporary values within it. We can then ask students, what are those values?

A second area of rhetorical studies we may use to deepen students' understanding of plain language is the relationship between written text and audience. In "A Humanistic Rationale for Technical Writing," Carolyn R. Miller (1979) troubles the windowpane myth of language, the notion that "language provides a view out onto the real world, a view which may be clear or obfuscated" (p. 611). This approach treats style and content as discrete categories, implying that content is independent, and the goal of the writer is simply to reveal it in transparent text. Miller states, "We have not said anything very useful about the writer-reader relationship when we say the purpose of technical communication is to be clear" (p. 615). With a windowpane approach, one would only consider "the relationship between the reader and reality (and whether the reader is mentally adequate to the reality)" (p. 615).

Readability formulas and limited, rule-governed iterations of plain language might be said to rely on this windowpane theory, suggesting that a particular metric of clarity should ensure comprehension. But more robust approaches to plain language engage audiences and their tasks in much more nuanced ways, raising questions of communication design, users' goals and histories, written style. A discussion of rhetorical audience can help students approach plain language as a highly contextualized, reflexive, and, as Willerton (2015) suggests, dialogic strategy to prioritize audience.

Lastly, I see the recent move in plain language toward building user trust (Schriver, 2017) as a way to acknowledge and investigate the ways plain language is persuasive. Framing plain language in the terms of persuasion can reveal more clearly the stakeholders, socio-political implications, and assumptions about language and clarity undergirding the movement. This move is important to keep present, particularly in light of the social justice work that plain language may potentially support.

Potential Resistance to Plain Language

A key reason I engage plain-language guidelines and the movement in technical communication courses is that they are already firmly anchored in industry, government, and elsewhere. Put differently, they are a *starting point* where public attention to language and communication is already established and integrated into policies, practices, and industries. This opportunity cannot be underestimated. However, the realities of plain-language application can be fraught, surface-level, rule-based, decontextualized, and exclusively reliant on readability scores. These kinds of applications are unfortunate and can deter instructors from engaging with plain language at all. But I believe that the plain-language guidelines promoted by national organizations and the federal government reflect a public investment in rhetorically grounded strategies to revise communication and prioritize users. We can prepare our students with resources and tools to recognize and—I hope—challenge weak applications of plain language. By teaching them stronger ways to define and apply plain language, we help bolster against weak applications in the future while harnessing the public buy-in that currently exists for the movement.

Instructors may also resist plain language because it doesn't widely consider its effects on speakers of other languages. Some research shows that while speakers of Germanic languages tend to appreciate English plain-language documents, those who speak Latinate languages (French, in the cited study) may not, due to the elimination of longer Latinate words and the use of phrasal verbs (Thrush, 2001). This kind of objection also prompts us to consider what other audiences are quietly obscured by plain language and, as I mention in the section on social justice above, how plain language offers both opportunities and potential risks for marginalized groups. These concerns should be made visible in the plain-language movement as we grapple with them in our classes. These concerns also help us emphasize that involving users is a crucial step to any plain-language work.

Other objections to plain language have been routinely levied and well-addressed, such as its oversimplification of material. I encourage instructors to read through the exchanges about these objections in the resources and references in this chapter, particularly Beth Mazur (2000) and any texts of Joseph Kimble.

Applying Plain Language in Technical Communication Courses

In previous sections, I offer introductory information and examples that can be used to situate and frame plain language in technical communication courses. In this section, I offer five strategies for incorporating this material into a syllabus through low-stakes assignments that can parallel existing syllabus materials. Effective plain-language guidelines tend to parallel the common goals of introductory technical communication classes already, so it can require little work to use them to support existing syllabi. In brief, instructors can introduce the plain-language movement in the beginning of the semester, then connect each existing unit to relevant aspects of plain-language guidelines. A few applied points of connection include audience analysis, content organization, effective use of headings and document design, and usability or user experience. These skills all can be linked to steps of existing plain-language guidelines and training tools (see Example 1). The remaining examples provide guidance on other ways to take up plain language as a theoretical, interdisciplinary, and international platform.

Plain-language materials can encourage student buy-in—especially if the examples are from plain-language resources in fields students have stakes in. These materials can also help students to see where their technical communication expertise can extend plain-language practices. In other words, students can make sense of themselves as practitioners who have highly marketable plain-language skills, and they can also lean to speak confidently about how and where they offer even more as technical communication experts.

Example 1: Applying Plain-Language Strategies

Assignment Context: In conjunction with some of the assignments above, I recommend that instructors prompt students to apply plain-language strategies to real texts. These kinds of write/rewrite or before/after assignments can be done as in-class activities, more extensive high-stakes assignments, or can be incorporated into the writing of existing projects. These kinds of applied tasks not only give students practice composing and revising texts with real contexts and audiences, but they lend themselves to student portfolios later. Nearly every plain-language framework included at the end of this chapter includes practice and before-and-after examples that instructors can introduce and use in the classroom. Instructors can provide the "before" version to students and work through various guidelines, including user testing in the classroom when possible, then introduce the "after" version along with students' revisions.

Below, I offer guidance on how an instructor might use the examples from earlier in the chapter (see Figures 3.1 and 3.2), to introduce application exercises. Note that while user involvement is a crucial portion of plain-language strategies, short, in-class application exercises often preclude effective user-tests; however, students in class can still brainstorm and prepare for user tests throughout the revision process. If higher-stakes assignments take up plain language, then teachers should encourage students to engage users to whatever extent is possible in the class context.

Exercise: Plain-Language Application—Water Meter Upgrade Letter

1. In groups of 2–3, you will receive a hard copy of a water meter upgrade letter sent out to eligible U.S. residents before. Take a

few minutes to become familiar with this letter and brainstorm with your groups any readability issues that are evident to you as technical communicators. What do you notice straight away?
2. Next, read over the plain-language guidelines provided to you by your instructor. Where do the issues you identified fit in the guidelines?
3. Using the strategies, develop a plan to revise the letter in plain language. Your plan should consider content, style, organization, and document design. Feel free to take some artistic license: if you believe something is missing, invent the content you believe is necessary. You may type your revised letter or sketch the layout and (rough) text on the blank paper provided. Be prepared to explain your decisions and your reasons.
4. With the class, discuss your group's decisions, reasons, and drafted product.
5. As a class, strategize how you would involve users to test this document at various stages of development and product. By what metrics would you measure success?

These in-class applications can be used in more targeted ways as well. For instance, plain-language guidelines and style textbooks both tend to recommend strong subjects and strong verbs, so targeted sentence-level work can map on well to these exercises. "Before" examples can offer real-life contexts for written style practice in relation to real user needs or actions. Alternatively, students can focus solely on other areas, such as content organization or use of headings/subheadings. These targeted exercises also help students discover that in real communication, such areas are not actually so neat or discrete.

Example 2: Making Sense of the Disciplinary Intersections in Technical Communication

Assignment Context: Technical communication intersects many other areas of expertise, including design, usability and user experience, content strategy, web design, and others. As such, students can struggle to make sense of themselves in the job market and even within the university. I suggest here that researching and practicing plain language can help ground students in these intersections of our field with others, helping them to see the common goals and the collaborative work possibilities. In the 1990s, technical communication largely abandoned plain language due to the limited ways it was being put into practice; plain language was understood in many cases to "dumm[y] down" texts (Schriver, 1997, p. 26). Much of the work of plain language at the time was taken up in fields like information design (Mazur, 2000). Technical communication recently stepped back into the ring of plain language, but other sister fields, such as usability, have grown considerably too, and they have also developed stakes in plain language. It has become a multi-disciplinary movement. I recommend class discussions that use plain language as a microcosm to make sense of the productive intersections and overlaps between technical communication and other fields.

Exercise: Plain Language: An Interdisciplinary Platform

Please read Beth Mazur's (2000) article titled "Revisiting Plain Language" before class. Compose answers to the following prompts. Be prepared to discuss them as a class.

1. Identify each common critique of plain language and Mazur's responses.
2. Discuss the ways Mazur positions plain language within and across disciplines. How do those disciplinary boundaries seem to be constructed?
3. Develop your own rationale for why technical and professional communication offers a strong foundation for developing and applying plain-language strategies.

Example 3: Plain Language in Specific Fields

Assignment Context: As I've mapped out earlier in the chapter, plain language has been taken up in both general ways (for instance, the Center for Plain Language's five steps), as well as in specific field contexts (health, law, business, web writing, etc.). An opportunity for engaging students who have other disciplinary bases—perhaps students who are majoring in something else but minoring in technical communication—is to prompt them in low-stakes activities to explore the plain-language resources in their own fields. Asking students to apply their field-specific plain-language resources to assignments can increase their benefit and buy-in while maintaining a relatively consistent class-wide assignment for the instructor to manage. Further, asking students in an in-class activity to compare guidelines across each other's fields can yield productive discussion about unique audience needs and tasks as well as the disciplinary cultures of different fields.

To prepare for this assignment, ask students to identify their disciplines (or anticipated disciplines). Place them in loose disciplinary groupings to the extent possible.

Exercise: Plain Language in Your Discipline

First, individually perform basic web research on any plain-language activity or requirements in your field. Start with basic Googling, then focus in on specific professional institutions or specific field expectations.

Second, come together as a group to discuss, compare, and compile what you found.

Each group should post two deliverables to our class-wide discussion board:

1. Introduce your field/sub-fields and post an annotated list of resources, including any relevant links and information.
2. Short Answer: What unique strategies, content, or considerations do the field-specific plain-language resources offer compared to the more general guidelines offered by plainlanguage.gov or Center for Plain Language?

Your instructor will compile the posts and provide students a "Plain Language Across Disciplines" resource for future professional use or reference.

Example 4: Approaching International Technical Communication through the Plain-Language Movement

Assignment Context: In introductory courses, I find it challenging to discuss issues of international and intercultural technical communication with adequate depth. I offer here a way to use plain language as a touchstone to engage with a global conversation about technical communication. Dozens of countries around the world have taken up plain-language initiatives in quite different ways. Instructors can ask students to explore and compare approaches from different nations and international organizations, looking for core values and strategies. This kind of assignment can also highlight multi-lingual students in the class, as they can investigate the strategies of non-English speaking countries as well. Two resources that can support this kind of work are the Plain Language Association International, which networks over 30 countries seeking to develop plain-language policy, and Clarity International, which publishes a regular journal, *Clarity*, that focuses primarily on law. Both are included in this chapter's list of resources.

Exercise: International Technical Communication and Plain Language

To complete this discussion post, first spend 5–10 minutes exploring the Plain Language Association International (https://plainlanguage-network.org) website to gain a sense of the breadth of plain-language movements across the globe. Next, select one country's plain-language resources to explore more deeply by following the links provided on the site and by searching for others yourself. You have some flexibility in terms of the scope of your exploration. (If you are able to navigate information in a language other than English, you are encouraged to do so!)

Provide an overview of what you have found by answering the following questions:

1. What nation's plain-language resources have you selected for this activity?
2. In your selected nation, what institutions or organizations have you found that support plain-language communication?
3. What do they claim plain language accomplishes? Provide evidence (in the form of cited quotes or screenshots of the websites you explored) and explanation.
4. How does their approach to or definition of plain language compare to some of the guidelines and goals of plain language in a U.S. context, such as those we have reviewed in class? Be specific in your answers.
5. Do they provide examples? If so, please include a screenshot of at least one example.
6. Is there anything else you noticed or would like to discuss about what you found?

To aid your readers, embed all relevant links in answers to questions 1–6.

The goal of this discussion post is for students to 1) explore a specific plain-language movement outside the US, 2) develop technical communication knowledge that is relevant outside the U.S. context, and 3) collectively archive a range of approaches to plain language for students' potential future use.

Example 5: Investigating Power and "Plainness"

Assignment Context: As I describe above, the plain-language movement is connected to prioritizing and protecting users. In this way, it is an important strategy to shift power to groups who are historically marginalized. However, we can't lose sight of the fact that being "clear" or "plain" can often be conflated with concepts like "standard edited English," which is bound up with linguistic, racial, and class privilege and a long, complex history of systemic inequalities, especially in education. We need to push our students to interrogate "plain language" as an evolving framework that should be continually (re)directed to challenge long-held power structures like these. Below, I offer some in-class discussion prompts that may help students begin to think through these issues, and I recommend referring back to some of the sources listed in the "Plain Language as a Strategy to Advance Social Justice" section.

Discussion 1

The concepts of "clear" or "plain" are not objective or neutral. As a class, come up with examples that show how these terms (and other concepts commonly associated with them, like "standard" or "proper" English) are non-neutral. Instructors can bring in examples or research from our field or others that help to make this point. I've found success with excerpts from the College Conference on Composition and Communication's (CCCC) statement, Students' Right to Their Own Language (https://cccc.ncte.org/cccc/resources/positions/srtolsummary). Using each example, discuss how a robust plain-language framework guards against (or should guard against, if it doesn't) participating in a "neutral" conception of language. The goal here is to not only think about how language is non-neutral but to think about how we act on that knowledge.

Discussion 2

Using one or two sets of guidelines, identify the dimensions of communication that plain language includes (or should include) beyond surface words. Discuss in specific terms how each dimension speaks to its ability to empower marginalized users. For example, a current tenet is involving real audiences. How can this involvement happen in a deep, collaborative way that allows for empowerment and moves beyond simply testing for effectiveness? The goal here is to again use plain language as an opportunity to point to specific actions or guidelines that enact or enable the values we want to support.

Discussion 3

One of the reasons plain language is anchored so strongly across fields is that prioritizing audiences sells itself—literally. Using plain language has proven to be very resource-effective in industry, government, and elsewhere. As a group, brainstorm scenarios where interests conflict and truly prioritizing audiences could create tensions for businesses or institutions. In these important cases of slippage, what can we learn? How does (or how should) plain language navigate such slippage to maintain its integrity?

■ Conclusion

In this chapter, I have attempted to provide instructors an overview of plain language, a preliminary map of its traction in technical communication across fields, and practical strategies and resources for incorporating plain language in courses.

These serve as launching points to help instructors see plain language as a recognizable, marketable skill for graduates as well as a platform to help students grapple with technical communication's theories, practices, and values in the world.

■ References

American Geophysical Union. (n.d.). *Sharing Science*. https://sharingscience.agu.org/.

Assy, R. (2011). Can the law speak directly to its subjects? The limitation of plain language. *Journal of Law and Society, 38*, 376–404.

Bonk, R. J. (2015). *Writing for today's healthcare audiences*. Broadview Press.

Center for Plain Language. (2015). *Privacy policy analysis*. https://centerforplainlanguage.org/wp-content/uploads/2016/11/TIME-privacy-policy-analysis-report.pdf.

Center for Plain Language. (2019). *Five steps to plain language*. https://centerforplainlanguage.org/learning-training/five-steps-plain-language/.

Cheung, I. W. (2017). Plain language to minimize cognitive load: A social justice perspective. *IEEE Transactions on Professional Communication, 60*(4), 448–457.

Derthick, K., Jones, N., Dowell, R., McDavid, J., Mattern, D. & Spyridakis, J. (2009). Effectiveness of plain language in environmental policy documentation for the general public. In *Professional Communication Conference proceedings* (pp. 1–5). IEEE International.

Dreher, K. (2017). Plain language revision and insider audiences: A qualitative case study of a revised city charter. *IEEE: Transactions on Professional Communication, 60*(4), 430–447.

Greer, R. R. (2012). Introducing plain language principles to business communication students. *Business Communication Quarterly, 75*(2), 136–152.

Grene, M., Cleary, Y. & Marcus-Quinn, A. (2017). Use of plain-language guidelines to promote health literacy. *IEEE Transactions on Professional Communication, 60*(4), 384–400.

Halloran, S. M. & Whitburn, M. D. (1982). Ciceronian rhetoric and the rise of science: The plain style reconsidered. In J. Murphy (Ed.), *The rhetorical tradition and modern writing* (pp. 58–72). MLA.

Jones, N. N. (2016). The technical communicator as advocate: Integrating a social justice approach in technical communication. *Journal of Technical Writing and Communication, 46*(3), 342–361.

Jones, N. N. & Williams, M. F. (2017). The social justice impact of plain language: A critical approach to plain-language analysis. *IEEE Transactions on Professional Communication, 60*(4), 412–429.

Jones, N. N., McDavid, J., Derthick, K., Dowell, R. & Spyridakis, J. (2012). Plain language in environmental policy documents: An assessment of reader comprehension and perceptions. *Journal of Technical Writing and Communication, 42*, 331–371.

Kimble, J. (1992). Plain English: Charter for clear writing. *Thomas M. Cooley Law Review, 9*(1), 1–58.

Kimble, J. (2012). *Writing for dollars, writing to please: The case for plain language in business, government, and law*. Carolina Academic Press.

Kunze, J. T. (2008). Regulating virtual worlds optimally: The model end user license agreement. *Northwestern Technology & Intellectual Property, 7*, 102.

Longo, B. (2004). Toward an informed citizenry: Readability formulas as cultural artifacts. *Journal of Technical Writing and Communication, 34*(3), 165–172.
Mazur, B. (2000). Revisiting plain language. *Technical Communication, 2*, 205–211.
Miller, C. R. (1979). A humanistic rationale for technical writing. *College English, 40*(6), 610–617.
Plain Language Action and Information Network. (n.d.). *History and timeline*. PlainLanguage.gov. https://www.plainlanguage.gov/about/history/.
Ramchandani, T., Chisnell, D. & Quesenbery, W. (2017). *Designing election systems for language access*. Center for Civic Design. https://civicdesign.org/wp-content/uploads/2017/07/Language-Access-V10-17-0710b.pdf.
Redish, J. C. (2000). Readability formulas have even more limitations than Klare discusses. *ACM Journal of Computer Documentation, 24*(3), 132–137.
Ross, D. G. (2015). Monkeywrenching plain language: Ecodefense, ethics, and the technical communication of ecotage. *IEEE Transactions on Professional Communication, 58*(2), 154–175.
Schriver, K. A. (1997). *Dynamics of document design: Creating texts for readers*. Wiley.
Schriver, K. A. (2000). Readability formulas: What's the use? *ACM Journal of Computer Documentation, 24*(3), 138–140.
Schriver, K. A. (2017). Plain language in the US gains momentum: 1940–2015. *IEEE Transactions on Professional Communication, 60*(4), 343–383.
Selzer, J. (1983). What constitutes a "readable" technical style? In P. Anderson, R. J. Brockman & C. R. Miller (Eds.), *New essays in technical and scientific communication: Research, theory, practice* (pp. 71–89). Baywood Publishing Company.
Steinberg, E. (1991). Introduction. In E. Steinberg (Ed.), *Plain language: Principles and practice*, 1–18. Wayne State University Press.
Summers, K., Chisnell, D., Davies, D., Alton, N. & McKeever, M. (2014). Making voting accessible: Designing digital ballot marking for people with low literacy and mild cognitive disabilities. *USENIX Journal of Election Technology and Systems, 2*(2), 11–33.
Tebeaux, E. (2004). Pillaging the tombs of noncanonical texts: Technical writing and the evolution of English style. *Journal of Business and Technical Communication, 18*(2), 165–197.
Thrush, E. A. (2001). Plain English? A study of plain English vocabulary and international audiences. *Technical Communication, 48*, 289–296.
US House, 111th Congress. *Plain writing act of 2010*. http://www.gpo.gov/fdsys/pkg/PLAW-111publ274/pdf/PLAW-111publ274.pdf.
Willerton, R. (2015). *Plain language and ethical action: A dialogic approach to technical content in the 21st century*. Routledge.
Williams, M. F. (2010). *From black codes to recodification: Removing the veil from regulatory writing*. Baywood Publishing Company.

■ Appendix. Web Resources for Plain Language

Alan Alda Center for Communicating Science: https://www.aldacenter.org/.
Center for Civic Design: https://civicdesign.org/.
Center for Civic Design—Language Access: https://civicdesign.org/projects/language/.
Center for Plain Language: https://centerforplainlanguage.org/.

Clarity International: https://clarity-international.net/.
Compiled Style Guides for Government Administrations and Agencies: https://www.plainlanguage.gov/resources/guides/.
"Health Literacy" (Office of Disease Prevention and Health Promotion): https://health.gov/our-work/health-literacy.
NASA Headquarters Library: https://www.hq.nasa.gov/office/hqlibrary/pathfinders/edusci.htm#web.
NIH Online Plain Language Training: https://www.nih.gov/institutes-nih/nih-office-director/office-communications-public-liaison/clear-communication/plain-language/training.
Plain Language Action and Information Network (PLAIN): https://plainlanguage.gov.
The Plain English Campaign (UK): http://www.plainenglish.co.uk/.
Plain Language Association International: https://plainlanguagenetwork.org/.
Plain Writing at the National Archives: https://www.archives.gov/open/plain-writing.

4. (Teaching) Ethics and Technical Communication

Derek G. Ross
AUBURN UNIVERSITY

Abstract: Ethics helps us make supportable decisions and explain those decisions to others. In this chapter, I discuss the role the study of ethics and ethical models play in helping us get at the ways ethical decision-making can inform our thought processes, thereby offering support for decision-making and consideration of the ways that the decision-making process shapes actions and outcomes. I discuss models such as Aristotelian, Kantian, utilitarianism, feminist, and ethics of care approaches. I consider how we might approach teaching and discussion of ethics in the classroom and offer an overview of many different approaches to ethics, including environmental ethics, different feminist approaches, and social justice models. The chapter uses a central scenario as a way to look at how different models enable different ways of problem solving and decision-making, ultimately arguing that an understanding of ethics opens the possibility of finding new ways of thinking and knowing in the classroom, in the workplace, or in research.

Keywords: ethics, decision-making, ethics of care, feminist ethics, social justice

Key Takeaways:

- Understanding how and why we make decisions allows us to more effectively communicate our decisions to others.
- Ethics-based decision-making is not a way to find a "right" answer but instead helps us to define "right" based on agent, action, recipient, and consequence.
- Ethics-based decision-making gives us a way to creatively solve problems and explore different possible outcomes and consequences.

As a field of study, technical and professional communication engages with ethics in deeply meaningful ways.[1] To teach ethics in the technical communication classroom, however, is no easy feat, nor is applying ethics in the workplace. We can teach or apply certain ethical moves, such as writing with inclusive language or considering accessibility in design, but getting at the complexity of ethics as it relates to the way we make decisions, and how those decisions might change as our ethical thinking changes, can be difficult.

---

1. Pieces of this chapter have been published in *Intercom Magazine* (Ross, 2017b) and *Mother Pelican* (Ross, 2012).

In this chapter, my hope is to help teachers of technical and professional communication help their students (and help students help themselves) start to get at the complexity and value of ethical thought in an accessible manner. While other chapters and books on ethics in technical communication often move quickly into rhetorical and theoretical complexity, my goal here is, instead, to get at the ways ethical decision-making can inform our thought processes, and thereby offer some support for decision-making and consideration of the ways that the decision-making process shapes actions and outcomes, whether in the classroom, in the workplace, or in research.

We have a growing body of scholarship on ethics in technical and professional communication that can help us navigate the complexity of ethics-based decision-making. Scott P. Sanders' (1997) chapter, "Technical Communication and Ethics," in Katherine Staples and Cezar M. Ornatowski's *Foundations of Teaching Technical Communication*, the spiritual predecessor to this volume, for example, offers a general overview of types of ethics. Sanders argues for three models: practical, philosophical, and rhetorical. He associates practical ethics with rules-based business ethics; philosophical ethics with a general, theoretical, understanding-problems approach; and rhetorical ethics with a postmodern model mixing construction and presentation of ethos, understanding of audience, and use of ethics, in general, as a model for analysis-writ-large. Texts like Paul M. Dombrowski's (2000a) *Ethics in Technical Communication* and Mike Markel's (2001) *Ethics in Technical Communication* address various theories of ethics and cases to which we might apply ethical thought to come to consensus with others on what we might consider "right" action, and, at this point, most, if not all, of our technical communication textbooks address ethics in some way.

The role of the technical communicator is increasingly expanding, and as roles expand, the decisions we make, or even now have the ability to make, take on more ethical weight. From transmitters of information to articulators of information (Slack et al., 1993), from information designers (Carliner, 2001; Redish, 2000) and information architects to experience architects (Potts & Salvo, 2017; Salvo, 2014), technical and professional communication is diverse, and how we identify ourselves and our profession is ever-changing. We identify as writers, editors, authors, teachers, researchers, user-experience experts, and more, and the methods we use to conduct our work are similarly diverse. We rarely work alone, however, and, as many authors have pointed out, ours is a profession that calls for collaboration (see, for example, Frith, 2014). Because working with information involves so many variables, such as determining origins of information, intent of the communicated information, and the impact information has on society, on top of the job of negotiating others' roles and involvement, an understanding of ethical theories, principles, and practices is increasingly important. We have more productive communication when we can see another's point of view, and we can produce more ethical communication (working alone or in groups) when we can clearly articulate our reasoning and desired outcomes. Ethics helps us make supportable decisions and explain those decisions to others.

■ The Basics of Ethical Decision-Making

Clear understanding of our actions allows us to communicate our reasoning to others. Following both Dombrowski's (2000a) and Markel's (2001) focus on decision-making—both authors begin their books by discussing how ethics ultimately shapes the way we make decisions—I argue that if we can teach nothing else about ethics in the technical communication classroom, we should at least show how a firm understanding of *why* we make decisions allows us to support our reasoning to both ourselves and others, which ultimately can make us more effective, insightful communicators. If we ourselves do not fully understand how we come to decisions, we are unlikely to be able to convince others to support our decisions or judgements in similar situations (Dombrowski, 2000a; Markel, 2001). This focus on setting standards (and defending them) means that when we make ethical decisions, we are making *normative decisions*.

A normative decision is one which makes an argument towards how things ought to be. Normative decisions guide our actions and seek agreement from others. So, given a simple situation, I might make an ethical judgement that I suspect most of us can agree with and say that "stealing is wrong." Rephrased, I can make an action-guiding statement and say, "Do not steal." Rephrased again, I can seek your agreement: "I think we can all agree that we should not steal." I have now made an ethical (normative) decision—not stealing, and agreeing that we should all not steal, becomes an action-guiding, agreement-seeking ethical principle.

Ethical situations generally involve four components: a moral agent, an action or series of actions, a recipient, and consequences. The agent takes action, the recipient receives consequences. Ethics comes into play when we consider what actions are appropriate to take in given circumstances and what consequences are justifiable for recipients of actions—even, in many cases, who or what we will even consider as a recipient for action.

Different ethical approaches privilege different elements of this decision-making equation. Virtue ethics, for example, relate to the agent's (or action-taker's) moral character. Deontological ethics refer to ethics that consider an agent's duties or obligations in any given scenario, and consequentialist ethics focus on the consequences of action.

Who or what is considered a viable recipient-of-action in any ethical equation also matters. In anthropocentric ethics, only humans have moral standing. In non-anthropocentric ethics, non-humans can be a part of that agent-action-recipient-consequence chain. Non-anthropocentric ethics takes at least three basic forms: zoocentric ethics assigns moral standing to all animals; biocentric ethics assigns moral standing to all living things, including plants; and ecocentric ethics assigns moral standing to ecosystems (communities of organisms in conjunction with non-living components like soil, air, and water). "Moral standing," then, becomes an important part of the way we think about ethics. If we agree to consider something in any part of an ethical equation, we have granted it (a fellow human, a dog, a tree, the air we breathe) moral standing. Andrew Kernohan, author of *Environmental Ethics*, succinctly defines moral standing by arguing that "if we must consider [a thing] or its interests for its own sake when we are making an ethical judgement," then we can consider that thing "morally considerable" (2012, p. 8).

Designating something as "morally considerable" is an important part of ethical decision-making because doing so means that we have agreed to build that morally-considerable thing into the fabric of our decision-making, agreed to make that morally-considerable thing an integral part of society (which can begin to envelop non-human components under various ethical models).

Morals are different from, though inextricably related to, ethics. In short, morals are concerned with how one situates oneself within society. Markel notes, for example, that "morality refers to a society's set of beliefs and mores about appropriate conduct" (2001, p. 28). Put another way, we can all agree that there are set expectations surrounding us regarding the way we conduct ourselves in public, in the workplace, in particular social settings, and more. Those always-surrounding-us belief systems are morals. "A person," Markel argues, "does not formulate his or her own morality; the morality of the society or culture already exists when that person is born, and that morality does not await the individual's approval or disapproval" (2001, p. 28).

Morals are societal. Ethics, on the other hand, are individual, though they may be socially constructed and agreed upon, and may lead to social action—a society's code of ethics, for example, such as that offered by the Society for Technical Communication (STC, 2020), offers guidelines for individual action and decision-making within the context of a larger organization. If society's morals suggest a particular course of action, following that course of action does not generally take much conscious thought. I wake up, eat breakfast, then leave the house to go to work. In all the things I do in the morning, I do not stop off at the store and steal a loaf of bread and some cheese for lunch. Not stealing, being a societal agreement, is part of our society's shared morality. If, however, I am starving, and my family is starving, and I have no immediate means of compensation and do not know where to turn for help, I might decide to steal that loaf of bread and some cheese. Such a choice falls under the purview of ethics, as there is now a situation (agent, action, recipient, and consequence) that conflicts with morality, but might, individually, be supportable. That we are all not likely to agree on the "right" choice of action without further argument and positional support works to highlight this scenario as one based in ethical decision-making. Any situation that involves agent, action, recipient, and consequence could potentially be an ethical situation. Stealing is an obvious example to work with when we start to think about ethics, but this easily translates into situations more in keeping with what technical communicators might encounter on the job: issues of copyright infringement and plagiarism, for example, which, really, are still just about stealing (theft of intellectual property).

Once we have established the basics of an ethical scenario, we might wish to begin to add complexity. For example, we might consider the agent's (the action-taker's) duty in an ethical equation by looking at indirect and direct duties. An indirect duty to a nonhuman is a duty owed to a human, and a direct duty to a nonhuman is a duty directly owed to that nonhuman. Put simply, if I can pollute your lake (let us say my company is directly upstream from you) but do not because you do not want me to and I have told you I will not, I am following an indirect duty. I did not pollute the lake because of the way I feel about the lake but because of our human-human agreement. If I can pollute your lake but do not, even though you have told me I can (perhaps because I think the lake is better off unpolluted), I am following a direct duty. It does not matter what you (another human) say. If I believe that I have a direct duty to a nonhuman (a tree, a lake, the environment-writ-large), I have assigned it moral standing, written it into the complexity of our society. Knowing where duties lie—and being able to articulate that knowing to others—allows an agent to make supportable, duty-based decisions. If I believe a lake has moral standing and I owe it a direct duty, then I can tell someone that I refuse to engage in actions that pollute the lake, even though our company might profit. Duties, as a decision-making heuristic, of course, extend far beyond the environmental. If I believe that all intellectual property is valuable, then I might decide that I have a direct duty to that concept and then can always support my decision not to plagiarize another's writing, music, art, photography, etc., even if that intellectual property is owned by a company I do not value or agree with. Assessment of where duty lies allows me to make (and support) an ethics-based decision.

This leads us directly into issues of value: when I make decisions based on action and consequence, I might consider something's instrumental value (its ability to cause value either through trade, sale, negotiation, etc.) or its intrinsic value (the belief that whatever I am considering has value no matter what I do with it). All of this—and much more—is why any theoretical discussion of ethical principles and values can get complex quickly. These elements, our consideration of agent, action, recipient, and consequence; our consideration of to whom or what we assign moral standing; our consideration of duty, or perceived duty; and our consideration of value, and how we assign it in any given instance, offer us complex ways to address problems. When confronted with a difficult situation at work (a co-worker who takes credit for your work, for example) or when thinking of how to solve difficult issues in the world (pollution, immigration policy, gun control, etc.), even the practice of building these ethical equations can start to help us interrogate how and why we are reaching decisions and making conclusions. Our action-guiding, agreement-seeking, normative decisions become potentially more supportable because we can work through the complexities of the decision-making process with some detail, and because we can clearly identify the components of any given argument.

■ An Introduction to Ethical Models

Ethics-based decision-making asks us to apply ethical models to ethical scenarios in order to establish a supportable course of action. These models are heuristics: ways to approach a problem that suggest courses of action without guaranteeing optimal results. They are not ways to find a universal "right" answer, as, arguably, such a thing does not exist. "Right" action is action designated as "right" given context. Instead, ethics-based decision-making in some ways *defines* "right" *through* audience, purpose, and context by working through the agent(s), action(s), recipient(s), and consequence(s) of a scenario.

I have used a variation of the following scenario for years as a way to get at the complexities of different ethical models, and, as simple as it may be, it has the benefit of letting us see how various components in any given situation work. Here is the situation: I am a university professor, and I am on my way to teach a class. I am running late and am forced to park across a busy street from the building where I meet with my students. There are 20 students in the class, all of whom have busy life/school schedules. By school policy, they are mandated to wait 15 minutes for me to show up, then they are free to leave. By social construction, they will most likely wait until one brave soul packs up and leaves, then everyone else will leave. I have roughly two minutes to get to class by the time I park my car. Given no obstacles, I can make it to my classroom within a minute or two of the official start time. So here is the situation: As I run up to the intersection to cross the street, I see an older woman with her arms full of bags also getting ready to head across. Do I help her across the road?

First, we need to establish how even the perceptual components of an ethics-based scenario work. Please understand that the scenario construction here is deliberate: perceived age, gender-identification, race, religion, political affiliation, ability, and more often play into the way we interact with each other, sometimes subconsciously. One of the strengths of ethics-based decision-making, particularly in a field dedicated to understanding how interlocutors and multiple publics interact, is considering how perception impacts action. So, here, my (the agent's) identification as "male" and my perception of the recipient's identification as "older" and "female" have a place in the way these models play out. When I teach this scenario in the classroom, I move from model to model, showing how each model creates different ethical tensions and, ultimately, ethics-based decisions. The models move from *Aristotelian* to *Kantian* to *utilitarian* to *feminist*, then into *ethics of care*, ultimately moving to then discuss other models and how they might shape the decision-making process as well. In each case, I remind students of the general scenario, then we apply that model's decision-process to the scenario.

Aristotelian Ethics

Aristotelian ethics are generally considered as virtue-driven and rule-based and are derivative of Aristotle's (384–322 BC) predecessors Socrates' and Plato's models. In this system of thought, the decision-maker's perspective is concerned with such concepts as goodness, truth, justice, and rightness. In virtue-driven, rule-based decision-making, one determines the most virtuous of possibilities from decision-making options and then chooses that outcome, regardless of outcome or personal backlash. Virtue—according to Aristotle—is "concerned with emotions and actions, and it is only voluntary actions for which praise and blame are given" (1975, p. 117). Once virtue in a given situation has been established, a personal ethical rule is created. Should a similar decision-making choice arise in the future, the decision-maker can simply follow the previously created rule.

In *Nicomachean Ethics*, Book II, Chapter VII, for example, Aristotle lists 12 individual virtues of character: Courage, also called bravery; Temperance; Liberality, also called generosity; Magnificence; Greatness of Soul, also called magnanimity; a nameless virtue concerned with appropriate concern for honor, defined in excess as ambition and in deficit as unambitious, where the virtue lies in the middle; Gentleness, also called mildness; Truthfulness; Wittiness; Friendliness; Modesty, or proneness to shame; and Proper, or righteous, Indignation (Aristotle, 1975, pp. 97–105). If I view myself as being virtuous of character and associate "friendliness" with being of good character, I might decide that the appropriate, friendly thing to do in our road-crossing scenario is to offer help. When I stop and offer help, I set precedent (create a rule). In the future, I need not stop to weigh the various components of this type of perceived ethical situation. I have established a rule that helping someone across the street that I read as needing my help is the right thing to do. That is the important catch here, however, and one we will come back to: I have established a virtue-based rule determined against my own internal perceptions of who or what is deserving of help without taking any other steps.

Kantian Ethics

Kantian ethics (from Immanuel Kant, 1724–1804) is an extension of Aristotelian ethics we can mark as situational, rule-based, motive-driven decision-making. Kant's decision-making process is governed by his overarching categorical imperative: that, simply put, one is duty-driven to base actions in relation to universal rightness and goodwill. Dombrowski sums up Kant's imperative as follows: "Act in such a way that, if you had your way, the principle guiding your actions would become a universally binding law that everyone must act in accordance with (in relation to you), applying to everyone, everywhere, and always, without exception" (2000a, p. 49). Kant's process differs from the Aristotelian approach in that both situation and guiding principles play a significant role in the decision-making process. If a choice appears in an ethical question where, given the situation, one can maintain pure motives (not acting out of greed, for example), regardless of the apparent good of the action itself, then that should be the decision-maker's choice.

In our scenario, I might make a similar choice to that made under Aristotelian ethics. Since under Kantian ethics one is duty-driven to act in good-will toward others, I could choose to help the woman cross the road—unless, of course, my motives are impure, or there is no real need. If my choice to help her cross the street is motivated by my knowledge that there is a group of students watching, I might decide that my actions could be entirely self-serving, therefore not universally-binding, thereby unethical. Or, simply, there might be no traffic. The situation might not warrant action. If there is traffic, and I determine my motives to be pure, however, off we go.

At this point you should be asking an important question: Namely, what if the recipient in our scenario, described here as an "older woman" does not want my help? How do elements like perceived gender identity, age, ability, and more figure into our decision-making process? What about the other part of the equation, namely, the students I mentioned I was on my way to teach? The next ethical models begin to get at these elements, leading us to question how culture and context fit into ethics-based decision-making.

Utilitarian Ethics

Utilitarianism, which can be traced to the writings of Jeremy Bentham (1748–1832) and John Stuart Mill (1806–1873), is often described as seeking the greatest good for the greatest number and is often referred to as cost-benefit analysis. This approach seeks to quantitatively assess—to the extent such a thing is possible—"good" vs. "bad" decision outcomes in relation to the number of elements involved. One problem here, of course, is that many views consider only the number of humans involved, a view with particular ethical connotations when we attempt to use cost-benefit analysis to assess ethical choices in relation to human vs. environment situations. In completing a cost-benefit analysis, value must be assigned to inputs and outcomes. As Claire Andre and Manuel Velasquez note, however, "it's often difficult, if not impossible, to measure and compare the values of certain benefits and costs" (2014, para. 8). How much is time worth? What is the value of a life? As opposed to the Aristotelian and Kantian models, which are concerned with the moral validity of a choice itself (a deontological approach, from the Greek "obligation" or "duty"), utilitarianism is primarily concerned with outcomes, with the consequences of any given action.

As Andrew Kernohan (2012) explains, utilitarian ethics have four aspects: They *cause* the *maximum total utility*. That is, causation is concerned with consequence, and consequences are considered in terms of total consequences counted for all affected recipients with regards to the consequence's utility, taken in aggregate. That means that we do not consider value to one side as being of greater or lesser importance over value to another side, just total utility gained or lost as it applies to all considered. Note that this is not best possible outcomes for all recipients but instead a computation of total utility.

As with all the models presented here, there are more in-depth explanations that get into complex issues of definition. In this case, for example, given that utilitarianism's goal is maximum total utility, how we define "utility" is important. If we define "utility" as "pleasure," for example, we are working with "hedonistic utilitarianism" and are concerned with achieving maximum pleasure for the maximum number of people. If we define "utility" as a satisfaction of wants and desires, however, we are working with what is commonly referred to as "preference-satisfaction utilitarianism." Our goal becomes working out how best to achieve a model of life that leaves the least number of people unsatisfied. In general, however, the model addressed here, of "greatest good for the greatest number," works to show how the way we think about those affected by an ethical scenario shapes our decision-making.

In our street-crossing scenario as viewed through the lens of a general utilitarian ethic, I might stop to ask myself who potentially benefits from my actions and what the potential costs might be. If I help the older woman across the street, she benefits. My students, on the other hand, all 20 of them, might leave before I could then make it to class. They would be out of a class, and our class would get behind schedule. There are two ways to think about this. Under the first model, we might assume that students care about the money they spend, or the money spent on them, to attend class. They would have wasted their time travelling to class on this day, and they would not be getting their money's worth. At my institution, according to our 2019–2020 cost of attendance tables, resident undergraduates can expect to pay roughly $10,000 for 12 hours of tuition and fees, plus books, transportation, supplies, and miscellaneous expenses. That comes out to roughly $833.33 per class hour, or $2,500.00 per class. In the fall, we are generally scheduled to meet 29 times, so each class costs approximately $86.20 per student per meeting. At 20 students per class, I've wasted $1,724.00 if I am late, and they leave—more, if my class includes non-resident students. Calculated this way, the greatest good for the greatest number lies in me ignoring the older woman who may need help crossing the road and running to class. If I do so, I maintain class momentum and protect student investments.

Under the second model, however, we might assume that a student's happiness will be increased by an unexpected day off and that this unexpected happiness-boost far outweighs the hypothetical $86.20 being spent for each class session. Under this model, the woman benefits if I stop to help, *as do all 20 of my students*. *This* may be the greatest good for the greatest number. How I define "utility," then, becomes a critical factor in assessing the ethicality of my actions and choices.

Feminist Ethics

Feminist ethics offers an alternative to the (White) male-dominated discourse which comprises the bulk of the history of ethical interrogation. Constructed as an alternative approach to male-dominated academic and scientific discourse, third wave (and beyond) feminist ethics asks us to consider our decision-making in relation to repercussions and perceived social hierarchies. Under a feminist consideration of ethics, we should avoid making decisions based solely on traditional models of authority, the desire for control or subjugation, or gendered stereotypes. Additionally, decisions should be based in an awareness of how our actions ultimately ripple outward to others. While this model of ethics often seems quite complicated, Gesa Kirsch notes that

> Ultimately, we have to learn to make political and ethical choices. These choices always entail risks—risks clarified by postmodern, postcolonial theories. We risk misrepresenting others (it is not a question of whether, but how much), we risk speaking for those who do not wish to be spoken for, and we risk speaking in voices that silence others. All this despite our best intentions. . . . But let me stress that such risks should not lead to intellectual paralysis. (1999, p. 63)

Under a feminist model of ethics, we strive to more carefully relate our decisions to our perceptions of virtues and outcomes, and an awareness of how our choices affect others. We should be very aware of how power is ascribed to us by society, by place, and by position, and, not conversely, but synergistically, how power is ascribed to those we consider of moral value. And, to add to that, equally aware of how and why moral value is assigned in the first place, and aware, if not hyper-aware, of gaps in the assignation of such value.

Under a feminist ethic, I should be aware of decision-making repercussions and social hierarchies, both real and perceived. Quite simply, considering culture and being Southern, the first question I should ask in our scenario might simply be "Excuse me, Ma'am, do you need help getting across the road?" If I ask *because I identify as male* and *because she is an older woman*, however, I'm already in a difficult situation. In fact, my typification of her as potentially needing help already creates a situation where I have removed power. I could start to remedy the situation, then, by rephrasing my question to "Excuse me, Ma'am, *would you like* help getting across the road?"

Under a feminist ethic, I need to think outside of stereotyped roles, particularly those which establish male/female power discrepancies. If I remove all outside elements, and my fellow human needs, and wants, my help, then off we go. Even though I have 20 students paying money for my time, a feminist ethic asks that I consider repercussions of my choices as well as my reasoning. For example, is there any decision-making calculus which warrants leaving a fellow human in potential danger? A feminist perspective also asks me to move outside of my preconceptions. What if she is in no danger at all but would like help? What if I would like to help her? Kindness, putting another human's needs before my own, might supersede any other expectations of this scenario. I might also bring my students back into the equation as well: we have an important relationship here, one where obligations—professional, personal, and institutional—are at play (I emphasize obligation and constructed relationships here, as status-oriented decision-making is problematic in this case). The key point here, as it relates to the overarching lesson of thinking through decision-making strategies and the way we explain them to others, is that a feminist approach does not follow the rules-based, hierarchical, often patriarchal, models established by so many other models. Instead, decision-making should engage participants as complex humans, not artifacts. Regardless, any decision-making in a feminist ethic should be based on communication and, many would argue, care, which leads to the last model I consider here.

Ethic of Care

An ethic of care, which has also been referred to as "feminine," or "feminist," ethics, further complicates feminist reconsiderations of repercussion and hierarchy by asking decision-makers to show caring concern for all involved parties. An ethic of care is not rule-bound. Unlike Kant's Categorical Imperative, each action must be context-based, and contexts are immense and multi-faceted.

Though the ethics of care contains many voices, those most often associated with this approach are authors such as Carol Gilligan, whose *In a Different Voice* (1982) drew attention to differences between masculine and feminine approaches to problem solving, and Nel Noddings, who argues for a one-caring/cared-for relationship where one "reaches out to the other and grows in response to the other" (2003, p. 81). As ethicist Ruth Groenhout describes in her synopsis of care ethics, it is a model of ethics built from feminist ethics but concerned "not so much by innate or essential gender differences but by the different social location of women in the particular social and historical circumstances found in contemporary American life" (2003, p. 3). She notes that

> One of the central strengths of care theory is its ability to identify gaps in traditional accounts of ethics that may be partially caused by the social location of the theorists who have traditionally done philosophy. When theorists who are largely male, upper-class, and single think about their own ethical experience, they do not note the extent to which they are located within caring relationships. (2003, p. 6)

Groenhout continues, noting that "care theory emphasizes the extent to which we are all dependent on communal and social structures for our existence and our lives, and also emphasizes the extent to which we cannot leave this dependence of our analyses of ethical issues" (2003, p. 24). Under this ethical model, the decision-maker does not privilege the virtue of a decision over the outcome, or weigh costs and benefits, but strives to act in a way which shows caring concern to all involved parties—no one "wins," no one "loses." Instead, the decision-maker ("one-caring," in this model) explores alternative pathways which potentially ameliorate majority/minority, win/loss structures.

Groenhout argues that an ethic of care can be likened to Martin Buber's argument for relationships which value the other, what he terms as "I-thou" relationships, and notes that "human lives . . . are not the lives of disconnected, discrete rational egos, but rather the lives of fundamentally interconnected social beings" (2003, p. 17). Under an ethic of care, I make decisions based on context, circumstance, and the participants in any given ethical scenario by considering how the participants (agent[s] and recipient[s], now contextualized as one-caring and cared-for) relate, or could relate, to each other.

For our scenario, an ethic of care would build on the decision-making scenario established through consideration of feminist ethics. If everything about the situation suggests that the person I see about to cross the road truly needs help, but I truly can't spare the time, then I look—quickly—for alternatives. Simply stopping another passerby to ask if they can help might be an option, as might offering to carry the other's load, so that we both make it across the street, her safely, though perhaps without my full attention, myself perhaps more slowly than usual, but still while helping my fellow human. It asks, once again, that I really interrogate those labels identifying gender as a reason to make decisions and start making decisions based on deeper considerations of care for an(other).

■ Ethics in Research and Application

As I hope the previous sections show, how we make and justify decisions forms the backbone of ethics-based decision-making. In the justification of our decisions, we are setting and defending standards that we hope others will follow. In the workplace, this sort of ethics-informed decision-making process can lend credence to our actions, helping us to model desired behavior, argue for social justice, and explain how design choices and rhetorical structuring influence user behavior and ability. In the design and conduct of research, ethics-based thinking allows us to think through research questions to get at complex levels of participant/observer engagement (that often then influence the way we operate in the workplace). Using different ethical models to think about the different ways agent(s), action(s), recipient(s), and consequence(s) interact, or *might* interact, offers technical communication researchers powerful ways to discover and describe our world.

In *Plain Language and Ethical Action* (2015), for example, Russell Willerton develops and applies what he terms a BUROC (Bureaucratic, Unfamiliar, Rights-Oriented, and Critical) model to identify and analyze "situations in which plain language supports ethical action" (p. xv). His model draws heavily from dialogic communication ethics, which considers Martin Buber's depiction of the "narrow ridge," a "place from which people in a dialogue genuinely listen to each other and remain open to the others' persuasion" (p. 44), as a way to establish both research method and process. Similarly, Jared S. Colton and Steve Holmes (2018) re-envision, and, in many ways, re-invigorate, virtue ethics as a research tool in their book *Rhetoric, Technology, and the Virtues* by updating Aristotle's framework of *hexeis*, "the cultivated bases for orienting oneself toward virtuous activity in varied circumstances" (p. 12) to consider the ways we engage with digital technologies. Willerton's book looks at issues like civic design, federal rules of evidence, and the way complex communication organizations work together, and Colton and Holmes' book considers such issues as digital sampling and remixing, and generosity in social media. Both are firmly based in ethics, not just as a way of thinking, but literally as a way of framing problems and researching solutions.

The field of technical and professional communication has embraced ethics as a systematic model of program development, both in the way we think about our research and in the ways we think about each other. Consider, for example, the conference proceeding titled "Social Justice in UX: Centering Marginalized Users" (2018). This proceeding serves as a valuable artifact for those of us interested in the way ethics-based thinking shapes research, as it places nine scholars in technical communication—Emma Rose, Avery Edenfield, Rebecca Walton, Laura Gonzales, Ann Shivers McNair, Tetyana Zhvotovska, Natasha N. Jones, Genevieve I. Garcia de Mueller, and Kristen Moore—in conversation about the way human-centered design may "intentionally or unintentionally" push "certain types of people" to the margins. It evolves *from* ethics-based decision-making because it argues for a way of thinking that we should all adopt. It seeks normative agreement on deeply important human-rights issues.

This same sort of agreement-seeking can be seen in much of our scholarship. Rebecca Walton and Sarah-Beth Hopton (2018) argue for consideration of non-Western rhetorics and the value of unity-seeking in "'All Vietnamese Men are Brothers': Rhetorical Strategies and Community Engagement Practices Used to Support Victims of Agent Orange"; Derek G. Ross, Brett Oppegaard, and Russell Willerton (2019) argue for a model of ethical thinking for technical and professional communicators which hybridizes Aldo Leopold's land ethic, Martin Buber's narrow ridge, and anticipatory technology ethics in "Principles of Place: Developing a Place-Based Ethic for Discussing, Debating, and Anticipating Technical Communication Concerns"; and Jared S. Colton, Steve Holmes, and Josephine Walwema (2017) reinvigorate care ethics by foregrounding Adriana Caverero's concept of vulnerability in their examination of documents produced by the collective Anonymous in "From NoobGuides to #OpKKK: Ethics of Anonymous' Tactical Technical Communication."

It is not my intent here to produce a full literature review of work on ethics in technical and professional communication, but it is worth noting the breadth of scholarship our field has produced on ethics, as these pieces have shaped not only our field, but any organization that hires our students. Our scholarship includes such pieces as Steven Katz's (1992) examination of technical documentation, expediency, and the Holocaust; Wanda Martin and Scott Sanders' (1994) consideration of ethics and public policy in the classroom; Nancy Allen's (1996) consideration of how electronic technologies allow us to mediate truth (which Jonathan Buehl extends in his own consideration of ethical rhetorics of scientific image-making in 2014); Sam Dragga's (1999) examination of Confucian ethics; Brenton Faber's (1999) critique of intuition in the role of ethical decision-making; and Paul Dombrowski's (2000b) rich synthesis of approaches to ethical thought. Sam Dragga and Dan Voss' (2001, 2003) work on ethics in visuals remain a staple in many of our classes, and Mark Ward's (2010) work on information design and the Holocaust extends many of Katz's ideas to account for "naturalized authority" (p. 60). My own work has included considerations of ethics and plain language (Ross, 2015); the role of ethics, culture, and artistry in scientific illustration (Ross, 2017a); and, with Marion Parks, mutual respect in an ethic of care (Ross & Parks, 2018), along with the piece I briefly described earlier on a hybrid place-based ethic for technical communicators (Ross et al., 2019).

The ideas discussed here can serve as jumping-off points for discussion in the classroom, and activities engaging the various ethical models result in often robust (and in some cases, impassioned) discussion. For example, in my own classrooms, both undergraduate and graduate, I often use a version of Lawrence Kohlberg's Heinz Dilemma (Gilligan, 1982; Kohlberg, 1971) to set up discussion of how perception alters potential action. The exercise works as follows. I first introduce students to the Heinz Dilemma. In short, the dilemma is a scenario where a man's wife is dying, and the chemist of their small town has a potential cure. Unfortunately, however, the druggist is asking more for the potential cure than the man (Heinz) can afford, so Heinz breaks into the chemist's office and steals the drug. The question is then asked of the class, "Should Heinz have done this?" Initially, it's a fairly simple set-up, but I ask students to commit to an answer, then we tally the vote to determine how many students in the class think "yes, Heinz should steal the drug," and "no, Heinz should not." We then discuss the justifications for their choice. Our initial discussions focus on issues of legality (Is stealing ethical, if not legal, when a life is at stake?), fairness (Shouldn't the chemist just charge less for the drug?), and even love (Should family always come first in all things?). Even this initial discussion can go on for quite some time, and leads us into issues of capitalism, profit, well-being, community, and more. Then, however, the dilemma begins to change. As with Kohlberg's original version, which he used to assess moral development, we start to add variations and ask questions: Would it matter if Heinz had been cheating on his wife? If she had been cheating on him? If the chemist was independently wealthy? If Heinz is a member of the police force? If the chemist is a member of the clergy? If, if, if, and so on. Variations can include everything from social status, gender and sexual identification, and religious issues to elements that get at capitalism vs. socialism and more. Each time, we take a new vote and tally responses, noting along the way how sometimes seemingly simple perceptual differences can lead to very different perceptions of "right" and "wrong."

My use of the Heinz dilemma in class is not unique—I know many who use it, and a simple online search shows many variations on what I have described above. It is an effective introduction to ethical thinking, however, and I have found that by tailoring the questions to class intent (in some classes we focus more on policy and politics, in others more on personal morals, in others more on social norms and societal expectation), we can get into rich discussions of ethical issues on any given subject in often passionate, well-considered ways. Having students think about the scenario from multiple ethical viewpoints also adds a layer of complexity that facilitates rich, engaged discussion and (potentially) writing. For example, considering the scenario from a utilitarian vs. ethics of care perspective can yield interesting contrasts, and often, I have found, result in conversations that come back into play throughout the semester. In fact, I have even had students bring in materials later in the semester from other classes that they found to be relevant to our discussion of the Heinz dilemma: Ursula K. Le Guin's (1975) "Those Who Walk Away From Omelas," which gets at issues of happiness at another's expense (see Olivia Burgess' [2019] "Stand Where You Stand on Omelas" for one potential teaching ethics activity), has come up several times, for example, as have news stories dealing with theft-for-a-good-cause, or even discussions of personal experiences with ethical-conundrum components.

This idea of approaching a scenario from multiple ethical viewpoints is, perhaps, one of the most powerful teaching strategies I have encountered when teaching ethics. Our field is full of case studies we might ask our students to engage with—ethics textbooks in technical and professional communication and engineering often contain scenarios, and our periodic publications regularly feature ethics sections. *Intercom Magazine: The Magazine of the Society for Technical Communication*, for example, regularly runs an ethics column which offers insights, discussions, and cases on ethics. Past cases include issues of use of inferior materials (Ross, 2013), Facebook use in the workplace (Hockenhull et al., 2013), business startups (Everett, 2014), insurance claims (Gosser, 2015), expediency (O'Neil & Cooney, 2015), edutainment (Lambert, 2016), group work (Grisham, 2016), use of common knowledge (Gehrke, 2016), conflicts of interest (Bippes, 2017), implementation of care ethics in style guides (Karr, 2017), creative messaging (Generaux, 2018), and more. Including even just one of these cases in an extended writing and discussion session in class can be valuable. For example, in Jessie Lambert's (2016) "That's [Unethical?] Edutainment!" we are presented with the scenario of a scientific illustrator being asked to alter drawings for a textbook to make them more entertaining. Lambert introduces us to the general scenario, then discusses "edutainment" and its role and impact on culture, then leaves us with a series of questions: What should the illustrator do? How will their choices alter their relationship with the client, or shape the way others interact with the work? What does authorship look like in technical communication? And so on.

In discussing these issues in class, as presented in the scenario, we already have ways to get into ethical issues related to the way communicators create, modify, and publish content. But we can then revisit this piece from any variety of angles: What does the illustrator's decision-making tree look like if we take on this case from a purely utilitarian point of view? How might a feminist ethics approach to the scenario alter the outcome? What (looking ahead to the end of this chapter) might the illustrator's decision-making process and potential outcomes look like in indigenous models, or through the lens of Black womanist ethics? After all, the content we create shapes others' perceptions of the world, so creating visuals and describing findings has important implications for whose work is seen, for how those around us are seen, even for who and what is allowed to be seen.

Last, of course, having students engage with the ethical principles of their organizations is an excellent way to get into conversations about what constitutes right action on the job, and of how being able to articulate our decision-making process to employers, clients, and co-workers empowers us. In 2017, for example, I gave a webinar offered by the Society for Technical Communication that I later wrote up for *Intercom* (Ross, 2017b), in which I unpacked our ethical principles of legality, honesty, confidentiality, quality, fairness, and professionalism (STC, 2020). Having students address the concepts as written, then work though what makes those concepts normative (remember that a normative decision is one that makes an argument towards how things ought to be) is powerful. Having students then interrogate what model the principles are assumed to operate under adds even greater understanding of how ethical decision-making works, and, finally, having students write their own ethical principles based on specific ethical models adds yet another layer.

Having students develop their own set of ethical principles early on in class, that they then agree to abide by for the remainder of the semester (and question when necessary as new information becomes available), is also quite powerful. When I teach classes specifically related to ethics, we do this on day one. We began an upper-level undergraduate class on ethics, communication, and society, for example, with a five-part, simple, yet powerful, entirely student-created set of guidelines:

1. No passive aggressive attacks.
2. No malice.
3. No attacks on character.
4. No degrading your classmate's point(s): All views are worth hearing.
5. Discussion and debate will remain civil and academic.

In a graduate-level class on ethics and technical communication, my students created an entirely different set of rules (though you can certainly see shared concerns):

1. Listen to comprehend, not respond.
2. Learn to interrupt respectfully.
3. Consider every option as valuable. Some opinions are based in moral outrage, some on education, but a fairly offered opinion should be fairly considered.
4. Be able to entertain a thought without necessarily agreeing to it.
5. Avoid ad hominem attacks, and do not make assumptions about each other based on our in-class discussions.
6. Class disagreements end at the doorway—take the ideas away, not the outrage.
7. If it's a personal story told to make a point (or ask a question), leave it in the classroom.

These ethical principles gave us a way to not only discuss ethics, but openly self-moderate often intense discussions. Because the class created them, not the teacher, they became a powerful unifying tool that we could use in multiple ways throughout the semester: to mediate discussion, of course, but also to discuss how ethical principles shape professional spaces, to look at how principles enable (or prohibit) types of discussions, to make arguments about the way organizational policies shape behavior, and more.

In conclusion, it is my hope that this chapter offers ways to reinvigorate your thought when it comes to ethics, perhaps moving away from the model of ethics as a way to somehow "do the right thing" (though, please, do the right thing), and instead starting to get at ethics as a deeply complex, yet immensely valuable, system of thought that can inform many different aspects of life. The models of ethics I present here are only a few of many powerful heuristics available. You may instead find yourself drawn to one or more of many different models: In *Ethical Theory* (2018), for example, Heimir Geirsson and Margaret R. Holmgren offer overviews of divine command theory, egoism, consequentialism, deontology, moral pluralism, virtue ethics, and feminist ethics, and, in each section, consider different models or approaches for each type. If you are interested in environmental approaches, you might consider Tormod V. Burkey's (2017) *Ethics for a Full World*. For an animal rights perspective, try Peter Singer's (2009) *Animal Liberation*. For an indigenous perspective, consider Daniel R. Wildcat's (2009) *Red Alert! Saving the Planet with Indigenous Knowledge*.

Social justice is certainly a component of ethics, and Rebecca Walton, Kristen Moore, and Natasha N. Jones' (2019) *Technical Communication After the Social Justice Turn* gets us into ways of thinking about oppression and justice, even taking us partially into ethics in the Global South—an area which, at the time of this writing, would benefit greatly from increased attention dedicated specifically to looking at ethical models outside of the Western canon. Work on localization (Agboka, 2013), politics (Dorpenyo, 2019), environmental action (Walwema, 2020), and more all get at ethical issues of justice in the Global South, but there are relatively few pieces which *specifically* target ethical models. We need more work here—to echo Gerald Savage and Godwin Y. Agboka (2016), "research studies and activities involving professional communication scholars in the Global South offer some of the most important and interesting, and the least investigated work, to be done in our field" (p. 6). We've come a long way in a few years, but there is so much more to be done.

Within the scope of feminist approaches, there are many models: Carol J. Adams' (2002) *The Sexual Politics of Meat* challenges not only male-centric models of ethics, but the way we eat and the impact these choices have on society; Donna Haraway's (2006) essay "A Cyborg Manifesto" de- (and re-) constructs our bodies and relationships; Adriana Cavarero's (2009) *Horrorism: Naming Contemporary Violence* addresses images of violence and issues of vulnerability; and Katie G. Cannon's (2006) *Black Womanist Ethics* offers a vastly different approach to feminism from the Black woman's perspective that stands to change the way we conceptualize ourselves, our bodies, and our interactions by asking us to consider lived experience and states of suffering. First published in 1988, Cannon's model has not been widely covered in technical communication, though I believe that may soon change. This might be another way for you to think about the ethical models you choose: What are people marking as "important" ethical models, and who is doing the marking? What models are getting ignored, and why? Our research and teaching of ethics often begin with what we commonly discuss as the traditional models—the Aristotelian, Kantian, etc., models I have discussed here. You might, however, completely change the way we think about our decision-making by starting not with the traditional canon, as, admittedly, I have done, but by starting with indigenous ethical models. Robert Begay's (2001) "Doo Dilzin Da: Abuse of the Natural World," for example, offers insight into the Navajo view of the natural world as sacred, and working within this ethic dramatically changes any environment-related, ethics-based decision-making when coming from the Western utilitarian model. Whatever you choose, by making ethics the starting point of research and workplace decision-making, rather than an end-of-the-day note, I argue that we open the possibility of finding new ways of thinking and knowing.

■ References

Adams, C. J. (2002). *The sexual politics of meat: A feminist-vegetarian critical theory*. The Continuum Publishing Company.

Agboka, G. Y. (2013). Participatory localization: A social justice approach to navigating unenfranchised/disenfranchised cultural sites. *Technical Communication Quarterly*, 22(1), 28–49. https://doi.org/10.1080/10572252.2013.730966.

Allen, N. (1996). Ethics and visual rhetorics: Seeing's not believing anymore. *Technical Communication Quarterly*, 5(1), 87–105. https://doi.org/10.1207/s15427625tcq0501_6.

Andre, C. & Velasquez, M. (2014). *Calculating consequences: The utilitarian approach to ethics*. Markkula Center for Applied Ethics. http://www.scu.edu/ethics/publications/iie/v2n1/calculating.html.

Aristotle. (1975). *The Nicomachean ethics* (H. Rackham, Trans.). Harvard University Press.
Begay, R. (2001). Doo dilzin da: Abuse of the natural world. *American Indian Quarterly, 25*(1), 21–27. http://dx.doi.org/10.1353/aiq.2001.0001.
Bippes, B. (2017, February). Recognizing and disclosing conflicts of interest. *Intercom,* 33–35.
Buehl, J. (2014). Toward an ethical rhetoric of the digital scientific image: Learning from the era when science met Photoshop. *Technical Communication Quarterly, 23*(3), 184–206. https://doi.org/10.1080/10572252.2014.914783.
Burgess, O. (2019). Stand where you stand on Omelas: An activity for teaching ethics with science fiction. *Teaching Ethics, 19*(1), 63–70. https://doi.org/10.5840/tej202022570.
Burkey, T. V. (2017). *Ethics for a full world.* Clairview Books, LTD.
Cannon, K. G. (2006). *Black womanist ethics.* Wipf and Stock Publishers.
Carliner, S. (2001). Emerging skills in technical communication: The information designer's place in a new career path for technical communicators. *Technical Communication, 48*(2), 156–175.
Cavarero, A. (2009). *Horrorism: Naming contemporary violence* (W. McCuaig, Trans.). Columbia University Press.
Colton, J. S. & Holmes, S. (2018). *Rhetoric, technology, and the virtues.* Utah State University Press.
Colton, J. S., Holmes, S. & Walwema, J. (2017). From NoobGuides to # OpKKK: Ethics of Anonymous' tactical technical communication, *Technical Communication Quarterly, 26*(1), 59–75. https://doi.org/10.1080/10572252.2016.1257743.
Dombrowski, P. (2000a). *Ethics in technical communication.* Allyn and Bacon.
Dombrowski, P. M. (2000b). Ethics and technical communication: The past quarter century. *Journal of Technical Writing and Communication, 30*(1), 3–29. https://doi.org/10.2190%2F3YBY-TYNY-EQG8–N9FC.
Dorpenyo, I. K. (2019). Risky election, vulnerable technology: Localizing biometric use in elections for the sake of justice. *Technical Communication Quarterly, 28*(4), 361–375. https://doi.org/10.1080/10572252.2019.1610502.
Dragga, S. (1999). Ethical intercultural technical communication: Looking through the lens of Confucian ethics. *Technical Communication, 8*(4), 365–381. https://doi.org/10.1080/10572259909364675.
Dragga, S. & Voss, D. (2001). Cruel pies: The inhumanity of technical illustrations. *Technical Communication, 48*(3), 265–274.
Dragga, S. & Voss, D. (2003). Hiding humanity: Verbal and visual ethics in accident reports. *Technical Communication, 50*(1), 61–82.
Everett, H. I. (2014, September). Holiday madness: Tough totes shoulders (un)successful startup. *Intercom,* 29–31.
Faber, B. (1999). Intuitive ethics: Understanding and critiquing the role of intuition in ethical decisions. *Technical Communication Quarterly, 8*(2), 189–202. https://doi.org/10.1080/10572259909364659.
Frith, J. (2014). Social network analysis and professional practice: Exploring new methods for researching technical communication. *Technical Communication Quarterly, 23*(4), 288–302. https://doi.org/10.1080/10572252.2014.942467.
Gehrke, M. (2016, July/August). The "cover up" of common knowledge. *Intercom,* 32–34.
Geirsson, H. & Holmgren, M. R. (Eds.). (2018). *Ethical theory: A concise anthology* (3rd ed.). Broadview Press.

Generaux, V. (2018, May/June). The elephant in the room. *Intercom*, 33–34.
Gilligan, C. (1982). *In a different voice: Psychological theory and women's development*. Harvard University Press.
Gosser, R. (2015, February). Hurricane Sandy: The perfect storm of ethical dilemmas. *Intercom*, 29–31.
Grisham, J. (2016, May). It is what it is: Deadlines, groupwork, and cherry-picked data. *Intercom*, 42–44.
Groenhout, R. (2003). Theological echoes in an ethic of care. *Occasional Papers of the Erasmus Institute* 2003–2, Notre Dame University.
Haraway, D. (2006). A cyborg manifesto: Science, technology, and socialist-feminism in the late 20th century. In J. Weiss, J. Nolan, J. Hunsinger & P. Trifonas (Eds.), *The international handbook of virtual learning environments* (pp. 117–158). Springer.
Hockenhull, D., Martin, A., Mayhall, V. & Stude, S. (2013, March). Ethics scenario: Facebook use in the workplace. *Intercom*, 45–46.
Karr, D. (2017, September). It's all about the reader: How to implement care ethics in your company style guide. *Intercom*, 26–28.
Katz, S. B. (1992). The ethic of expediency: Classical rhetoric, technology, and the Holocaust. *College English*, *54*(3), 255–275.
Kernohan, A. (2012). *Environmental ethics: An interactive introduction*. Broadview Press.
Kirsch, G. (1999). *Ethical dilemmas in feminist research: The politics of location, interpretation, and publication*. State University of New York Press.
Kohlberg, L. (1971). Stages of moral development. *Moral education*, *1*(51), 23–92.
Lambert, J. (2016, January). That's [unethical?] edutainment! *Intercom*, 25–27.
Le Guin, U. K. (1975). Those who walk away from Omelas. In U. K. Le Guin, *The wind's twelve quarters* (pp. 254–262). Orion Publishing.
Markel, M. (2001). *Ethics in Technical Communication: A Critique and Synthesis*. Praeger.
Martin, W. & Sanders, S. (1994). Ethics, audience, and the writing process: Bringing public policy issues into the classroom. *Technical Communication Quarterly*, *3*(2), 147–163. https://doi.org/10.1080/10572259409364563.
Noddings, N. (2003). *Caring: A feminine approach to ethics and moral education* (2nd ed.). University of California Press.
O'Neil, F. & Cooney, J. (2015, September). Out of time: A double feature on expeditious ethics. *Intercom*, 26–27.
Potts, L. & Salvo, M. J. (2017). *Rhetoric and experience architecture*. Parlor Press.
Redish, J. C. (2000). What is information design? *Technical Communication*, *47*(2), 163–166.
Rose, E. J., Edenfield, A., Walton, R., Gonzales, L., McNair, A. S., Zhvotovska, T., Jones, N., Mueller, G. I. G. D. & Moore, K. (2018, August). Social justice in UX: Centering marginalized users. In *Proceedings of the 36th ACM International Conference on the Design of Communication—SIGDOC '18* (pp.1–2). https://doi.org/10.1145/3233756.3233931.
Ross, D. G. (2012). Why ethics?: Can doing the right thing really change the world? *Mother Pelican*, *8*(11). http://www.pelicanweb.org/solisustvo8n11page9.html.
Ross, D. G. (2013, January). Ethics scenario: Chinese drywall. *Intercom*, 37–38.
Ross, D. G. (2015). Monkeywrenching plain language: Ecodefense, ethics, and the technical communication of ecotage. *IEEE Transactions on Professional Communication*, *58*(2), 154 –-175. https://doi.org/10.1109/TPC.2015.2425135.

Ross, D. G. (2017a). The role of ethics, culture, and artistry in scientific illustration. *Technical Communication Quarterly, 26*(2), 145–172. https://doi.org/10.1080/10572252.2017.1287376.

Ross, D. G. (2017b). Why Ethics?: Interpreting "Ethics" and What STC's Ethical Principles (Can) Do. *Intercom Magazine, 64*(10), 29–32.

Ross, D. G., Oppegaard, B. & Willerton, R. (2019). Principles of place: Developing a place-based ethic for discussing, debating, and anticipating technical communication concerns. *IEEE Transactions on Professional Communication, 62*(1), 4–26. https://doi.org/10.1109/TPC.2018.2867179.

Ross, D. G. & Parks, M. (2018). Mutual respect in an ethic of care. *Teaching Ethics, 18*(1), 1–15. https://doi.org/10.5840/tej2018112156.

Salvo, M. (2014). What's in a name? Experience architecture rearticulates the Humanities. *Communication Design Quarterly, 2*(3), 6–9.

Sanders, S. P. (1997). Technical communication and ethics. In K. Staples & C. Ornatowski (Eds.), *Foundations for teaching technical communication: Theory, practice, and program design* (pp. 99–11). Ablex Publishing Corporation.

Savage, G. & Agboka, G. (2016). Guest editor's introduction to special issue: Professional communication, social justice, and the global south. *Connexions, 4*(1), 3–17.

Singer, P. (2009). *Animal liberation*. Harper Perennial.

Slack, J. D., Miller, D. J. & Doak, J. (1993). The technical communicator as author: Meaning, power, authority. *Journal of Business and Technical Communication, 7*(1), 12–36. http://dx.doi.org/10.1177/1050651993007001002.

Society for Technical Communication. (2020). *Ethical principles*. https://www.stc.org/about-stc/ethical-principles/.

Walton, R. & Hopton, S. B. (2018). "All Vietnamese men are brothers": Rhetorical strategies and community engagement practices used to support victims of Agent Orange. *Technical Communication, 65*(3), 309–329.

Walton, R., Moore, K. & Jones, N. (2019). *Technical communication after the social justice turn: Building coalitions for action*. Routledge.

Walwema, J. (2020). Rhetoric and Cape Town's campaign to defeat Day Zero. *Journal of Technical Writing and Communication*. https://doi.org/10.1177/0047281620906128.

Ward, M. (2010). The ethic of exigence: Information design, postmodern ethics, and the Holocaust. *Journal of Business and Technical Communication, 24*(1), 60–90. https://doi.org/10.1177/1050651909346932.

Wildcat, D. R. (2009). *Red alert!: Saving the planet with indigenous knowledge*. Fulcrum Publishing.

Willerton, R. (2015). *Plain language and ethical action: A dialogic approach to technical content in the 21st century*. Routledge.

Part Two: Shaping Curriculum

5. Confronting Methodological Stasis: Re-Examining Approaches to Technical Communication Pedagogical Literacy Frameworks

Halcyon M. Lawrence
Towson University

Liz Hutter
University of Dayton

Abstract: The layered literacies pedagogical framework has been a dominant model in the field of technical communication for the discussion of literacies and their interrelatedness. Although the field has regularly applied the framework to course and curriculum planning, in the 20 years since its development, there has been limited examination of the assumptions that form the framework's foundation. The under-theorization of the framework has led to what we term methodological stasis. We examine the field's prevalent patterns of engagement with literacy frameworks—checklisting, adding, deepening, and stacking—and discuss the ways that these patterns reinforce the unchallenged assumptions of the framework. As an alternative method for naming and categorizing technical communication skills and knowledge, we demonstrate an iterative, inductive method of examining classroom activities. This method is centered on classroom activities and makes visible a more complex, inter-related set of writing practices. The outcome is a set of literacy themes which provide a rich set of descriptors of student skills and knowledge. We end our chapter by proposing questions to guide the field in the development of responsive, multidimensional, and sustainable pedagogical literacy frameworks for the twenty-first century.

Keywords: pedagogical literacy frameworks, layered literacies, methodological stasis, literacy categories

Key Takeaways:

- There has been limited critical examination of technical communication pedagogical literacy frameworks, leading to methodological stasis.
- There is a need for our pedagogical literacy frameworks to demonstrate the qualities of responsiveness, multidimensionality, and sustainability.
- An iterative, inductive method for identifying and understanding technical communication literacies has the potential to make complex skills and knowledge more conspicuous.

Kelli Cargile Cook's (2002) seminal work on layered literacies advocated for "a more integrative frame that incorporates all of the literacies, into a single articulation of technical communication pedagogical goals" (p. 8). This pedagogical framework provided the field with both a nomenclature to help conceptualize and a structure to help organize the skills and knowledge important to our field's research, pedagogy, and program development. In the nearly 20 years since Cargile Cook conceptualized this framework, our field has undergone significant changes in terms of competencies demanded by the workplace, recruitment and training of instructors who teach technical communication, increased demand for the service course, and an expansion of professional writing and technical communication programs. Despite the significant changes experienced in the field, there has been limited critical examination of the layered literacies framework, in particular the assumptions that underlie the framework's application and the method through which the literacies are identified, named, and organized. We term this limited critical examination of the framework *methodological stasis*.

While we do not want to undermine the impact of the literacies framework on the field, we are interested in drawing attention to the fact that since the proposal of the layered literacies framework and the field's subsequent engagement with this and other pedagogical literacies frameworks, there has been minimal reexamination of how pedagogical frameworks ought to be developed and expanded. As the field has responded to a range of new workplace contexts, technological innovations, and shifting institutional requirements, we believe a reexamination is necessary to ensure that our pedagogical frameworks are sustainable, responsive, and multidimensional in the face of the field's growth and change.

To examine the problem of methodological stasis, we use a new thematic-analytic approach. This approach uncovers assumptions that underpin the layered literacies framework. Additionally, the approach yields a set of themes that provide more nuanced descriptions of classroom practices and lends more insight into the complex interrelationships of technical communication classroom activities.

In the sections that follow, we present a review of the field's engagement with the layered literacies framework, critique its distinctive characteristics, and offer reflection on the framework's limitations. Finally, we describe an inductive method for identifying skills and knowledge, followed by a discussion of how this method provides critical re-thinking of the deductive methodology used in the layered literacies model. We conclude by raising questions that might shape the field's future research on the development and application of technical communication pedagogical literacy frameworks.

Layered Literacies Pedagogical Framework for Technical Communication

Our interest in pedagogical literacy frameworks for technical communication derives from our past seven years as instructors in the field. During this time, we have collectively taught at six different institutions (research and teaching). At these institutions, we have held positions of graduate instructors, post-docs, visiting assistant professors, and most recently, tenure-track assistant professors. In these positions, we have taught upper-level courses in technical and business communication, and most often the service course, which has been directed to majors in computer science, health science, engineering, and environmental science. In none of these scenarios indicated above have we taught in an undergraduate technical communication program.

As we reflected on our collective teaching experience in the field, we came to realize a few shared realities as it informs our interest in pedagogical literacy frameworks. First, because for many students there is no writing course requirement after first-year composition, service courses have often uncovered that our students have varying degrees of readiness for upper-level technical and professional writing. Second, our colleagues don't readily understand what we teach. We have regularly inherited curricula and program learning objectives developed by colleagues outside of the field of technical and professional communication. Consequently, we found it problematic to design classes around inherited syllabi or programmatic outcomes that have been designed by colleagues unfamiliar with our field's curricular requirements. Furthermore, assessment criteria, like learning outcomes, are often not derived from technical practice. As a result, we perceived that many of our course learning outcomes have been genericized in order to be taught by any instructor, especially those who have not been trained in the field.

As a result of these realities, we felt that we needed to produce "multiply literate students in one semester" (Cargile Cook, 2002, p. 8) to meet the curricular learning outcomes. For example, in our classes, it was not uncommon to ask students to complete assignments such as writing cover letters and resumes—a conventional genre in the business and technical communication classrooms. However, this foundational assignment uncovered a host of skills and prerequisite literacies that students had not acquired in previous classes. In fact, much of our time was spent "unteaching" the academic essay before our students could begin to engage with other types of writing. Consequently, we were concerned that students were leaving our classrooms without adequate workplace literacies. As we tried to make sense of these constraints and sought to find creative pedagogical solutions, we talked almost daily about what literacies we could reasonably expect our students to demonstrate having completed our classes. In addition, we wrestled with how those literacies could be scaffolded within a single semester. Our interest in literacy frameworks, therefore, was an organic outcome from the issues we were grappling with in our classrooms.

In the following section, we introduce the layered literacies pedagogical framework that, on the one hand, has offered potential answers to the questions we were asking, yet on the other hand, has also demonstrated the practical limitations of working with a framework that integrates a comprehensive set of pedagogical goals.

Overview and Structure of the Layered Literacies Pedagogical Framework

In the period up to Cargile Cook's work, Katherine Staples (1999) characterizes the field of technical communication as going through major changes in the late 1990s. These changes included

1. an expanding field characterized by newly emerging specializations in areas like international communication, document design, and usability, to name a few;
2. new research agendas connected to the expansion of the field's interests and new venues at which to present research;
3. an increased demand for trained technical communication practitioners; and
4. the growth, both in number and complexity, of technical communication programs.

The layered literacies framework conceptualized by Cargile Cook consists of two key characteristics: discrete, static categories and the principle of layeredness. First, the framework identifies six discrete literacy categories—basic, rhetorical, social, technological, ethical, and critical. For each of the six categories, Cargile Cook offers an explanation of the range of skills and knowledge that comprise each literacy. Additionally, for each of these categories, she suggests how each literacy might be taught by instructors, demonstrated by students, or assessed within a curriculum or program by administrators. A second characteristic of the layered literacies framework is the interrelationships—the layering—among one or more literacies. Drawing on Wahlstrom (1997), Cargile Cook emphasizes that literacies "are not isolated but integrated and situated through a complex of classroom goals and activities" (2002, p. 6).

The framework has helped legitimize the work of our field. In the application of the framework, the studies we highlight below have collectively (1) accounted for skills and knowledge that are important to the field, (2) connected technical communication theory to practice, (3) developed a robust research agenda, and perhaps most importantly, (4) provided a link between classroom literacies and workplace competencies. Since the publication of Cargile Cook's work, scholars have engaged with both the literacy categories, to be able to account for new or newly-valued, yet under-examined, literacies, and the layeredness model presented in Cargile Cook's work. While not exhaustive, Table 5.1 provides a sampling of studies to show the range of ways in which scholars have engaged with the layered literacies framework.

Table 5.1. Selected sample of studies that have engaged with the layered literacies framework

Author	Literacy Category	Description of Engagement
Classroom pedagogy: Studies in this section extend the discussion around literacy categories established in the layered literacies framework.		
Kienzler & David (2003)	Ethical	Demonstrates how integrating ethics into the professional communication curriculum "provides students with experience in ethical problem-solving that requires that they consider not only immediate but also long-term consequences of their decisions" (p. 487).
Swarts (2011)	Technological	Expands understanding of technological literacy and its interrelationship with social literacies to include "network-building."
Bacabac (2013)	All six literacy categories	Uses a teaching case to present an e-portfolio assignment and demonstrate the presence of Cargile Cook's six literacy categories in the assignment.
Hovde & Renguette (2017)	Technological	Synthesizes the field's definitions of technological literacy to propose "a four-level technological literacy framework that can guide curricular decisions" (p. 396).
Classroom pedagogy: Studies in this section add new literacies not previously mentioned in the layered literacies framework		
Portewig (2004)	Visual	Engages directly with the layered literacies framework, making an argument that visual literacy should not be subsumed in the basic category but rather recognized as "a literacy that we must teach, research, and practice" (p. 32).
Starke-Meyerring (2005)	Global	Identifies themes in the discourse of globalization and draws on Cargile Cook's plural literacies to propose a framework for a global literacy.
Hannah (2010)	Legal	Argues for the establishment of a legal literacy and indicates that the layered literacies framework is a useful starting point to describe how technical communicators and students should see the law in their work.
Chong (2016)	Usability	Establishes the need for a usability literacy and makes the point that although the layered literacies framework does not explicitly mention usability as a literacy, a usability-centered approach is implied in Cargile Cook's work.

Author	Literacy Category	Description of Engagement
Colombini & Hum (2017)	Quantitative	Argues for the inclusion of quantitative literacy in the technical writing classroom as a site where multiple literacies convene.
Swacha (2018)	Embodied	Proposes the addition of embodied literacy as a "seventh" literacy; explores how embodied experiences interrelate with each of the six literacies.
Programmatic assessment: Studies in this section demonstrate how the layered literacies model has informed the development of programmatic assessment.		
Thomas & McShane (2007)	Programmatic assessment	Presents a case study of a rubric developed for the assessment of a technical communication program at Weber State.
Henschel & Melonçon (2014)	Assessment	Presents a program-focused method of assessment by combining Cargile Cook's layered literacies and Reich's (1992) symbolic analytic discussion of twenty-first century skills.
New workplace contexts: Studies in this section demonstrate the use of the layered literacies framework applied to new workplace contexts.		
Bivens et al. (2018)	Multisensory	Draws on the concept of layered literacies to argue that health literacy is an embodied, multisensory experience, invariably mediated by healthcare technologies.
Angeli (2020)	Embodied/ Multisensory	Demonstrates the presence of Cargile Cook's six literacy categories as well as multisensory and embodied literacies across a range of writing contexts for Emergency Medical Services (EMS) trainees.

Patterns of Engagement with the Framework

In this section, we define and illustrate four patterns of engagement to demonstrate how the framework is applied and expanded in classroom, program, and workplace contexts. These patterns include *checklisting*, *adding*, *deepening*, and *stacking*. *Checklisting* is a means to register the absence or presence of a literacy in the design of an assignment or curriculum. *Adding* and *deepening* are the ways in which the layered literacies framework is expanded. The addition of a literacy category to the framework occurs when new pedagogical or workplace contexts are identified; on the other hand, the deepening of an existing literacy category occurs when there is a need to recognize nuanced demonstrations of the literacy or to recognize other characterizations of the literacy not previously described in the framework. Finally, patterns of *stacking* refer to the description of layering of the literacy categories, which is limited to their co-existence rather than how they are interconnected.

We provide examples of each of these patterns of engagement below and examine their significance and implications. Finally, we show how understanding the patterns of engagement with the framework belies key assumptions about how a framework is structured as well as its purpose and development.

Checklisting

One form of engagement we observe is that the framework's reliance on discrete, static categories (i.e., basic, rhetorical, etc.) allows the framework to be used as a kind of rubric. This leads to a checking off of whether or not the six literacy categories are demonstrated in an assignment or in course objectives. An example of a checklist approach to the design of a course is in Cargile Cook's application of the framework to the curriculum for a technical communication capstone course. She notes that the course assignments did not reflect the six literacies, and in an attempt to align the course objectives with the assignments, she states, "[i]n order to incorporate more instruction in the other literacies, new assignments were added to the course content" (2002, p. 19). The checklist approach we note here is Cargile Cook's decision to add more instruction to address the missing literacy so that all six literacies are reflected in the course objectives.

Another expression of a checklist pattern is seen in the desire to demonstrate the presence of the six literacy categories in an artifact. Florence E. Bacabac (2013) offers an example with regards to the students' assessment of an e-portfolio assignment. She ascertains the presence of the six literacy categories in students' reflections as "evidence that this assignment series helped them to develop Cargile Cook's (2002) layered literacies" (p. 106). Checklisting is also demonstrated in studies focused on assessment of curricula or programs (Henschel & Melonçon, 2014; Thomas & McShane, 2007).

While the process of checklisting may be helpful for an approximation of the skills and knowledge an assignment asks students to demonstrate, the complexity of skills and knowledge embedded within a literacy category can become obscured. Furthermore, an understanding of what other pedagogical activities might be needed for students to demonstrate a particular literacy and an understanding of the extent to which the literacy is performed or developed is also not recognized by a checklist approach. In the case of technological literacy, Bacabac (2013) uses the following quotation from a student reflection to prove the existence of the literacy category in an e-portfolio assignment:

> [On technological literacy] I feel that the development of my professional eportfolio allowed me reach [sic] one of the course learning outcomes of conceptualizing, designing, planning, and critiquing an informational project. This was my favorite project of the entire semester because I feel it enabled me to use my TECHNOLOGICAL AND critical writing skills, but also my creative side to create a site that fulfilled its purpose. (p. 107)

While it is not clear if the student or the instructor put the word "TECH-NOLOGICAL" in all caps, what is clear is that Bacabac is matching the named category to the key term used in the student's reflection as evidence of the presence of the literacy. In this example, the student reflection identified several activities around which the technological skills were demonstrated. However, in the process of matching the terms to the demonstration of technological literacy, there is no opportunity for a nuanced discussion about other skills and knowledge supportive of the students' acquisition and demonstration of technological literacy.

The significance of the checklisting approach used in the application of the framework is the potential reductive effect on how we understand students' acquisition, development, and understanding of particular literacies. An implication of the reductive effect we see is an unresponsiveness of the framework to new contexts. The unresponsiveness of the layered literacies framework is evident when a new literacy cannot be meaningfully incorporated by the framework, as Elizabeth Angeli demonstrates in her case study later in this collection. Although Angeli does not add literacies to the framework, she acknowledges that the six categories of the layered literacies framework do not accommodate the new context of technical communication training that happens in the Emergency Medical Services (EMS) workplace. She states that "despite the wide range of studies that apply the layered literacies framework, less explored is how this framework translates into training courses outside of the university."

Adding and Deepening

There are many examples of the field's approach to adding new literacies to the existing framework over time. When the framework does not align with what is being taught in classrooms or when the workplace presents new skills and knowledge, there is a need to recognize a new literacy, such as in the case of Kristin M. Bivens and colleagues (2018), who argue for the inclusion of a multi-sensory literacy to account for different skills and knowledges practiced in a biomedical healthcare context. Additional examples of work that add to the existing framework include the work of Tiffany C. Portewig (2004), who argues for the inclusion of visual literacy, distinct from Cargile Cook's basic literacy, and Mark A. Hannah (2010), who argues that technical communicators would be well served to develop a legal literacy that reflects a "complex understanding of how their work intersects with legal concerns" (p. 5). Doreen Starke-Meyerring (2005) proposes a framework for global literacies consisting of "plural literacies" and drawn on themes she observes in the discourse of globalization (p. 470). More recently, Kathryn Y. Swacha (2018) asserts the necessity for an embodied literacy as "a distinct seventh literacy" (p. 262). Felicia Chong (2016) examines usability as core skills and knowledge in the technical communication classroom. Although Chong does not explicitly acknowledge usability as a discrete category, her focused examination of usability in technical communication textbooks and syllabi makes explicit "a user-centered approach" she sees implied in the layered literacies framework (p. 12). Similarly, Crystal B. Colombini and Sue Hum (2017) advocate for a more systematic integration of quantitative literacy in the technical communication classroom, noting its lack of inclusion in literacy frameworks, despite their emphasis on multiple literacies (p. 381).

Another way to expand the framework is through the process of deepening a discrete literacy category. Donna Kienzler and Carol David (2003), for example, agree with Cargile Cook's argument of the necessity to cultivate students' ethical awareness by having them "identify and explain ethical choices they made in their classroom projects . . ." (p. 16). In their work, they deepen the demonstration of Cargile Cook's category when they describe how students will learn to identify and explain their choices through exposure to ethical theories and ethical vocabulary along with the case studies and other learning activities that Cargile Cook recommends. Other studies engage with the framework by deepening the range of skills and knowledge that can be demonstrated in a particular literacy category. Jason Swarts (2011) extends technological literacy to include "the behind-the-scenes work of gathering information, collaborating, distributing labor, and gaining buy in" (pp. 274–275), whereas Marjorie R. Hovde and Corrine C. Renguette (2017) identify four levels of technological literacy: functional, conceptual, evaluative, and critical.

At issue with the methods of adding to and deepening of the discrete literacy category is the framework's ability to maintain a manageable scope. Although these methods provide a more nuanced understanding of a literacy, the ability of the framework to absorb the repeated contributions of the field to provide ever richer understandings of a literacy has the potential to allow a category to become a catchall. As such, the literacy category over time risks encompassing too wide a scope, thereby losing its descriptive power and ultimately its usefulness (e.g., too many activities could be labeled as "rhetorical" or "technological").

We speculate that over time, the catchall effect will become unsustainable. As the field continues to grow and as new workplace contexts and their relevance to technical communication are identified, the continued practice of adding to and deepening of the categories makes it harder to identify what is critical to the field. If we were to follow Swacha's example of naming a distinct seventh literacy, what restrains the field from simply adding additional literacies (eighth, ninth, and so on) to the framework? As the framework does not provide any threshold that limits this method, the processes of adding and deepening could continue infinitely. These current practices are actively working against the field's ability to intentionally define and organize the skills and knowledge we teach; this could in turn compromise the field's disciplinary identity over time.

Stacking

The framework has the potential to think about how a set of skills and knowledge associated with a specific task or series of tasks fosters students' understanding; however, in its application, in many of the examples of scholarship we reviewed, we see little demonstration of the complexity with which literacies are interrelated. A static, linear relationship emerges rather than one that is more interactive.

As a result of the linear relationship between the literacies, the pattern of stacking (like a sandwich) emerges, in which one literacy is stacked upon another and the sequencing of the literacy layers is interchangeable. Jennifer L. Bay and Samantha Blackmon (2016) also observe this tendency in literacy frameworks and suggest that while frameworks such as Cargile Cook's conceptually should demonstrate the interrelatedness of the literacies, in reality, "they are still divided into discrete blocks of skills that can be combined and recombined in a variety of ways" (p. 213). An example of stacking is found in Cargile Cook's (2002) own work. Her example is limited to showing the literacies' co-existence, rather than how they are (inter)connected. On the one hand, she conceptualizes the interrelatedness of the six literacies by showing how one literacy draws upon and incorporates other skills and knowledge. For example, she explains that basic literacies associated with reading, writing, and document design are not simply rules or templates, but rather demonstrations of the student's rhetorical ability to "mak[e] informed decisions about usage, grammar, mechanics, styles, and graphic representations based on knowledge of readers and writing situations" (2002, p. 9). On the other hand, in her demonstration of interrelatedness for a sample interview assignment, she lists each literacy and discusses how the students demonstrate that literacy in the assignment. Because the framework is premised on discrete literacy categories, it emphasizes the coexistence of literacies more than it promotes an understanding of any inter-relationship among the literacies.

The (re)stacking of literacy categories over time has the potential to flatten what is in practice a more multidimensional and integrated relationship in the classroom to a relationship that is more static and linear. This linear conceptualization is counterintuitive to pedagogical scaffolding and understanding how literacies develop and mature over time.

We summarize our discussion of the layered literacies pedagogical framework and the implications of the critiques we raise in Figure 5.1. Given our analysis, the field needs to rethink how to develop a framework to account for technical communication literacies and what we conceptualize so that it can be more responsive to new contexts, accommodate multidimensional relationships among literacies, and ultimately be more sustainable.

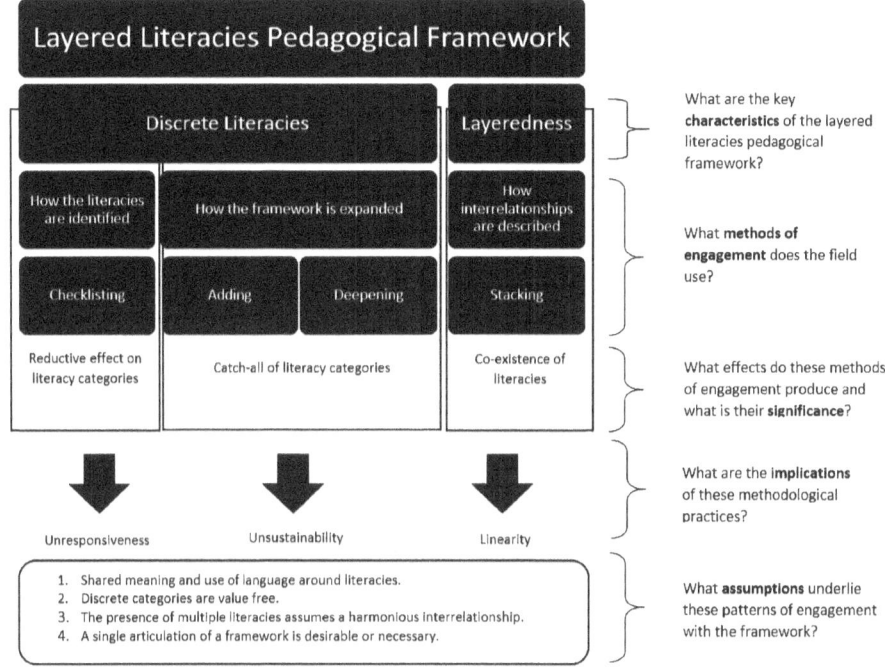

Figure 5.1. Layered literacies framework analytical model.

Rethinking Method: An Inductive Approach to Establishing Pedagogical Literacy Themes

In this section, we demonstrate an alternative approach to thinking about technical communication skills and knowledge. First, we describe the general design and the rationale of an ongoing study that examines what teachers ask students to do in their technical and professional writing courses. We are not presenting the results of this study, but rather using one of the study's questions to demonstrate what emerges when we shift or approach literacies with a different methodological lens. Next, we explain the four stages of iteration which our inductive approach takes and what emerges at the end of each iteration. As we go along, we highlight how our method is a departure from existing literacy framework methods. Finally, we discuss four assumptions about literacy frameworks that are revealed because of the shift in method, and we discuss the implication of these assumptions for the field.

Modeling an Inductive Approach

To demonstrate an inductive method that can lead to technical communication pedagogical literacy themes, we draw data from a corpus from our ongoing study in technical communication pedagogy. In this study, 65 instructors of technical communication took a 26-question survey.[1] We draw from one question where respondents were asked to name a traditional technical communication and professional writing assignment that they ask their students to complete. Additionally, respondents were asked to identify all the activities that students were required to *do* to complete the assignment they named. In the survey, we defined a traditional assignment as one that asks students to engage with conventional tools, approaches, processes, or all three in the completion of that assignment. We further indicated that traditional could also refer to an assignment that, to the instructor's knowledge, is standard or conventional in the field's mainstream pedagogy and practice.

The responses to this question generated a rich collection of verbs demonstrating an array of activities performed in a technical communication classroom. These verbs formed the basis and therefore the unit of analysis of our iterative, inductive coding method. While many of the studies we reviewed used the classroom as the level of analysis, the verb (as a measure of student activity) provides a departure from other studies that begin their analysis at the unit of the assignment and/or course objectives (e.g., Cargile Cook, 2002; Swacha, 2018; and Bacabac, 2013).

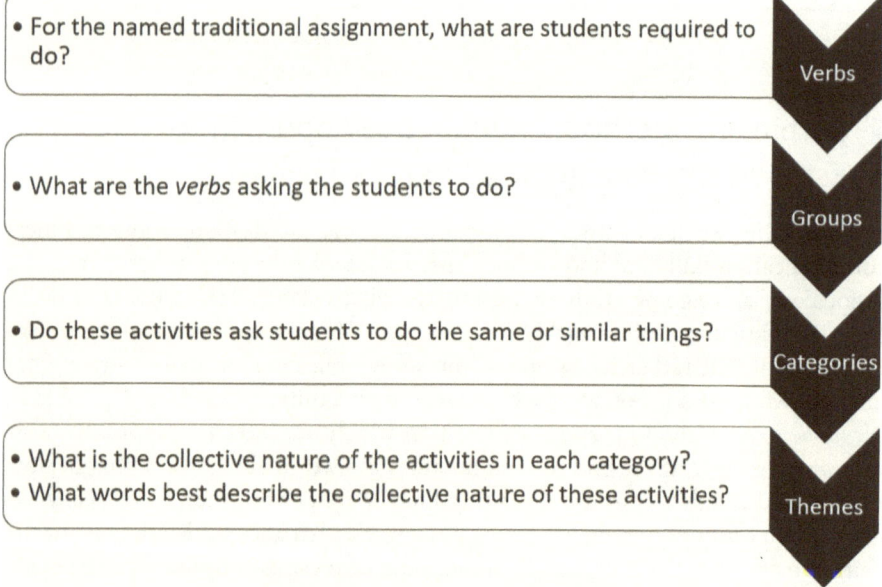

Figure 5.2. Questions asked in the inductive method to move from verbs to themes.

---

1. The study received IRB approval from the University of Dayton on March 26, 2019.

The inductive process we describe was facilitated by a series of questions to guide how we coded and sorted the data. Each question led us to a new stage of the process, moving from verbs to themes (see Figure 5.2). Our iterative, inductive process demonstrates what is perhaps the most significant departure from existing methods used in the layered literacies framework as the nomenclature (literacy themes) is established at the end of the iterative process rather than at the beginning.

We began our sorting process by writing each activity (e.g., "conduct research using journalistic and scholarly sources, including trade journals") on a sticky note. As we asked a question and at each of the four stages, we moved the sticky notes around when we came to a consensus to an answer to our question. We expand upon the inductive process in the following paragraphs, drawing on select examples from our data to demonstrate the kinds of patterns we observed.

Iteration 1. In the first iteration, we returned to the following question asked in the survey: "For the named traditional assignment, please indicate what students are required to do." The use of the verb is significant in this iteration of our method as it allowed us to focus on the multiple doings associated with completion of a single assignment (see Table 5.2, column 2). At this stage, we were interested in understanding what types of activities made up the named traditional assignment.

Table 5.2. Example of first iteration of indicative method

Assignment description	What students were asked to do to complete assignment
Write a professional report with a presentation.	1. **Conduct** research using journalistic and scholarly sources, including trade journals. 2. **Use** citation correctly. 3. **Write** multiple drafts. 4. **Use** professional formatting.

Iteration 2. Once we had each step written on a sticky note, we asked the second question, "What are the verbs asking students to do?" We understood that the verbs instructors use to describe what students do might embed many more actions than the verb on the surface conveys. According to the respondent (Table 5.2), completing a professional report and presentation assignment entailed conducting research using journalistic and scholarly sources, using citations correctly, writing multiple drafts, and using professional formatting. However, when we asked what the verb "use" is asking students to do, we were prompted to think more carefully about other skills and knowledge a student needed to draw on in order to use a citation style correctly. Using citation correctly entails that a student not only applies a professional convention, but also understands what the legal, ethical, and/or professional standards are that inform the convention. The verb "use" in the context of citation conventions also entails lower order processes such as listing, correcting, and ensuring consistency. Furthermore, this example demonstrates the importance of context in uncovering other tasks associated with what a verb is asking stu-

dents to do. The example "Use professional formatting" uncovers other variations of "use" because the context of use is different. In this context, using professional formatting requires knowledge of document design, visual design, and perhaps even knowledge of software to accurately apply the formatting. Uncovering these implicit skills and knowledge seemed not only to be a step toward acknowledging these under-recognized skills and knowledge, but also a way to understand their interrelationship. The consensus we came to regarding what the verbs were asking students to do allowed us to sort the sticky notes with similar verbs to form groups.

Iteration 3. In the next step, we asked the question "Do these verbs ask students to do the same or similar things?" We recognized that although our groups at this stage were made up of similar verbs, the activities weren't necessarily asking students to do similar tasks; conversely, we also found that different verbs were indeed asking students to engage with similar activities. For example, a set of activities from different respondents produced a group like "Agree on a proposal idea for groups; Create a team contract; Peer review a document; Hold team meetings." While all these activities are represented by different verbs, in essence, they all ask students to engage in some act of collaboration. Answering this question allowed us to move from groups to categories.

Iteration 4. Finally, we asked the question "What is the collective nature of the activities in each category and what words best describe the collective nature of these activities?" In this final iteration, we were interested in making observations and finding labels that best described the range of lower order (e.g., "measuring against a standard") and higher order (e.g., "engaging with information, research, data, and range of sources") activities to plan, develop, and produce an artifact (Table 5.3, column 3). Answering this question allowed us to identify a list of themes that characterized the range of skills and knowledge described in our data. These themes are noticeably more descriptive than the label of a literacy category (e.g., technological, social). The usefulness of a more descriptive literacy category is its concreteness, which makes it easier to recognize in spaces outside the classroom and more likely for students to transfer to another space (e.g., collaboration describes a more transferable skill than the literacy category "social"). Additionally, these themes are likely more recognizable and connected to workplace competencies.

The inductive approach to coding these activities exposed a wider set of skills and knowledge than a discrete literacy category suggests. Table 5.4 demonstrates an example of how a literacy category can oversimplify, or even erase, the complexity of an activity. For example, we took the list of activities for a named traditional assignment and coded each verb according to whether it was a demonstration of one of the six layered literacy categories. Doing so, our conversation around the verb began to conform to the category (i.e., a basic, rhetorical, social, technological, ethical, or critical literacy) so that it could "fit." As a result, we focused less on what skills and knowledge the activity was engaging and more on matching the characteristics of the activity to align with the category.

Table 5.3. An example of themes that emerge from the inductive method

Assignment description	What students were asked to do to complete assignment	Themes which emerged from inductive method
Write a professional report with a presentation.	1. **Conduct** research using journalistic and scholarly sources, including trade journals. 2. **Use** citation correctly. 3. **Write** multiple drafts. 4. **Use** professional formatting.	1. **Engaging** with information, research, data, and range of sources 2. **Recognizing** and **understanding** and applying standards and conventions (genre, legal, academic, professional) as they relate to artifacts; attention to routine and regularized activities 3. **Measuring** against a standard of quality through iteration; working with peers/users/constituents/audiences to refine product 4. **Producing** final or culminating artifact

On a larger scale, this process of fitting activities into a literacy category sets up the potential for a literacy category to become a catch-all. In the example coded in Table 5.4 for instance, three of the four activities can be coded as "basic." The breadth of literacy themes uncovered through an inductive process foregrounds more explicitly the interdependence of literacies across different stages of planning, developing, and producing an artifact. The richness of these interrelationships, in turn, brings a different orientation to or raises questions that allow us to think beyond a checklist (i.e., whether a literacy is absent or present) and allows us to think more about development, interdependence, and scaffolding of skills and knowledge across an assignment or assignment sequence, for instance.

We outlined this process to demonstrate how asking a series of questions begins to move us out of methodological stasis. As we disrupt methodological stasis, we are prompted to confront some assumptions that undergird the framework. Four of these unchallenged assumptions about literacy pedagogical frameworks include the following:

Assumption 1. A single, comprehensive framework is desirable or necessary. An assumption that a single, comprehensive framework is necessary means that we are constantly adding or expanding to make our activities fit a framework, instead of exploring other models that might better define and describe our pedagogical practices. A single articulation of a framework may not be sufficiently flexible to respond to the changing and specific cultural and institutional contexts in which we teach technical communication; in fact, it may be constraining the field's ability to respond to new and emerging contexts, practices, and workplaces.

Table 5.4. A comparative example between the layered literacy categories and the sample themes from the inductive method

Assignment description	What students were asked to do to complete assignment	Themes which emerged from inductive method	Layered literacy categories
Write a professional report with a presentation.	1. **Conduct** research using journalistic and scholarly sources, including trade journals. 2. **Use** citation correctly. 3. **Write** multiple drafts. 4. **Use** professional formatting.	1. **Engaging** with information, research, data, and range of sources 2. **Recognizing** and understanding and applying standards and conventions (genre, legal, academic, professional) as they relate to artifacts; attention to routine and regularized activities 3. **Measuring** against a standard of quality through iteration; working with peers/users/constituents/audiences to refine product 4. **Producing** final or culminating artifact	1. Rhetorical (**conduct** research) 2. Basic (**use** citations) 3. Basic (**write** drafts) 4. Basic (**use** professional formatting)

Assumption 2. Frameworks and literacy categories are value free. Assignments or course objectives designed with the six literacy categories in mind may promote the assumption that the framework and its literacy categories are "neutral" and "context-less" (Wysocki & Johnson-Eilola, 1999, p. 355). The field's recent scholarship on social justice, diversity, and inclusion (Jones, 2016; Melonçon, 2017; Walton & Jones, 2013) demonstrates that there are other urgent contexts within and beyond the workplace that require our students to think critically about how power and accessibility impact communication behaviors and practices. Thus, we must recognize that what our students do in our classrooms is often a reflection of the particularities of our classroom contexts, such as the skills and knowledge our students may already have or the community and institutional structures in which we teach. The responsiveness of an inductive approach can be leveraged to recognize the ways in which skills and knowledge that we deem necessary and thus teach are also products of institutional or cultural ideologies.

Assumption 3. The discrete literacy category should be the foundation of a framework. Much of our scholarship engages with the discrete literacy category as an unchallenged, foundational characteristic of the framework. We have not challenged the discrete literacy categories at the level of their definition (e.g., what is the meaning of "technological" literacy?) nor at the level of their inclusion (e.g., why these six literacies?). As a result, the meaning of the categories and their value to the framework are deemed self-evident. This assumption perpetuates the field's patterns of engagement around adding to, deepening, and stacking of the literacy categories, thereby limiting the usefulness of the framework to guide pedagogical and programmatic choices. The inductive approach provides other ways of identifying and explaining skills and knowledge other than by using a single term (e.g., basic, ethical, etc.). For example, the themes that have emerged from the inductive method are more descriptive and allow for a more sustainable expansion of the framework.

Assumption 4. The presence of multiple literacies assumes a harmonious interrelationship. Our inductive approach shows that the interrelationships between literacies are more complex and intentioned than the stacking of the layered framework suggests. The inductive approach exposes the fact that too often, even if multiple literacies are present in an assignment, there is not sufficient demonstration of their layeredness. As a field, we have not done enough work to explore what the interrelationship among literacies looks like, how they might exist in tension with one another (Angeli, this collection), and why these tensions might be valuable. The interactive process of the inductive method offers us more insight into the complexity and interrelatedness of technical communication activities, and moves us away from flat, linear thinking about classroom activities. Upending this assumption might also push the field to consider if metaphors other than layeredness are needed to interrogate these multidimensional interrelationships.

Pedagogical Literacy Frameworks for the Twenty-First Century

Our field continues to experience the growth of technical communication courses and programs, particularly the technical communication service course. Given this growth, we need a framework that is responsive, multidimensional, and sustainable. Therefore, it is timely for the field to ask such questions as the following about our pedagogical literacy frameworks and our engagement with them:

1. Articulation of a framework

    Is a single articulation of a framework necessary? Would multiple articulations of a framework erode our field's identity or strengthen it?

If we as a field explored other frameworks and literacy models, what types of assignments, course designs, and program objectives might emerge or be recognized as a result?

2. Methodology

Who and by what mechanism (i.e., methodology) does our field determine technical communication literacies?

At what point in our field's evolution of practices do we begin to recognize a new literacy (with its own unique demonstrations of skills), and not simply a new demonstration of an existing literacy? In other words, how do we account for emerging literacies?

Can and should our framework account for a hierarchy of literacies? If so, what would be the benefit to the field of a hierarchization of literacies? How would such a hierarchization be organized and rationalized?

3. Assessment

How do we as a field balance the need for assessment with a responsive, sustainable framework, so that the framework does not become "a handy shortcut for covering a wide range of skills, procedure, and practices" (Wysocki & Johnson-Eilola, 1999, p. 360)?

We encourage those in the field—graduate students, instructors, administrators, practitioners, and advisory boards—to consider and adopt these questions as part of an active field-wide research agenda. Doing so can serve to strengthen our disciplinary identity and enrich our pedagogical practices.

■ Acknowledgments

We would like to thank the participants of the Squirrel Hill Scholars Workshop, August 8–11, 2019, in Pittsburgh, PA, especially Brian Larson, Ph.D. and Sue Provenzano, J.D., for their feedback and support in the early stages of this project's conceptualization. We also thank Elizabeth Angeli, Ph.D. for her comments as we refined our paper in its later stages.

■ References

Angeli, E. L. (2021). Technical communication pedagogy and layered literacies in workplace training courses. In M. J. Klein (Ed.), *Effective teaching of technical communication: Theory, practice, and application* (pp. 287–302). The WAC Clearinghouse; University Press of Colorado.

Bacabac, F. E. (2013). Creating professional eportfolios in technical writing. *Journal of Business and Technical Communication, 27*(1), 91–110. https://doi.org/10.1177/1050651912458921.

Bay, J. L. & Blackmon, S. (2016). Inhabiting professional writing: Exploring rhetoric, play, and community in Second Life. In J. DeWinter & R. M. Moeller (Eds.), *Computer games and technical communication: Critical methods and applications at the intersection* (pp. 211–232). Ashgate.

Bivens, K. M., Arduser, L., Welhausen, C. A. & Faris, M. J. (2018). A multisensory literacy approach to biomedical healthcare technologies: Aural, tactile, and visual layered health literacies. *Kairos, 22*(2). http://kairos.technorhetoric.net/22.2/topoi/bivens-et-al/index.html#1.

Cargile Cook, K. (2002). Layered literacies: A theoretical frame for technical communication pedagogy. *Technical Communication Quarterly, 11*(1), 5–29. https://doi.org/10.1207/s15427625tcq1101_1.

Chong, F. (2016). The pedagogy of usability: An analysis of technical communication textbooks, anthologies, and course syllabi and descriptions. *Technical Communication Quarterly, 25*(1), 12–28. https://doi.org/10.1080/10572252.2016.1113073.

Colombini, C.B. & Hum, S. (2017). Integrating quantitative literacy into technical writing instruction. *Technical Communication Quarterly, 26*(4), 379–394. https://doi.org/10.1080/10572252.2017.1382259.

Hannah, M. A. (2010). Legal literacy: Co-producing the law in technical communication. *Technical Communication Quarterly, 20*(1), 5–24. https://doi.org/10.1080/10572252.2011.528343.

Henschel, S. & Melonçon, L. (2014). Of horsemen and layered literacies: Assessment instruments for aligning technical and professional communication undergraduate curricula with professional expectations. *Programmatic Perspectives, 6*(1), 3–26.

Hovde, M. R. & Renguette, C. C. (2017). Technological literacy: A framework for teaching technical communication software tools. *Technical Communication Quarterly, 26*(4), 395–411. https://doi.org/10.1080/10572252.2017.1385998.

Jones, N.N. (2016). The technical communicator as advocate: Integrating a social justice approach in technical communication. *Journal of Technical Writing and Communication, 46*(3), 342–61. doi:10.1177/0047281616639472.

Kienzler, D. & David, C. (2003). After Enron: Integrating ethics into the professional communication curriculum. *Journal of Business and Technical Communication, 17*(4), 474–489. https://doi.org/10.1177/1050651903255418.

Melonçon, L. (Ed.). (2017). *Rhetorical accessibility: At the intersection of technical communication and disability studies*. Routledge.

Portewig, T. C. (2004). Making sense of the visual in technical communication: A visual literacy approach to pedagogy. *Journal of Technical Writing and Communication, 34*(1), 31–42. https://doi.org/10.2190/FGJ6-UETB-9CA6-5PC3.

Reich, R. R. (1992). *The work of nations: Preparing ourselves for 21st century capitalism*. Vintage.

Staples, K. (1999). Technical communication from 1950–1998: Where are we now? *Technical Communication Quarterly, 8*(2), 153–164. https://doi.org/10.1080/10572259909364656.

Starke-Meyerring, D. (2005). Meeting the challenges of globalization: A framework for global literacies in professional communication programs. *Journal of Business and Technical Communication, 19*(4), 468–499. https://doi.org/10.1177/1050651905278033.

Swacha, K.Y. (2018). "Bridging the gap between food pantries and the kitchen table": Teaching embodied literacy in the technical communication classroom. *Technical Communication Quarterly, 27*(3), 261–282. https://doi.org/10.1080/10572252.2018.1476589.

Swarts, J. (2011). Technological literacy as network building. *Technical Communication Quarterly*, 20(3), 274–302. https://doi.org/10.1080/10572252.2011.578239.

Thomas, S. & McShane, B. J. (2007). Skills and literacies for the 21st century: Assessing an undergraduate professional and technical writing program. *Technical Communication*, 54(4), 412–423.

Wahlstrom, B. (1997). Teaching and learning communities: Locating literacy, agency, and authority in a digital domain. In S. A. Selber (Ed.), *Computers and technical communication: Pedagogical and programmatic perspectives* (Vol. 3, pp. 129–146). Ablex.

Walton, R. & Jones, N. N. (2013). Navigating increasingly cross-cultural, cross-disciplinary, and cross-organizational contexts to support social justice. *Communication Design Quarterly*, 1(4), 31–35. https://doi.org/10.1145/2524248.2524257.

Wysocki, A. F. & Johnson-Eilola, J. (1999). Blinded by the letter: Why are we using literacy as a metaphor for everything else? In G. E. Hawisher & C. E. Selfe (Eds.), *Passions pedagogies and 21st century technologies* (pp. 349–368). Utah State University Press.

6. Trial and Error: Designing an Introductory Course to Technical Communication

Chen Chen
WINTHROP UNIVERSITY

Abstract: This chapter presents the experience of an undergraduate introduction to technical communication (TC) class from design to execution in a four-year public university without a technical communication degree program. The chapter contributes to scholarship on technical and professional communication (TPC) pedagogies and curricular design by sharing reflexive narratives from the instructor and students on what happened in the classroom in an institutional context often not represented in established scholarship. I argue that the challenges of the class are to maintain a good balance and connection between theory and practice to help students begin to develop core conceptual skills of TC and facilitate transfer. Through trial and error, students gained some conceptual skills but might have gained a limited view of technical communication in this first iteration. Upon that reflection, I discuss the changes in the second iteration and offer suggestions for designing the class with a problem-solving perspective and social justice orientation in an institutional context without TPC programmatic structures and learning outcomes, using more scenarios and examples to help students see how technical communicators can be advocates for change and to facilitate transfer. I also argue for adaptive, flexible, socially just pedagogical practices and discuss implications for classroom research and professional development practices.

Keywords: classroom research, intro to technical communication, theory and practice, social justice pedagogy

Key Takeaways:

- Curriculum development and course design must be in tune with disciplinary trends as well as accommodating and adaptable to local institutional contexts.
- An introduction to technical communication (TC) course can take a problem-solving and social justice-orientation that asks students to explore fundamental TC concepts and connect them with practical scenarios in class activities and assignments.
- When designing new courses, we should enact equitable, inclusive, and flexible pedagogical approaches by maintaining an open dialogue with students throughout the course, enacting a human-centered pedagogy.

Researchers in technical and professional communication (TPC) have long been studying our pedagogical practices, curricular design, and program administration (Cargile Cook, 2002; Melonçon & Henschel, 2013; Staples & Ornatowski, 1997; Thatcher & St.Amant, 2011; Walton et al., 2016). Some of the scholarship focuses on the varieties of the multi-major professional writing course (Breuch & Sadler, 2016; Read & Michaud, 2018), while others focus more on technical communication degree programs that prepare future practitioners and scholars (Melonçon & Henschel, 2013; Melonçon & Schreiber, 2018). More recently, we have seen new collections on specific pedagogical theories and practices (Bridgeford, 2018; Haas & Eble, 2018). Joining this new trend of pedagogical research on technical communication, this chapter reports on a research project that responds to Lisa Melonçon and Sally Henschel's (2013) call for more research about TPC program design and development, particularly "what occurs in the classroom" (p. 60). Specifically, this chapter reports my experiences of designing and teaching an introductory technical communication course in the context of an institution without a TPC program. This research will be especially useful for instructors in similar institutional contexts who are faced with the challenges of developing courses and/or programs from the ground up by reminding us of the importance of contextualizing curriculum and programmatic research in local situations (Cooper, 1991) and by emphasizing the lived experiences of an instructor and students in the classroom.

I teach in a four-year public university with a liberal arts focus. Our institutional context is unique but also reflective of liberal arts institutions with traditional English departments trying to build and/or enhance their curriculum in technical and professional communication. Recently, we shifted the ways we want to prepare our majors by rearticulating the objectives and learning outcomes of our B.A. in English degree program, transforming it from a content-oriented degree with concentrations in literature and language or writing into a more skill-oriented major without concentrations but highlighting skills such as "critical reading and research, as well as strategic, creative, and critical communication" (Winthrop University English department, 2016). This revision of the major was driven by the department's desire to empower students to recognize the value of an English degree and the transferrable skills they would gain from this program that would prepare them for a variety of careers. This change also resulted in a renewed emphasis and interest in writing and rhetoric courses.

I joined the department in the fall of 2018, expected to contribute to the rhetoric and professional writing curriculum and coordinate internships. I was very excited to learn that I would be teaching a 300–level technical communication (TC) course in my second semester. But very little direction or record was given to me about what this course was meant to be, largely due to the lack of a TPC programmatic structure; thus I had the freedom to design it however I wanted it to be. While I enjoyed this freedom and was thrilled to design a new course, I also immediately recognized the challenges as I began to conceptualize this course and its learning objectives. Designing and teaching the course proved to be challenging due to a variety of factors, the first of which was the limited number of courses we offer on technical communication, thus positioning my class as an introductory window to a vast field with knowledges and practices that might not always be later explored in more depth in other courses. Further, the lack of a TPC program and the changes of personnel provided limited infrastructure for sharing resources and developing interpersonal relationships among instructors in a consistent and sustainable manner. In fact, I only discovered after I finished the semester that this course was taught remotely online three years earlier by an adjunct instructor. I was able to obtain his syllabus and course website, which could have been helpful in my planning process the previous semester.

Therefore, when designing the class, I was faced with an incredibly challenging question: If this was the first class where students would learn about technical communication or even the only class where they would learn about some of the fundamental concepts and theories of TC, what should they get out of this course? I had to situate this question in scholarly perspectives on TPC curriculum design.

"Basic" and "Intro" Courses in Technical Communication

In their research on U.S. TPC undergraduate degree programs, Melonçon and Henschel (2013) categorized courses in curricula they studied by course description and purpose. Two categories are especially relevant to my project here: "basic" and "intro." The basic category refers to "introductory courses to the *practice* of technical and professional writing and communication" (Melonçon & Henschel, 2013, p. 51). The authors also mentioned that often this course is also "the 'service course' for other departments" (p. 51). The intro category refers to "[c]ourses that are an introduction to the *field* of TPC. Unlike the basic course, the intro course establishes the history and theories of the field, and then prepares students to produce or create professional documents" (p. 52). It seems that the distinctions here are driven by the different perceptions of technical communication as a discipline versus a profession. While the basic course focuses more on what a technical communicator does, the intro course provides students with more disciplinary and theoretical content about technical communication as a knowledge-producing discipline.

By comparing their study to Sandi Harner and Anne Rich's (2005) survey of undergraduate curriculum in scientific and technical programs in the US, Melonçon and Henschel (2013) showed a curricular trend where the number of courses in the basic category had decreased, which was partly attributable to the diversifying of the degree programs. In fact, Harner and Rich (2005) only used "technical communication" as a category but did not distinguish "basic" from "intro" as Melonçon and Henschel did. Nancy Allen and Steven T. Benninghoff (2004) had already identified a variety of central topics covered by most courses in TPC programs: "audience, genre, visual rhetoric, document design, rhetorical analysis, collaboration, ethics, user-centered design, project management" (p. 165). By December 2011, when Melonçon and Henschel (2013) verified their programmatic data collection, some of these topics already had their own designated courses: "The decrease in the percentage representation [of the 'basic' course category] also could be attributed to the diversity of recent course offerings and to the fact that many of the basic skills covered in this course could be divided and completed in other courses" (p. 57). Melonçon and Henschel (2013) thus concluded that the field of TPC had become more defined and more mature. However, as the field has become more mature, some institutional contexts may not yet align with the trends of the field. Therefore, the local challenge becomes how to develop new curricula that reflect the current trends of the field but also accommodate the needs of student populations and respond to institutional constraints.

While Melonçon and Henschel's categorization can be useful in investigating curricular trends in the field, local institutional contexts may often require more nuanced understandings and adaptation of courses. As I described before, my institutional context without a TPC program determined that I must design a course without guiding programmatic objectives and learning outcomes. However, my institution does have a minor in writing that allows students to focus on professional writing or creative writing, in which my course is an option. Within the minor's requirements are other writing courses that do cover some of the core topics of technical and professional communication:

- WRIT300 Rhetorical Theory;
- WRIT367 Editing for Professionals (this course has not been taught in a while);
- WRIT43X: Academic Internship in Writing;
- WRIT465 Preparation of Written and Oral Reports (the service course);
- WRIT501 Writing for New Media;
- WRIT502: Digital English Studies: Literature, Rhetoric, and Technology;
- WRIT566 Writing for Sciences and Technology; and
- two other special topics courses at the 300 and 500 level on rhetoric and writing.

By just looking at the course titles, one might discern that these courses would cover some of the topics on Allen and Benninghoff's (2004) list, such as audience, genre, rhetorical analysis, visual rhetoric, document design, collaboration, and ethics. Nonetheless, explicit curricular efforts across these courses that are driven by shared programmatic outcomes for professional and technical communication are lacking, which could have facilitated students' transfer experiences. On the other hand, being one of the two rhetoric and writing faculty in the department, I have another disciplinary partner to bounce ideas off. Since these courses are mostly taught by the two of us, we have been discussing our course design, and pedagogical practices so that we might informally build some continuity across the courses for students and inadvertently facilitate some knowledge transfer.

It is also important to point out that our service course has been primarily serving business students; thus, it has a stronger emphasis on professional communication more broadly than technical communication. My course WRIT366 can take on the role of both the basic and intro categories as it introduces students to technical communication both as a field and as a profession, in theory and in practice. Within our majors, many students were not familiar with technical communication. On the other hand, there was also hope that this course would draw other students into our English major and/or professional writing minor. The purposes of this course were certainly multifold.

■ Designing WRIT366: Technical Communication

With this complex role and multiple purposes of this course in mind, I began to determine the important topics the course should cover and to conceptualize how to introduce students to these topics via readings and course assignments. Below is the course description I came up with:

> This course introduces you to the field and profession of technical communication. Technical communication refers to activities of preparing and delivering written and oral documents that present specialized information in a way that allows non-specialists to understand the information and use it to perform tasks. For example, a software company needs technical writers to develop documentation for their software packages; a non-profit organization needs technical writers to develop and maintain content for their websites. Technical writers provide a bridge between technical experts and non-specialists. You will learn the theories of technical communication, how to conduct research to solve workplace communication problems, how to retrieve, evaluate, and present information for different types of audiences in different genres in ethical and legal ways. In turn, you will explore what it means to be a technical writer and develop an understanding of technical communication with a social justice perspective. (Chen, 2019)

This description aims to provide a straightforward explanation of technical communication as a field and a profession by using simple language and examples to help students conceptualize what this course might cover since many of them might not know what technical communication entails. At the same time, this description also mentions some of the important concepts this class would cover: ethics, social justice, genre, etc. From here, I developed a more detailed list of student learning objectives to show students that in this course, they would

- explore what it means to be a technical writer;
- develop a critical understanding of technical communication with a social justice perspective;
- understand the relationships among language, knowledge, and power, including social, cultural, historical, and economic issues related to information, writing, and technology;
- understand that writing is driven by specific purposes and audiences and rhetorical situations;
- understand that genres are socially and rhetorically constructed;
- develop skills to communicate technical information to non-specialists;
- gain practice in collective decision making, team building, and group projects;
- learn to conduct research to solve workplace communication problems;
- practice technical writing and editing and document design; and
- begin to develop a professional identity as a technical writer.

Here you may notice that not only did I include the traditionally core concepts, but I also foregrounded the social justice approach in this course. Recent scholarship has been drawing more attention to teaching technical communication with a social justice approach, signaling a "social justice turn" in the field (Jones, 2016; Jones et al., 2016; Walton et al., 2016). Angela M. Haas and Michelle F. Eble's (2018) edited collection offered us a number of ways to teach technical communication with a social justice-informed pedagogy. Responding to this disciplinary call for more social justice-oriented work, I decided to design the course with a social justice orientation, through both developing course content that pushes students to see how technical communication can be oppressive and empowering to marginalized populations, as well as engaging with social justice pedagogical practices that foster inclusivity and equality in the classroom.

These learning objectives were developed to cover the five core conceptual skills Sally Henschel and Lisa Melonçon (2014) developed: rhetorical proficiency, abstraction, social proficiency, experimentation, and critical system thinking. I will discuss more how these conceptual skills were realized through these learning objectives in my analysis of student reflections later. Rhetorical proficiency is the most fundamental conceptual skill in TPC curriculum; students need to understand how to write for different kinds of audiences and purposes and that the different genres of technical communication should be understood rhetorically and as social constructs. It is also important to understand first and foremost that existing systems and structures contribute to the rhetorical contexts of technical communication; therefore, it is important to analyze and critique those structures. From there, students can learn to research, write, and organize information and content rhetorically and ethically. Because of the central role technology plays in the field of technical communication, students should be able to enhance their technological literacies in this class as well. Finally, students need to recognize that technical communication, just like other writing and research practices, is a social endeavor and that technical communicators actually spend more of their working time communicating with others, such as subject matter experts or other tech writers and editors, rather than actually sitting at their desk writing alone.

Based on these course objectives and pedagogical goals, I selected Heather Graves and Roger B. Graves' *A Strategic Guide to Technical Communication* (2012) and Krista Van Laan's *The Insider's Guide to Technical Writing* (2012) as primary texts supplemented by other texts written by both scholars and practitioners of technical communication. Graves and Graves' text resembles similar professional and technical communication textbooks, with chapters on major genres as well as important topics such as document design, style, and presentation. Van Laan's book is written primarily for practitioners, offering useful descriptions of what it means to work as a tech writer, especially in the software industry, as well as practical advice and sources for people to jump-start their career in the field. I assigned a variety of scholarly articles and other resources that would introduce students to important concepts, such as social justice and feminism, as they relate to technical communication, as well as providing them with more in-depth guidance and resources on certain practices, such as technical editing.

Major assignments included a white paper and a software documentation project that required students to take a critical view of technology and ground their work in human experiences while practicing writing rhetorically about technology to non-expert audiences. They would also collaboratively write a research proposal to explore a workplace problem, which would help them develop a problem-solving view of technical and professional communication. Another major project enabled students to gain editing experience and skills by completing two editing reports that would require them to practice both comprehensive and copy editing and include an explanation of their editing objectives and justifications for editorial changes and comments. Finally, students would design an online portfolio where they could begin building their professional identity as a technical communicator and showcase their work. Students would also reflect on their learning processes from all these assignments. Along with these major assignments, students also completed 19 notecards throughout the semester with prompts that ranged from guided responses to readings, reflections on course activities and my teaching, to short writing exercises, etc.

To further enact a social justice pedagogy, I planned to regularly collect students' thoughts about the readings, discussions, and activities of the class and to use contract grading in this course to ensure a more equitable assessment of student work (Medina & Walker, 2018). These pedagogical decisions reflect some of the principles of the "apparent feminist pedagogy" Erin A. Frost (2018) laid out, such as ensuring students read a variety of materials and paying attention to their genres, creating space for students to reflect on their positionalities and their instructor's positionalities in this course, and leading students to consider the situatedness of the authors of technical documents.

Due to the institutional limitations I mentioned earlier, I perceived the needs for sustainable course development and curricular design practices. Therefore, I obtained Institutional Review Board (IRB) approval the semester before to collect student work and reflections to conduct research on this class, and also as a pilot study in preparation for later curricular development on technical and professional communication. I recruited students to participate in this research by obtaining permission to use their written work (excluding the online portfolio) and reflections on all major assignments in the research. Because I was also the teacher of the class, I made sure that students understood that their participation would not impact their grades in the class. And I waited until the class ended before I looked at the informed consent forms to learn who opted to participate. In the end, 14 students out of 16 enrolled consented to participate, and 13 of them provided written work (one student did not submit any major project). In the next section, I will share the lived experiences of this course: what went well and what did not. Using both my pedagogical narrative and reflections and the analysis of student reflections on major assignments to support my discussion, I hope to provide a comprehensive picture of my trial and error in teaching the first iteration of this course that will be helpful to instructors working in a similar institutional context.

■ Teaching WRIT366: Technical Communication

I had a grand plan for this class, but not everything worked out as well as I hoped it would. Throughout the semester, I kept a reflective journal on this course to note down my reflections of my teaching methods and things that I would change in future semesters based on how students reacted to the course. Thus, I will start with a brief teaching narrative.

One of the first tasks I performed at the beginning of the semester was to distinguish this class from the service course, WRIT465: Preparation of Written and Oral Reports. I made it clear that my class was not going to necessarily prepare students for communication in their respective fields; instead, it was aimed at introducing them to the field of technical communication and preparing them for the professional careers of technical communicators. This distinction was important to make, especially when there aren't explicit programmatic goals or narratives that would delineate the roles of a variety of courses.

Contract grading allowed students to plan and set their own goals for the class, which might have helped ease some concerns and anxieties students had learning about new concepts and practices. If students completed the four major assignments in good standing and did not miss more than four class notecards and four classes or violated my professional communication policy no more than four times, they would receive a B for the class. Completing major assignments in good standing meant that they had to turn in all components on time and meet assignment requirements with good efforts. To receive higher grades, students simply had to revise previous work in this class to put on their professional website. They could also complete an optional assignment that asked them to write a letter to future students taking this class; this letter served both as a reflection and review for them of everything they had learned in this class and as a great teaching tool for me when teaching this class again in order to support future student learning with lived experiences and perspectives of former students. Contract grading alleviated pressure from them on producing the "best" work so that they could make mistakes with these new genres they were learning; at the same time, it required students to be better at time management and focus more on the process of their writing and learning.

Two-thirds of the way into the semester, just as we were wrapping up the documentation project, I realized that there was very little time left to work on a group research proposal before the students had to develop the final online portfolio. Many students did not seem to be technologically savvy, and I suspected that they would feel overwhelmed with the final web design project. On top of that, learning about conducting workplace research and developing a research proposal required a shift back to the "academic" side that at the time might seem disconnected from the rest of the class to them. Therefore, I made the decision of cutting the group research proposal assignment in response to these concerns. This change also made me realize that students needed a lot more scaffolding and explicit transfer among assignments in this course; I had simply placed the research proposal at the wrong time in the schedule, and it would have required more scaffolding than time was allowing for.

I intentionally front-loaded the class with more theoretical readings, having students read scholarly articles about rhetoric, feminism, ethics, and social justice. At the same time, I also used both lectures and readings to illustrate to students what it meant to be a technical writer by presenting them with resources and reports from the industry. The quick introduction of both theory and practice seemed to work well, to the extent that students could quickly gain an understanding of what technical communication *was*. But I noticed that they were struggling to engage with theoretical concepts and the scholarly readings in their practices, especially when it came to ethical considerations and how tech writers could serve as advocates for social changes, which I suspected had to do with the fact that there was a stronger perceived emphasis on tech writing in the software industry, and perhaps not enough discussions and activities were given to critical understandings of technologies and the complexities of workplace dynamics and tensions. I certainly felt that I could have done a better job at threading social justice throughout course assignments and activities, especially from a perspective that's more action and change oriented, beyond just the accommodative practices.

I gained this impression both from students' discussions in class and from their reflections on all the major assignments. The reflections for all major assignments asked the following five questions:

- What have you learned from this assignment? What have you learned about [assignment name] as a genre? What have you learned about technical writing through this assignment?
- What was easy about this assignment and what was challenging?
- How did you overcome any challenges to complete this project?
- What ethical considerations did you have when writing this assignment?
- How will you transfer what you've learned from doing this project to other projects in this class and other contexts?

For this chapter, I coded students' reflections on four major assignments (white paper, documentation project, editing reports, and professional portfolio), first using an "evaluation coding method" (Saldaña, 2013, p. 119) with my learning objectives as codes in order to see if students had indeed achieved the objectives I built this class on. I then used the "descriptive coding method" (Saldaña, 2013, p. 87) to capture students' experiences from these assignments, such as what they enjoyed doing and what they found challenging, how they overcame these challenges, and any other significant experiences that they mentioned. I placed these codes into three categories: learning objectives, learning challenges, and perceptions of learning experiences. Because I completed the first draft of this chapter soon after teaching the class, I reviewed the coding and analysis process again during later revisions, having gained some distance from the class and those students. During this coding review, I also "shop talked" (Saldaña, 2013, p. 35) with my writing colleague about my analysis of the data to improve the validity of coding. Of course, these are not perfect measures to ensure research validity and reliability. But under the institutional limitations, this shop talk allowed me to improve my perception of student experiences and help me better situate my analysis in our institutional context. Additionally, it further strengthened the informal exchanges between us—the only two writing and rhetoric faculty—which will be valuable for future curriculum and programmatic development. Next, I will discuss some key insights from this analysis based on the following themes: what learning objectives students met, what students' perceptions of their learning experiences and transfer were, and how the challenges of balancing theory and practice manifested.

What Skills and Practices?

In the introduction, I mentioned the question I was faced with when beginning to design this class: If this was the first class where students would learn about technical communication or even the only class where they would learn about some of the fundamental concepts and theories of TC, what should they get out of this course? The coding shows that students did gain something along the lines of the conceptual skills Henschel and Melonçon (2014) developed (p. 8):

- Rhetorical Proficiency: compose content for a variety of audiences and purposes
- Abstraction: discover patterns and meaning, rearrange information in new ways
- Social Proficiency: collaborate, negotiate, and achieve consensus
- Experimentation: try new approaches and concepts
- Critical System Thinking: understand the processes by which parts are linked together; the ethical responsibility to consider ideological/power stances of those structures and critique when necessary

Out of the 49 reflections coded across four assignments, the top five learning objectives most frequently coded were understand writing is driven by specific purposes, audiences, and rhetorical situations (32); practice technical writing and editing and document design (29); explore what it means to be a technical writer (26); understand the relationships among language, knowledge, and power including social, cultural, historical, and economic issues related to information, writing, and technology (23); and understand that genres are socially and rhetorically constructed (16).

More students gained the conceptual skill of rhetorical proficiency and recognized that they practiced technical writing like a practitioner. However, fewer of them gained the other conceptual skills: abstraction, experimentation, social proficiency, and critical system thinking. But when they did discuss ethical considerations, they were often cognizant of their writing processes that reflect some critical system thinking and experimentation by talking about how they overcame style, design, and technical challenges in their consideration of rhetorical and technological contexts through problem solving. For example, one student said the following in their reflection on the white paper:

> The ethical challenge with creating this assignment was not showing bias towards the document format that I prefer when creating documents. I had to make sure that I included the limitations of the product as well the benefits of the other programs even if it might have showed the other product in a better light. This also required that I conduct a little bit of research to find out more about the programs I wanted to discuss in my white paper.

Another said in their reflection on the online portfolio,

> I spent the majority of the time trying to configure a website for this assignment. In the end, I had to manipulate the website for three different viewing format [sic]. On a laptop, parts of the website's content is [sic] cut out. I had to rearrange the information so that all of the relevant information appeared. In the process, this new arrangement left blank space when viewed on a desktop.

> I filled this space with a video about my beliefs on education. It took a lot of work, but it should be accessible on multiple formats.

By reflecting on these processes, students articulated the awareness that they had been practicing what technical writers and editors would be doing, and that they learned firsthand what it meant to be technical writers, including the complexities and nuances of the profession. For example, one student wrote in their reflection for the documentation project, "Technical writers for software documentation must be willing to adjust to the feedback received from their usability testing participants, but also make decisions that best benefits [sic] the end users, even if it goes against comments from their testing results." Here, it's important to note that students more frequently perceived they were exploring what it meant to be a technical writer in the documentation project and the editing report project, seeing those as more practically what technical writers do, resulting in a limited view of the profession. This might also be caused by my choice of Van Laan's textbook, which focused on the practitioner's perspective in the software industry.

Learning and Transfer

My coding of students' perceptions of their learning experiences also revealed what most supported their learning and transfer. While I cannot argue that students will successfully transfer what they have learned in this class to other contexts, some transfer did occur among assignments within the class, and students also recognized other explicit sites of transfer. So while our department offers a limited number of specialized technical communication courses, students could already see how this course might prepare them for a technical communicator job or for tasks in other contexts. As our writing internship coordinator regularly asking students to talk about knowledge transfer from their courses to internships, I have already seen student interns who have explicitly discussed how our writing and rhetoric classes prepared them for their internships, which is very gratifying.

Some students found peer review helpful in their learning processes. Although I had to cut the group project, thus ridding students of a collaboration opportunity, they were still able to gain some social learning experience by commenting on each other's work, learning from each other, and troubleshooting with each other. One student said in their white paper reflection,

> The feedback from the peer reviews was the best way to overcome most challenges I faced because it required others to be able to understand what you said and determine if you did an effective job of creating the document and communicating the information.

Another said in their professional portfolio reflection,

> The main thing I had to do to jump over some of the hurdles I found myself in was to just ask someone who I knew was also doing the same thing. Most of the time, they had encountered the same problems and they were able to show me how to fix it.

Peer review thus not only offered opportunities for collaborative learning but also more authentic situations for composing where students would interact with a suitable audience.

I saw explicit transfer and connection between assignments, especially in the final online professional portfolio assignment. One student said, "I think it was a good way to finish of [sic] the class with an assignment that would incorporate everything that we have learned about technical communication throughout the semester." Students also drew connections between what they did in this class and what they had done or would do in other contexts; here, transfer is a two-way street for them. One student said in their professional portfolio reflection, "Because I understood the relationship between written text, purpose, audience, usability and design from this assignment, I will be able to apply them everywhere and in different contexts." One student mentioned that they were familiar with usability testing because they had done it in their digital information design classes; another said the documentation project reminded them of something they had done for their broadcast concentration. While I think I did a fairly good job at giving students an explicit rhetorical education and by designing authentic writing opportunities—two principles Elon Research Seminar on Writing Transfer participants laid out to support writing transfer—I certainly could have done more in providing them with "strategies and tools to think about how writing functions in communities" as well as discovering more what dispositions would better afford their transferring experiences (Moore & Anson, 2017, p. 10). And those authentic writing opportunities should be enhanced more to help students see how technical communicators could be advocates and agents of change.

Theory Versus Practice?

The gap between theory and practice widened when students began to create technical documents without being able to explicitly apply a social justice perspective to the work they were doing, especially with a more action-oriented approach to diversity and inclusion of different cultures that Natasha Jones and Rebecca Walton (2018) argued for. It was more difficult for them to see how writing documentation for a software required a social justice perspective. For example, in their reflections, they talked mostly about ensuring document accessibility, using gender-neutral pronouns, and being objective about their products (to not exaggerate and to acknowledge limitations in their white papers), which are certainly very important, but more accommodation-driven and less advocacy-driven. They knew they had to make their writing accessible and inclusive, but they couldn't always see themselves as agents of social change in the practice of their writing for this class. I suspect that this was due to the limitations of the genres of the assignments and the lack of depth and breadth of discussions on social justice in class. As Rebecca Walton, Kristen Moore, and Natasha Jones (2019) reminded me, "[i]t's impossible to reject and replace injustices if you can't recognize them" (p. 133). While theoretically, we explored how technical communicators could be agents of social change by reading prominent scholarship by Steve Katz (1992), Melody Bowdon (2004), and Emily J. Petersen and Rebecca Walton (2018), not enough time was devoted to exploring these perspectives more in-depth in connection with more "real life" examples. When students created their own projects, the situations and topics they worked with only provided them with a more accommodative view of building accessible and inclusive content rather than an active change-oriented view for the writing decisions they made. For example, they recognized that they needed to provide captions for visuals to make them accessible for users who might be visually impaired. However, they might not necessarily recognize the structural and systemic inequities and oppressions that technical communication can enhance or combat.

Nevertheless, some students did gain an understanding that technical communicators, as argued by Johndan Johnson-Eilola (2004), do not hold just a supporting or auxiliary role to technologies or software developers but are crucial in creating user experiences and advocating for users. For example, one student wrote in their documentation project reflection, "Through this assignment I have learned that technical writing has a very big influence on people because it is technical writers that provide the information to users that they need to be able to use a product or service." The accommodative view is the first step for them to move toward a deeper reflection on social justice and technical communication. In order to push students for deeper reflections on the social justice perspective of technical communication, I need to provide them with more opportunities to expand their perception of what technical writers could do in various sectors, and practice and articulate the kinds of influences they could bring to diverse people's lives.

▮ Conclusion and Looking Forward

To conclude this chapter, I will offer some thoughts and questions on both the development of an intro to TC course and the pedagogical practices in such a course with respect to programmatic development or the lack thereof. Further, I will emphasize the values of informal exchanges and infrastructure for fostering inter- and intra- institutional connections in supporting this work and research, as well as how it should be acknowledged and recognized.

Thoughts on Course Development

When I designed this class, I was afraid that I was trying to do too much by trying to cover too many conceptual skills in just one course. Upon reflection at the end of the semester, I also worry that I didn't do enough. For example, student reflections showed that my course wasn't able to focus more on the conceptual skill of "social proficiency" (Henschel & Melonçon, 2014). Similarly, some of the important TC skills were not as explicitly emphasized in my course. For example, more knowledge of business operations, knowing how technical communicators fit in an organization and how to navigate organizational culture; improved interpersonal communication skills; and project management are cited to be useful for increasing the marketability of the students across scholarship (Kim & Tolley, 2004; Rainey et al., 2005; Whiteside, 2005). Should these other skills be incorporated in an introductory course? If so, how should we introduce students to these skills without overburdening them with extra course work? If not, what types of skills, both conceptually and practically, should be emphasized in an introductory course? Ultimately, how could I bring the critical system thinking more to the foreground in this course, especially without tethering to programmatic goals?

In a way, these questions are intimately linked with the challenge of balancing theory and practice in such an introductory course, which I discussed earlier. Jones and Walton (2018) showed us how to use narratives to teach students to develop a critical perspective on social justice issues and apply it to technical communication. Walton et al. (2016) proposed three strategies to help frame courses with a social justice perspective, which were formulated based on service-learning courses. Is service learning the answer to help students see technical communicators as advocates? Other than service-learning courses, can we offer students other learning opportunities by perhaps constructing "conditional rhetorical spaces" (Anson & Dannels, 2004) that allow students to apply theories to hypothetical scenarios?

With these questions in mind, in future iterations of the course, I planned to spend more time earlier in the semester exploring theories of technical communication with practical examples for students to analyze before moving on to more production-based work. Instead of having students practice several genres, as I did in this iteration, I might ask them to focus on one main genre, such as a documentation project. Moving away from a production-heavy format to a more balanced analytical and production model, I hoped students would be introduced to a larger variety of technical writing genres in order to ground the theories for them even if they don't get to practice writing many of them. At the same time, I hoped to offer scenarios that can inspire more authentic composing practices and broaden their view of what technical writers could do, especially as agents of change and advocacy. In the meantime, we could devote more time to discussions on other issues I wasn't able to cover this time that can be more beneficial in students' future workplaces, such as project management and how to navigate organizational cultures and business operations.

Consequently, I designed the second iteration of the course with a problem-solving perspective and social justice orientation. We spent more time exploring fundamental concepts and theories such as rhetoric, genre, information design, and ethics to develop rhetorical proficiency. For every concept, students worked with practical examples in homework assignments and in-class activities to connect theory with practice. I used scenarios that especially pushed them to think about how technical communicators could serve as user advocates in terms of social justice impacts with a more active perspective so that the social justice theme could be more foregrounded and threaded throughout the course. For example, in one class activity, I gave students a list of phenomena that took place in China during the early emergence of the COVID-19 outbreak and asked them to come up with best TC practices in crisis response that would explicitly actively address the possible oppressions inflicted on different populations. In the second half of the semester, students worked on a collaborative documentation project in groups for different campus clients, which strengthened their critical system thinking skills and allowed them to work in a more realistic professional setting. Short of a service-learning component, this client project at least helped students improve upon project management, collaboration, and interpersonal communication skills while practicing a major technical communication genre. While I have not analyzed this semester's data, my perception as an instructor is that this problem-solving and social justice-oriented course design with scenario-based practices can be a useful way to marry theory and practice together and offer students a good window into the field and discipline of technical communication.

Thoughts on Pedagogical Practices

One of the most beneficial parts of this experience teaching this class for the first time was my effort to create open dialogues with students as equitable pedagogical practices, especially in a class where students might be overwhelmed by the workload and challenging content. This should be done both in terms of having students communicate with the instructor on their learning experiences regularly and maintaining a good interpersonal rapport with students. Students need to consistently reflect on their learning experiences in the class, and instructors need to be reflexive with them as well.

I had suspected that my own positionality and identity as a woman of color might have an impact on how students would respond to my pedagogical practices. But I did not experience any challenges in this regard. On the other hand, I did experience some unexpected personal challenges. On top of having two new course preps in my second semester on the tenure track with a 4–4 teaching load, we lost a close family member to cancer. For a full month, I was consumed by grief and stress. In line with my equitable pedagogical approach, I told my students what I was going through, partly also to model a practice that I hoped my students would do with me. In fact, several students, from this class and others, told me their own struggles that were impacting their performance in class because they were encouraged by what I had done. This open dialogue was crucial in supporting student learning, especially in such a challenging course. Since we teach students human-centered technical communication, we must practice first seeing ourselves and our students as humans with real emotions and recognize the interdependence of our personal and professional lives.

Finally, when instructors are asked to develop a class like this, they must teach it with a great degree of flexibility and adaptability, for instance keeping open spots in the schedule and offering optional assignments to adapt to student needs and asking for student feedback on their ongoing learning experience. Maintaining flexibility is not only an important feminist approach to teaching but also useful in new curricular development situations. Because in a context where it is difficult to predict how students might respond to the course materials, it is all the more important to be flexible and adaptable and dialogic. Of course, this must be explicitly communicated to students early on as well. These equitable, inclusive, and flexible pedagogical approaches are just one small way to enact the social justice turn in technical communication pedagogy.

Thoughts on Professional Development

My course design and research process also indicate that in institutional contexts where rhetoric and writing curriculum is small and limited, instructors need to build professional networks with intra- and inter-institutional connections to help one another with curriculum development and pedagogical practices, especially when more formal programmatic structures are lacking. I know I could not have designed and taught this class without all the conversations with my fellow writing faculty in the department, and I certainly benefited from the larger TC professional community I'm attuned into on Twitter and various professional listservs. These support networks are crucial in our growth as teachers and researchers. Thus, it is important for us to advocate for such collaborative and supportive professional environments from within departments, institutions, and professional organizations, such as recognizing the values of collaboration and peer learning in faculty evaluation mechanisms like tenure and promotion guidelines.

Acknowledgment

I would like to thank Devon Ralston for her help with data analysis and for her continuous generous support as a colleague.

References

Allen, N. & Benninghoff, S. T. (2004). TPC program snapshots: Developing curricula and addressing challenges. *Technical Communication Quarterly, 13*(2), 157–185.

Anson, C. M. & Dannels, D. (2004). Writing and speaking in conditional rhetorical space. In E. Nagelhout & C. Rutz (Eds.), *Classroom space(s) and writing instruction* (pp. 55–70). Hampton Press.

Bowdon, M. A. (2004). Technical communication and the role of the public intellectual: A community HIV-prevention case study. *Technical Communication Quarterly, 13*(3), 325–340.

Breuch, L. K. & Sadler, V. (Eds.). (2016). Programmatic research [Special issue]. *Programmatic Perspectives, 8*(2).

Bridgeford, T. (Ed.). (2018). *Teaching professional and technical communication: A practicum in a book*. Utah State University Press.

Cargile Cook, K. (2002). Layered literacies: A theoretical frame for technical communication pedagogy. *Technical Communication Quarterly, 11*(1), 5–29. https://doi.org/10.1207/s15427625tcq1101_1.

Chen, C. (2019). WRIT366: Technical Communication course syllabus.

Cooper, M. (1991). Model(s) for educating professional communicators. In J. Zappen (Ed.), *The Council for Programs in Technical and Scientific Communication Proceedings: 1990* (pp. 1–12). Rensselaer Polytechnic Institute.

Frost, E. A. (2018). Apparent feminism and risk communication: Hazard, outrage, environment, and embodiment. In A. Haas & M. Eble (Eds.), *Key theoretical frameworks: Teaching technical communication in the twenty-first century* (pp. 23–45). Utah State University Press.

Graves, H. & Graves, R. (2012). *A strategic guide to technical communication*. Broadview Press.

Haas, A. M. & Eble, M. F. (Eds.). (2018). *Key theoretical frameworks: Teaching technical communication in the twenty-first century*. Utah State University Press.

Harner, S. & Rich, A. (2005). Trends in undergraduate curriculum in scientific and technical communication programs. *Technical Communication, 52*, 209–220.

Henschel, S. & Melonçon, L. (2014). Of horsemen and layered literacies: Assessment instruments for aligning technical and professional communication undergraduate curricula with professional expectations. *Programmatic Perspectives, 6*(1), 3–26.

Johnson-Eilola, J. (2004). Relocating the value of work: Technical communication in a post-industrial age. In J. Dubinsky (Ed.), *Teaching technical communication: Critical issues about the classroom* (pp. 573–594). Bedford/St. Martin's.

Jones, N. N. (2016). The technical communicator as advocate: Integrating a social justice approach in technical communication. *Journal of Technical Writing and Communication, 46*(3), 342–361. https://doi.org/10.1177/0047281616639472.

Jones, N. N., Moore, K. R. & Walton, R. (2016). Disrupting the past to disrupt the future: An antenarrative of technical communication. *Technical Communication Quarterly, 25*(4), 211–229. https://doi.org/10.1080/10572252.2016.1224655

Jones, N. N. & Walton, R. (2018). Using narratives to foster critical thinking about diversity and social justice. In A. Haas & M. Eble (Eds.), *Key theoretical frameworks: Teaching technical communication in the twenty-first century* (pp. 241–267). Utah State University Press.

Katz, S. B. (1992). The ethic of expediency: Classical rhetoric, technology, and the Holocaust. *College English, 54*(3), 255–275.

Kim, L. & Tolley, C. (2004). Fitting academic programs to workplace marketability: Career paths of five technical communicators. *Technical Communication, 51*(3), 376–386.

Medina, C. & Walker, K. (2018). Validating the consequences of a social justice pedagogy: Explicit values in course-based grading contracts. In A. Haas & M. Eble (Eds.), *Key theoretical frameworks: Teaching technical communication in the twenty-first century* (pp. 46–67). Utah State University Press.

Melonçon, L. & Henschel, S. (2013). Current state of U.S. undergraduate degree programs in technical and professional communication. *Technical Communication, 60*(1), 45–64.

Melonçon, L. & Schreiber, J. (2018). Advocating for sustainability: A report on and critique of the undergraduate capstone course. *Technical Communication Quarterly, 27*(4), 322–335. https://doi.org/10.1080/10572252.2018.1515407.

Moore, J. L. & Anson, C. M. (2017). Introduction. In C. M. Anson & J. L. Moore (Eds.), *Critical transitions: Writing and the question of transfer* (pp. 3–16). The WAC Clearinghouse; University Press of Colorado.

Petersen, E. J. & Walton, R. (2018). Bridging analysis and action: How feminist scholarship can inform the social justice turn. *Journal of Business and Technical Communication, 32*(4), 416–446. https://doi.org/10.1177/1050651918780192.

Rainey, K., Turner, R. & Dayton, D. (2005). Do curricula correspond to managerial expectations? Core competencies for technical communicators. *Technical Communication, 52*(3), 323–352.

Read, S. & Michaud, M. (2018). Hidden in plain sight: Findings from a survey on the multi-major professional writing course. *Technical Communication Quarterly, 27*(3), 227–248. https://doi.org/10.1080/10572252.2018.1479590.

Saldaña, J. (2013). *The coding manual for qualitative researchers.* SAGE.

Staples, K. & Ornatowski, C. M. (Eds.). (1997). *Foundations for teaching technical communication: Theory, practice, and program design.* Ablex.

Thatcher, B. & St. Amant, K. (Eds.). (2011). *Teaching intercultural rhetoric and technical communication: Theories, curriculum, pedagogies, and practices.* Routledge.

Van Laan, K. (2012). *The insider's guide to technical writing.* XML Press.

Walton, R., Colton, J. S., Wheatley-Boxx, R. & Gurko, K. (2016). Social justice across the curriculum: Research-based course design. *Programmatic Perspectives, 8*(2), 119–141.

Walton, R., Moore, K. R. & Jones, N. N. (2019). *Technical communication after the social justice turn: Building coalitions for action.* Routledge.

Whiteside, A. L. (2005). The skills that technical communicators need: An investigation of technical communication graduates, managers, and curricula. *Journal of Technical Writing and Communication, 33*(4), 303–318. https://doi.org/10.2190/3164-e4v0-bf7d-tdva.

Winthrop University English department (2016). *WUENGL program modification proposal.*

7. Regenerating a Once Fallow Ground: Theorizing Process and Product in Twenty-First-Century Technical Communication Ecologies

Adrienne Lamberti and David M. Grant
UNIVERSITY OF NORTHERN IOWA

Abstract: This chapter describes how one institution revised its professional and technical communication program to include more technology and community engagement experiences. The program originally was highly instrumental, focusing on document design skill sets (e.g., use of Adobe InDesign). Before they could evolve the curricula, program faculty needed to ready themselves to invoke technical communication scholarship's historically key talking points regarding theory, because one significant trait of the program's institutional context was a perceived irreconcilable split between theory and practice. Demonstrating to institutional stakeholders a more nuanced relationship between theory and practice justified the teachers' changes to their pedagogical practice. In addition, strengthening their fluency in scholarship's discussions about theory assisted the program faculty in settling upon the specific theoretical frameworks that the revised curriculum embodies: ecologies of practice and civility. Furthermore, increasing community engagement opportunities in the classroom revealed the benefits of incorporating into the curriculum theoretical content knowledge—but without connecting theory exclusively to one particular assignment or project.

Keywords: instrumentalism, theory, community engagement, ecologies of practice

Key Takeaways:

- Although technical communication scholarship now largely fuses theory and practice, the relationship between the two has not been a static one throughout the discipline's history.
- Technical communication pedagogy often privileges application, one result of the discipline's historical emphasis on instrumentalism.
- There are benefits to focusing on theory in technical communication curricula without explicitly attaching it to an application exercise or assignment.

When Katherine Staples and Cezar M. Ornatowski's *Foundations for Teaching Technical Communication* was published in 1997, it entered a disciplinary scene characterized by debates over theoretical frameworks' relationship to technical communication teaching and scholarship, appropriate locations for universities' technical writing programs, implications for technology-mediated communication in the professions (especially regarding distributed work teams), and best and best-for-now workplace practices. In the subsequent decades, although these topics have not quieted in the field, they obviously have altered, and to a degree that may be considered remarkable when compared to their presence in some other disciplines.

We are specifically intrigued by the role of theory in teaching technical communication. There is a traceable thread in our field's literature that discusses theory's place, with many corners of the discipline advocating for theory's existence in the classroom—just so long as it somehow is transformed into an application opportunity in which students can engage. It's been argued that examining theory in the classroom without also enacting it (see Turnley's [2007] discussion of service-learning assignments for an example of theory/practice fusion) contradicts the discipline's pragmatic and instrumental history (e.g., Moore, 1996). It's additionally been suggested that the technical communication field is made less distinctive from others when it is taught from a largely theoretical perspective (even though the field's disciplinary boundaries themselves often undergo redefinition [e.g., Henning & Bemer, 2016; Johnson et al., 2018; Kimball, 2016]). Further, theory is often seen as too universalized and inattentive to institutions', contexts', and places' local exigencies.

Inarguably, there is merit in striking a theory/application balance in technical communication curricula, and in fact this balance has so long been a disciplinary staple that it now may be considered an assumed value within the field. However, we write as technical communication teachers and scholars who nevertheless have continued to experience marked and ongoing contestation of institutional "turf" that is fueled largely by a persistent belief in a theory/application split. In our experience, practical application continues to be regarded by some as a-theoretical, whereby hands-on learning in some way sullies or oversimplifies intellectual effort. Theory-practice debate also muddies the lines between different values about writing, sometimes allowing others to co-opt what we seek to do in our particular technical communication program. Consequently, in our work to defend strongholds gained by the field within academic contexts and demonstrate its value without, we wonder if leaning so heavily on application and on curricula driven by product outcomes has itself become a disciplinary vulnerability.

Other disciplines, such as composition and writing studies, have more and more needed to justify their existence via tangible results, lest they simply concede to institutional forces beyond their control (Skinnell, 2016). Narrowly focused practical programs in communication are similarly feeling encroachment from fields that more explicitly embrace their theoretical legacies, especially because technology is blurring once clear lines among modes of communicative activity. We ultimately find ourselves needing to repeatedly return to this argument: pointedly theorizing both pedagogy and the purposes of technical communication does not have to squeeze out application in the classroom, but rather can enrich it.

The following describes how one university's professional and technical communication program increasingly incorporated visible theory into its curricula as a means of strengthening its institutional role and did so while retaining product deliverable-oriented assignments. Our program in professional writing, which began shortly after Staples and Ornatowski's publication of *Foundations*, initially was almost exclusively skills-based—the bulk of courses focused on teaching document design software. However beneficial this focus for students who would need functional skill sets upon graduation, this curricular content also operated during a time when the role of technology within technical communication was being questioned (e.g., Johndan Johnson-Eilola's [1996] call to reassess the importance placed on technological product-driven work).

Our program is housed in a literature-centered department that, with a few notable exceptions, has not addressed the shift from print-based literacy to other communicative modes. As a result, few literature majors were keen to enroll in professional writing courses, and as faculty, we found it difficult to incentivize enrollment through curricular reform. Luckily, we had allies in communication studies who, with the authors' involvement, founded a new interdisciplinary program, interactive digital studies (iDS). At a time when institutional enrollment had been falling for several years running, this cross-campus alliance benefited both programs with one of the most-enrolled optional "bundles" of the iDS program focused on digital writing.[1] While the professional writing program within the English curriculum remained stand-alone, iDS helped foster the exigency for teaching digital communication as a norm rather than an add-on in professional and technical contexts.

Close attention to technologically-mediated communication prompted us to revise assessment materials, professional development for instructors, and experimentation with potential courses and where concepts and practices might best fit. For example,

- Lamberti worked with staff members who taught the program's introductory course to generate assessment data that responded to their needs in the new landscape; Grant worked with rhetorically-minded allies across campus to provide opportunities for instructional staff to pedagogically respond to new ideas and modes.
- Both authors also reframed courses so that an experimental course in digital writing theory was wrapped into the professional writing pro-

---

1. Other bundles include Marketing, Activism, Digital Imaging, Sound, etc. Core courses are required at the beginning and end of the program, with courses in these areas in the middle.

gram's required course on theories of writing, and many of the courses were re-named "Applied Writing: _____" in order to signal consistency across the program as well as the ways in which students would be expected to use theoretical insights they gleaned across their coursework.

- Lastly, pedagogies were revised to include community engagement projects (students partner with organizational clients to compose workplace communications), allowing students to develop their own strategies to theorize, and to build on communicative strengths already possessed by most students.

Revision of our program needed to unfold carefully; as described later, a great deal of thinking-through had to occur regarding the technical communication discipline's historical discussions about theory/practice binaries, in view of our program's departmental and institutional contexts. Our consideration of the field's legacy was necessary before the revised curriculum could be focused down into an embodiment of particular theoretical frameworks (see Figure 7.1).

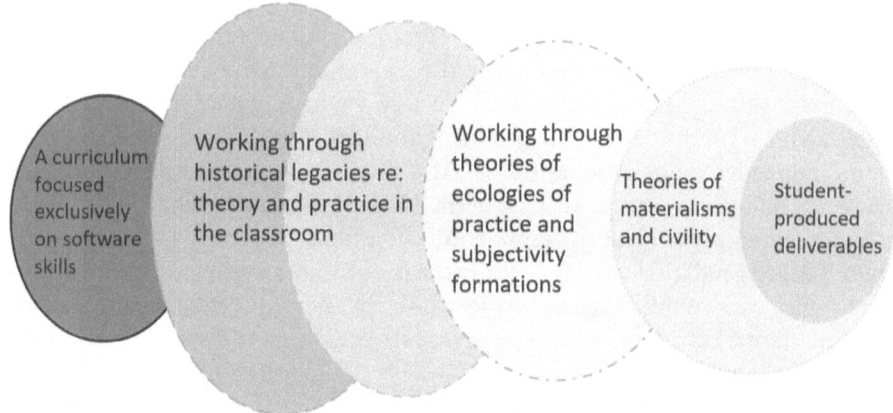

Figure 7.1. The process of revising a technical communication curriculum to more explicitly incorporate theory.

The program ultimately became more overtly theorized through a mindfulness of local ecologies of practice, or how, following Jenny Rice (2012), particular rhetorical practices lead to dynamic subjectivity formations across both private and public dimensions. In our case, how we teach—our own rhetorical practices—affects the ways both students and external stakeholders engage with or resist our curricular aims. The theoretical import which shapes our program derives its measures from the overall functioning and health and vitality of a techno-social web. We will offer examples showing how a theoretical focus specifically on materialisms and civility both evolved the curricula and made students' learning more lasting and robust.

■ In the Literature

Bearing in mind our institutional context, where theory and practice were still perceived in some corners as distinctive entities, we needed to equip ourselves with the historical stances regarding theory and practice that populate our discipline's discussion in order to successfully justify a dramatic change from a software-skills curriculum to a more theoretical, applied-writing pedagogy (see Figure 7.2).

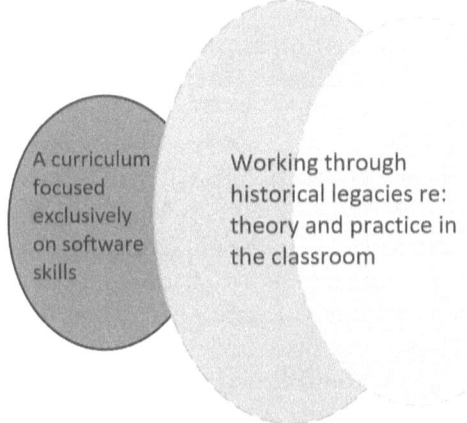

Figure 7.2. Revisiting technical communication's history of discussions re: theory's presence in the discipline.

The past several decades of technical communication disciplinary discussion about theory's role in the field have focused on implications for pedagogy and research, where programs in technical communication may most effectively "live" on a university campus, and how technology-mediated communication affects theories in non-academic professions and practices, among other issues. J. C. Mathes and Dwight W. Stevenson's (1976) definition of effective technical communication teaching and scholarship, and their relationship to theory, are attached to the argument that subject matter experts, such as engineers, are best poised to teach communication in that subject area. Such a claim is partially based on the criteria by which instructors and researchers are recognized for their work in cases of tenure and promotion; those in the English discipline who teach and publish about technical communication, the authors explain, likely would not be rewarded for what then was activity relegated to the boundaries of a literature-centric field. This reason is joined by others—including a quick reassurance that subject matter experts "could do basic research on communication theory" as a means of grounding their instruction in technical communication (p. 333). That this reference is the extent of any discussion of theory in the authors' article is representative of a moment in the discipline when instrumentality and practicality were urged as dominant values of technical communication pedagogy and research. Or, as Mathes and Stevenson put it, "[T]he design of a report [should] be seen as analogous to the design of an engineering system" (p. 333). In such a moment, communication is generalizable and universal, requiring quick study to understand, while the subject matter and context are exact, detailed, and of utmost practicality.

A similar approach to theory can be seen in Mathes and Stevenson's contemporaries' arguments as to where technical communication programs should be housed at universities. Robert J. Connors' (1982) review of technical communication's disciplinary history, instigated by his belief that "technical writing has been accepted as an important part of the discipline of English" (p. 329; this interestingly only six years after Mathes and Stevenson's article), tracks the field's migration across several locations within higher education architecture. From its early twentieth-century ascendancy as a response to institutions founded under the Morrill Act, to subsequent debates in English departments regarding "literature versus vocationalism," to the impact of post-WWII student-veteran populations upon university curricula (p. 341), the physical and philosophical place of technical communication in Connors' history reflects a trajectory of the field that, in its disciplinary theory, values the functional: instruction in technical writing should "increas[e] the efficiency of the work" of writing (p. 332). Yet, even this yen for functionality is cast as insufficiently practical. Connors describes early twentieth-century technical communication theory's focus on "'modes of discourse'" (Earle, 1911, as cited in Connors, 1982) as being too rhetorical, a focus also soon subsumed by a theoretical framework that prioritized genres and their respective—and, it could be said—prescriptive, conventions. Approximately a half-century later, theory moved back to a comparatively rhetorical focus, a shift concomitant with renewed discussions as to whether technical communication programs should live within English departments or elsewhere.

Perhaps surprisingly, it is beyond academic contexts that theory's role in technical communication even more so eschews pragmatism in favor of the less tangible. This especially is seen in technology-mediated professional practices. Wick's (2000) reconceptualization of knowledge management in the workplace—that it should be understood along a spectrum comprised of an organization's documents; technology; socio-cultural factors; and the capital accorded to specialized knowledge—moves philosophy of technical communication from being a product-driven enterprise to being a discernible body of expert knowledge. As it is enabled and supported by technology, particularly within mediated cross-functional work teams, a technical communicator's knowledge includes sophisticated rhetorical recognition as to how each communicative act is a unique sum of nuanced negotiations among these four considerations on the spectrum.

Other workplace practices in technical communication also encourage a subtler theorized approach, often in response to perceived restrictive ways in which the teaching of technical communication is theorized and exercised. In his chronicle of the Association of Teachers of Technical Writing's (ATTW) early years, Donald H. Cunningham (2004) reveals how theory in the professions evolved as the academic discipline moved away from using literature as its primary texts, commenting that his submission to College Composition and Communication of a bibliography that closely resembled his technical writing experience was received by an editor who was happy to see a piece "that actually might be of use to some readers" (p. 123). Cunningham's co-founding of ATTW's journal (now *Technical Communication Quarterly*) similarly was motivated by a dearth of systemic philosophy when it came to the ability for technical communication pedagogy to sufficiently prepare students for actual practice, i.e., work in locations that necessitate agile responsiveness to shifting rhetorical situations. (Indeed, when the Conference on College Composition and Communication demonstrated reluctance to make space for sessions on technical communication, citing a lack of relevance to the [then still mostly literary] manner in which writing pedagogy was theorized, the ATTW initiated its own conference [Cunningham, 2004]).

■ The Evolution from Practicality to Application

Upon scrutinizing how the history of technical communication theory is dotted by frequent moments of strong consensus in favor of the instrumental, we were able to better shepherd our curriculum's revision by explaining to institutional stakeholders how such instrumentality has evolved into forms of application within classroom contexts. For example, Teresa C. Kynell's (2000) account of a century of academic programs in engineering and their tense relationship with English curricula shows how changes in the engineering profession—especially the need for practitioners who could clearly communicate their expertise—eventually overrode a contempt for English curricula, which had been regarded as lacking application. This need created an opportunity for technical communication coursework that fused the humanistic dimensions of English study with an opportunity for engineering students to practice becoming rhetorically attentive to audience (Kynell, 2000).

We also kept in mind how, in addition to logistical need, a similar, perceived philosophical need for the tangible exercise of theory also was in operation. Specifically, studying theory without some form of attendant application was viewed as going against the field's historical identity, as Staples and Ornatowski (1997) themselves imply in Foundation's organizational structure. Their text begins with theoretical basics, but its bulk is devoted to application in practice, as professionalism, and in academic programming. Jeff Todd's (2003) review of the discipline's history, too, is an instance of the trajectory towards a preference for the application of knowledge, here, as the primary manner in which history may be used pedagogically. That is, Todd mentions early within a series of recommended guidelines that teachers look to "canonical works in the field" (p. 69) to maintain a reliable historical understanding of technical communication; these works are subsequently described in his piece as focusing on "technical" discourse, "technical" being used synonymously with "applied discourse" (p. 70).

Such identity formation, maintenance, and even protection are understandable missions for any discipline, particularly one connected to those humanities fields that at this time are enduring another wave of opposition in the U.S. North American socio-political landscape. Adhering to the visible, the countable, in technical communication, such as that offered by application-centered pedagogy, answers questions as to what the field does—actionable words that assist in defining disciplinary boundaries. Mark G. Cooper and John Marx's (2018) survey of the pushback against interdisciplinarity points to larger worries about blurry disciplinary borders as vulnerabilities prey to attack, especially within a higher education context driven increasingly by business models. External and internal forces upon academic and professional fields have prompted a doubling-down on their definition. Jane Tompkins (2018) echoes Cooper and Marx's piece with a cautionary example. Her experience when writing a deeply reflective and personal essay was followed by the sobering challenge of determining how this experience might fit within her pre-existing identity and work as a professor of literature, especially in the classroom. Or as Tompkins puts it, "[T]here's the departmental curriculum committee to conjure with." In the case of the technical communication field, flirting too strongly with the perceived vagueness of theory could be argued as muddying the field's integrity.

■ In the Classroom: Theory-Explicit Application

By tracing significant disciplinary arguments surrounding theory and practice as well as questions about the role of theory in technical communication pedagogy, we were able to subsequently make clear, both to ourselves and to our colleagues, our revised curriculum's focus (see Figure 7.3).

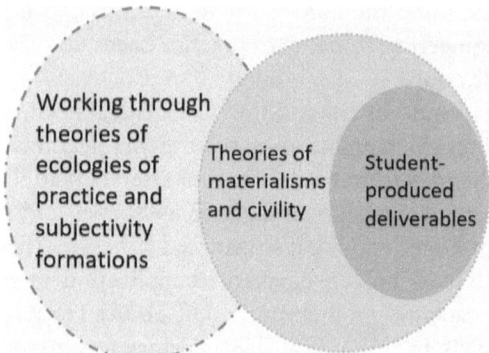

Figure 7.3. Focusing down our curriculum to particular theoretical frameworks.

We offer both our previously described process of exploring the history of theory and practice in the field's literature, as well as one classroom scenario resulting from our revised professional and technical writing program, to serve as examples for others in a situation similar to ours.

Our program's revised curriculum made increased use of theories in materialisms and public civility (Bennett, 2018; Keith & Mountford, 2014; Kynell & Tebeaux, 2009; Lueck, 2018; Robin, 2016) as a means to fortify its presence at our university and distinguish itself from newer programs yet continue to generate student-composed product deliverables. Specifically, after the professional writing program was comprehensively reconceptualized to incorporate community engagement assignments, students began to collaborate with local organizational clients to produce needed communications. Although it predated a larger institutional turn to student engagement, the program has been enhanced by institutional support, winning engagement grants, becoming recognizable in the wider community, and creating jobs in advertising and local industry.

We feel this civility-driven approach to applied communication projects meshes well with Rice's (2012) starting point in her "publics approach to place," which she details as a "look at the way . . . discourse helps to create particular kinds of *public subjectivities*" (p. 13, emphasis in original). In other words, rather than understanding students simply as private producers of texts, we view students as ecologically embodied subjectivities who can conduct themselves toward purposive ends, harnessing available energies and circumstances to achieve iteratively refined goals. That is, students solve the problems presented to them through an ecology and adopt a subjectivity of agential problem solver because they cannot be seen as "outside" the problem.

Also, the external stakeholders who partner with our students needed some theorization on our part in order for us to understand their role(s) and to grow our practices beyond regular skill-and-drill routines. In this sense, the program has struck a balance by not only incorporating and applying more universal theories, but also retaining specificity in its application assignments by focusing on local ecologies of practice (Fleckenstein, 2003; Wardle & Roozen, 2012). While theory's uncemented place in the technical communication field, as we have described it, can be attributed to anxiety that a theory-driven approach in the classroom cannot scope down to the uniqueness of a specific communicative situation in the same manner as application (Richardson & Liggett, 1993), our recognition of the potential for a theory-rich curriculum and our consequent programmatic revision suggest otherwise. Below, we detail classroom examples as to how this evolution was enabled by careful attention to particular theoretical concepts, without which, we believe, students' applied learning experiences would not be nearly as substantive or lasting.

In one of Lamberti's recent courses that collaborated with organizational clients, students working to produce a tourism video for a prisoner of war (POW) museum found themselves struggling ethically when the museum client demanded an exclusively positive "spin" on the video content. Had the students not been immersed in a theory-driven curriculum, their reaction likely would not have been as complex—or the resulting video as nuanced.

This particular class was an introductory course focusing on professional communication, populated largely by students coming from business communication and digital technology programs characterized by strongly instrumental curricula. The course included an assignment whereby a student team worked with a local museum that chronicled the lives of POWs in Iowa during WWII, to produce a film script for what eventually would become a video shown during guided tours for elementary school-aged children. As the students already possessed scriptwriting and video creation experience, their first impulse was to immediately begin production on the video itself, using content provided by the client (descriptions of POWs' daily schedule, work assignments, etc.).

The impulse to move directly into the creation of the actual product deliverable (video) arises not only from the assignment-and-deadline-driven structure normal in a classroom context, but also from a larger efficiency-and-product-deliverable-driven model of project management that characterizes Western workspaces and can operate at the expense of reflective practice (Lauren, 2018). Lamberti encouraged students in the class, however, to theorize their project management communications and their product deliverable's development process as having an ecological, symbiotic relationship (Fleckenstein et al., 2008); that is, to not assume their project management communications as being positioned in *response* to the development phases of their product deliverable, but rather that the communications and process mutually influenced and evolved one another.

Had theory not been deliberately introduced into class readings and discussion, any student's reflection upon their communications during the project, insofar as they facilitated the product deliverable's development process, likely would have only confirmed a project management efficiency paradigm—e.g., the development of a goal-oriented project plan and a map of a lock-step project lifecycle (Lauren, 2018). The comparative heightened complexity of a theorized relationship between communication and development process was especially noticeable during moments of conflict between what was expressed during student communications and how the students created their video.

As the project continued, students were able to spot the growing tension between the sociocultural consequences when composing project management communications and the consequences of their video production work. Using a theory-rich curriculum, Lamberti prompted the largely middle-class White students, who were accustomed to ready access to higher education and digital technology, to explore how the normalcy of their resources shaped their project management communications in a manner that (however unwittingly) confirmed their privilege. Rather than leaning especially on prior humanistic or social science learning, students were coming to understand how truly interdisciplinary technical communication can be. Meanwhile, the client was insisting that the students' video production decisions result in an exclusively positive depiction of POWs' lives. As revealed by the theorized, symbiotic interchanges between students' communications and their video work process, the chasm between the students' reality and that of the POWs became too irreconcilable to ignore.

Students' comments from one of their project management communications, an informal log maintained throughout the duration of the assignment, underscore the impact of the course's theory-driven curriculum. As one member of the student team wrote in his project log,

> We brainstormed how to make our script filled with information for children without watering down unflattering parts of America's past. There is a fine line to figure out how to portray the information we were given [by the client] to children[,] so they know that not everyone was treated well[, yet] without ruining [the children's] day. The story of an individual [POW] is our best route [as the focus of the video] because it can show what life was like for one man, who might not be the norm[. M]any people came through [the POW camp,] so not everyone was treated nearly as well as our individual. . . . It is difficult to inform kids about such a gritty aspect of human life during any war, but we cannot hide such things of American history. (C.W., December 13, 2018)

As another team member wrote,

> Our client . . . provided us with several articles from WWII detailing the lives of the POW members. Although the articles were informative, they still aired [sic] on the "Hakuna Matata" side of POW life, making us feel like we were watching an episode of Hogan's Heroes. The team and I knew that [that] was not the case, and we wanted to portray the harsh reality. This led us to include the following section in our final [video script]:
>
> NARRATOR
>
> By 7:30 a.m., we began the workday. Sounds WAY too early, right? You got used to it over time. Some of the men made their way to the field, while some stayed behind and worked in the camp; mowing, cleaning, and making it look nice.
>
> The work is hard, but it's better than sitting in a jail cell all day. Working with a large group of people makes the day go faster.
>
> Several shots of workers doing different tasks. (J.B., December 10, 2018)

Ultimately, students made the decision to rhetorically resist the image of the "contented prisoner" during their script production decisions. A curriculum that had been characterized by an overly efficient and pragmatic theory of technical communication, in contrast, would likely have replicated a sense of resignation and an apathetic subjectivity in the students. They probably would have dismissed their own participation and circumscribed themselves to a private transaction between them and their client. Instead, the students were affectively impacted; they were not just concerned with how their deliverable may influence its intended audience with a happy narrative of POW life, but they were simultaneously moved by the conflict between what they perceived to be true and what their client requested. Rice (2012), describing the importance of affect in the ecology of public rhetorical engagement, argues that a focus on affect is not "to revisit the old binary between feeling/rationality," but to "simultaneously affirm that feeling is one way we encounter and interact with others" (p. 59). Situating our students as participants in a public space enables them to experience on a complex level their wider roles in civic society.

Our program's pointed embrace of theory, as we have discussed and exemplified, allows an important shift for students and faculty by encouraging greater collaboration and fruitful exchange with creative writers, journalists, digital writers, and teachers—all significant occupational aspirations of local students. Such collaboration is the active doing of technical communication in context. This is a context that Carl G. Herndl and Lauren L. Cutlip (2013) argue entails a move "from analysis of science and its discourse to collaborating in the management of uncertainty." As seen in the example from one of our classes, students who were accustomed to the comforting clarity of a product deliverable's development process nonetheless were persuaded by a theorized curriculum to test unfamiliar waters, by rhetorically evading a client's problematic expectations. Like the broader field of rhetoric of technology, science, and medicine pointed to by Herndl and Cutlip, theory-staked technical communication builds ecologically through "interdisciplinary alliances, engages with our colleagues in science to help manage uncertainty and the threat of ecocide, and develops specific strategies and tools to put into practice our disciplinary intentions to make a difference." We maintain that seeing our students and ourselves as embodied, feeling, and decision-making beings within the ongoing developments of social and institutional ecologies achieves that outcome.

■ Final Thoughts: Theory's Implication of Faculty

As institutional members, we must pull out specific assessment measures for our own programmatic purposes. Still, rubrics for gauging affective public roles, such as those taken up by Lamberti's students, nonetheless are open to modulation and guidance. Assessment measures are not developed out of nothing, but as responsive and responsible conditions of our own context, which in turn is necessarily dependent on its institutional ecology. Kristie S. Fleckenstein (2003) notes how meaning in an ecology "is triadic, involving at least two organisms within an environment, all of which are mutually constitutive, mutually dependent" (p. 166). In our case, we cannot separate out our program or the students in our courses, because our institutional ecology has nestled us among digital studies, business communications, public relations, and other institutional needs. And we gladly accept this, for instance, by designing assignments that distinguish between technical communication roles of subject-matter expert (SME) and professional technical communicator.

Ecological assessment also allows us to consider the nature of freedoms balanced alongside public good. Democratic participation is not confined solely to the public sphere, but neither is limitless in private. Indeed, we are acutely aware of how public dissemination of technical knowledge is at a premium, a fact heightened even more by the COVID-19 pandemic. It was through Drew Harris' (Roberts, 2020) "flatten the curve" infographic that, a week before our university moved its classes to online-only, Grant's students were learning about the upcoming future, able to plan for, and accordingly adjust their own conduct.

When we assess our pedagogy and curricula, then, we value balancing democratic freedoms with public knowledge. Elizabeth Wardle and Kevin Roozen (2012) maintain that "ecological assessment recognizes and acts from the assumption that the breadth of students'... literate experiences—in and out of school—impacts their ability to 'do' academic literacy tasks" (p. 107), and we similarly recognize that technical writing education situates students at the nexus of school and workplace. As Staples and Ornatowski (1997) envision it, "The technical communicator emerges as an educated decision maker whose professional decisions are informed by critical thinking, skills, theory, application, ethics, communication ability and knowledge of and about technology" (p. xii). Meanwhile, William Keith and Roxanne Mountford's (2014) "Mt. Oread Manifesto" explicitly calls out the exigence to reunify communication modes under the umbrella of rhetoric, with attention to how "the civic dimension of the rhetorical tradition is plainly crucial to producing students with the communicative capabilities needed in this world" (p. 2). During assessment of our curricula, we respond to such expectations and calls by recognizing that students need autonomy when using theory to benefit their communities. That is, rather than choose sides between humanistic and technical training, rather than divorce outcomes between the logical precision of technical literacies and the passionate ethical orator, and rather than delineate particular areas where public persona stops to become private, we take the whole person as educated decision maker.

■ References

Bennett, E. (2018, April 13). Dear humanities profs: We are the problem. *The Chronicle of Higher Education.* https://www.chronicle.com/article/dear-humanities-profs-we-are-the-problem/.

Connors, R. J. (1982). The rise of technical writing instruction in America. *Journal of Technical Writing and Communication, 12*(4), 329–52.

Cooper, M. G. & Marx, J. (2018, November 9). Why we love to hate English professors. *The Chronicle of Higher Education.* https://www.chronicle.com/article/why-we-love-to-hate-english-professors/.

Cunningham, D. H. (2004). The founding of ATTW and its journal. *Technical Communication Quarterly, 13*(1), 121–30.

Fleckenstein, K. (2003). *Embodied literacies: Imageword and a poetics of teaching.* Southern Illinois University Press.

Fleckenstein, K. S., Spinuzzi, C., Rickly, R. J. & Papper, C. C. (2008). The importance of harmony: An ecological metaphor for writing research. *College Composition and Communication, 60*(2), 388–419.

Henning, T. & Bemer, A. (2016). Reconsidering power and legitimacy in technical communication: A case for enlarging the definition of technical communicator. *Journal of Business and Technical Communication, 7*(1), 12–36.

Herndl, C. & Cutlip, L. (2013). "How can we act?" A praxiographical program for the rhetoric of technology, science, and medicine. *Poroi, 9*(1). https://doi.org/10.13008/2151-2957.1163.

Johnson, M. A., Simmons, M. & Sullivan, P. (2018). *Lean technical communication: Toward sustainable program innovation.* Routledge.

Johnson-Eilola, J. (1996). Relocating the value of work: Technical communication in a post-industrial age. *Technical Communication Quarterly, 5*(3), 245–70.

Keith, W. & Mountford, R. (2014). Mt. Oread manifesto on rhetorical education 2013. *Rhetoric Society Quarterly, 44*(1), 1–5.

Kimball, M. A. (2016). The Golden Age of technical communication. *Journal of Technical Writing and Communication, 47*(3), 330–358.

Kynell, T. (2000). *Writing in a milieu of utility: The move to technical communication in American engineering programs, 1850–1950* (2nd ed.). Ablex.

Kynell, T. & Tebeaux, E. (2009). The Association of Teachers of Technical Writing: The emergence of professional identity. *Technical Communication Quarterly, 18*(2), 107–141.

Lauren, B. (2018). *Communicating project management: A participatory rhetoric for development teams.* Routledge.

Lueck, A. (2018, October 23). How high schools shaped American cities. *The Atlantic.* https://www.theatlantic.com/technology/archive/2018/10/how-high-schools-shaped-american-cities/573616/.

Mathes, J. C. & Stevenson, D. W. (1976). *Designing technical reports: Writing for audiences in organizations.* The Bobbs-Merrill Company, Inc.

Moore, P. (1996). Instrumental discourse is as humanistic as rhetoric. *Journal of Business and Technical Communication, 10*(1), 100–118.

Rice, J. (2012). *Distant publics: Development rhetoric and the subject of crisis.* University of Pittsburgh Press.

Richardson, M. & Liggett, S. (1993). Power relations, technical writing theory, and workplace writing. *Journal of Business and Technical Communication, 7*(1), 112–137.

Roberts, S. (2020, March 27). Flattening the coronavirus curve. *The New York Times.* https://www.nytimes.com/article/flatten-curve-coronavirus.html.

Robin, C. (2016, January 29). How intellectuals create a public. *The Chronicle of Higher Education.* https://www.chronicle.com/article/how-intellectuals-create-a-public/.

Skinnell, R. (2016). *Conceding composition: A crooked history of composition's institutional fortunes.* Utah State University Press.

Staples, K. & Ornatowski, C. (1997). *Foundations for teaching technical communication: Theory, practice, and program design.* Ablex Publishing Corporation.

Todd, J. (2003). Teaching the history of technical communication: A lesson with Franklin and Hoover. *Journal of Technical Writing and Communication, 33*(1), 65–81.

Tompkins, J. (2018, November 18). Speak, memory: A new way to read—and to teach—literature. *The Chronicle of Higher Education.* https://www.chronicle.com/article/speak-memory/.

Turnley, M. (2007). Integrating critical approaches to technology and service-learning projects. *Technical Communication Quarterly, 16*(1), 103–123.

Wardle, E. & Roozen, K. (2012). Addressing the complexity of writing development: Toward an ecological model of assessment. *Assessing Writing, 17*(2), 106–119.

Wick, C. (2000). Knowledge management and leadership opportunities for technical communicators. *Technical Communication, 47*(4), 515–529.

8. Creating the "Through-Line" by Engaging Industry Certification Standards in SLO Redesign for a Core Curriculum Technical Writing Course

Julianne Newmark
UNIVERSITY OF NEW MEXICO

Joseph Bartolotta
HOFSTRA UNIVERSITY

Abstract: This chapter describes one technical and professional communication program's (TPCs) revision of student learning outcomes (SLOs) in a sophomore-level technical writing course to engage industry standards and terminologies, specifically, the Society for Technical Communication's (STC's) nine areas of competency from the Foundation Certification Exam. These SLOs serve as an enculturative framework in the foundation-level technical writing course. This chapter also offers a discussion of how to navigate the challenges of implementing new SLOs, including getting buy-in from full-time and part-time faculty, especially when drawing upon industry-designed standards. Deriving from assessment data, this chapter argues that our program-specific adaptation of the STC's Foundational Exam competencies suggests effectiveness in setting the stage for a university-to-industry through-line that intends to benefit our students and reinforce the values of humanism, social justice, and user-centrism that figure as crucial emerging mandates in TPC today.

Keywords: student learning outcomes, assessment, program revision, technical writing, social justice, professionalization

Key Takeaways:

- Industry certification standards can be used to help shape technical writing course student learning outcomes (SLOs).
- Measurable changes reported via assessment, and qualitative indexes of improvement shared in faculty feedback, can suggest improvements in teaching and learning resulting from SLO redesign.
- Navigating the challenges of implementing new SLOs, especially when drawing upon industry-designed standards, necessitates getting buy-in from full- and part-time faculty.

Teachers in technical and professional communication (TPC) have long considered ways to more effectively bridge the gaps between their classrooms and the work of industry-situated technical communicators. To construct such a "bridging" curriculum (Blakeslee, 2001) for technical communication courses and larger curricula, TPC faculty have considered various interventions. Many such attempts exist in the form of genre assignments (proposals, procedures, feasibility studies, etc.) and others as discrete courses that are curricular requirements, such as a junior- or senior-level internship or service-learning course. Yet, are there ways to reformulate a curriculum, whole cloth, so that the founding premises of that curriculum have an eye towards students' eventual workplace realities? This question lies at the core of the curricular reform project studied in this chapter, a reform that began with a re-envisioning of our foundation-level technical writing course's student learning objectives (SLOs). We strove to determine how we could best endow our students with both the practical know-how that would inform their day-to-day duties as future professional writers along with providing them a theoretical basis in a university context that would prepare them to deal with emerging media, genres, ethical concerns, and audiences. This chapter offers a view into our strategies of meaningfully engaging such questions and making curricular changes as a result.

The interplay between what might be called practical and theoretical factors informs the work that TPC faculty do in the context of program development, curriculum design, and individual course planning. Teachers in TPC cannot design course plans and assignments that capture the dynamic nature and true diversity of writing situations that students may find themselves managing once they leave the classroom. Because it is likely not possible for TPC curricula to replicate or anticipate industry genre diversity and situational/compositional typologies in a "mirror image" fashion, we must re-think TPC curriculum development so as to foreground students' theoretical foundation as attached to "habits of mind" development, as relayed through carefully paced practices in standard conventions of multimodal TPC communication. What we share here is one such model, a model that foregrounds industry-situated terminologies—namely, the Society for Technical Communication's (STC's) Foundation Certification Exam's nine areas of competency—as an enculturative framework, a framework installed throughout one institution's TPC curriculum, beginning in its foundational, multi-major 200–level core curriculum course. Our model, then, is not so much a bridge as a *through-line*.

Our rationale for reinventing our curriculum with an eye towards training students as emerging communicators who are *already* imagining themselves as part of an industry and professional culture, and who are crucially doing the work that they can uniquely do in their university context, is that we well know that the definition of what exactly technical writing *is* is always in discussion. We include our students in this discussion from their first technical writing class. Our foundation-level course dedicates class time to practicing the conventions of specific communication outputs and probing the ethical dimensions of a technical communicator's work. But we also ensure that our instructors alert students to the slippery nature of what technical writing/communication *is* by way of perusing the STC job board, doing analyses of technical communicators' personal websites, and reading prominent tech writers' personal blogs. Students come to discover, then, what Eva Brumberger and Claire Lauer (2015) found in their research on nearly 1,000 job postings in technical communication. Brumberger and Lauer observed that the job postings displayed considerable diversity in both job title and desired skills. We knew that in redesigning SLOs, we should be attentive to the breadth of skills this research described.

Brumberger and Lauer identified five main categories for position titles: content developer/manager, grant/proposal writer, medical writer, social media writer, and technical editor/writer. Teaching the writing practices inclusive of all of these different writing exigencies would be difficult in one common course. However, Lisa Melonçon and Sally Henschel (2013) observe that a "basic" technical communication course, described as one that introduces students to the "*practice* of technical and professional writing" (p. 51), is the most common course required of majors in technical communication (along with a later capstone course). That such courses have become a curricular standard is important for us to consider. In the case we share here, we describe how we reinvented our "standard" or "basic" class to be one that we believe helps set the tone for a student's decision to more generally embrace a technical communication certificate, minor, or major—and career.

Foundation-level (or "standard," "basic," core curriculum, lower-division—there are many identifiers) technical writing courses often function as "service" courses for faculty, in that often students enrolled in the courses are from outside an institution's TPC major, minor, or certificate track; many students come from the sciences and engineering, business schools, or nursing programs. Thus, these courses serve several populations. Sarah Read and Michael J. Michaud (2015) describe these courses as "multimajor professional writing course[s]," or MMPWs, underscoring the challenging nature of being designed both as the most common sort of course TPC students take while also having to operate in a service obligation to students outside a major. So, MMPWs are populated by students with diverse interest areas, but they are also taught by a wide-ranging faculty staffing structure at many institutions. The focal institution of this chapter's discussion, the University of New Mexico (a flagship state university at the Research 1 designation), staffs its 200-level (sophomore-level) technical writing courses, of which there are on average 28 sections each Fall and Spring semester, with a mixture of full-time faculty, part-time instructors, and graduate teaching assistants. Thus, there is interest-area diversity among our instructors (as very few are explicitly specialists in TPC) within English studies. All instructors have been trained, however, using the newly developed curricular model described in the following pages.

In sum, such courses—often the foundation for a student's expanded university-level study of TPC and potential later entrance into the profession—are challenging on many levels. Several years ago, we decided to address some of these challenges directly in the form of a substantial redesign of this MMPW—"basic" or core curriculum—technical writing course. We wanted to create an effective course that could both serve the important work of preparing students for our certificate and minor TPC programs (we do not have a major concentration) while also being responsive to the needs of students from across the university who had to take the course as a requirement for a completely different sort of major.

We were also inspired by the social justice turn in the field (Jones, 2016) to more explicitly include language that embraced attention to diversity in its myriad forms. We were particularly interested in ways to draw student attention to issues as they pertain to race, gender, sexuality, language diversity, and (dis)ability in a professional context, so as to prepare them to think critically about historical practices in TPC when and if they enter the profession as practitioners. We were also eager to add nuance to a set of SLOs such that they would be responsive to our own particular university context. The University of New Mexico, as our Land Acknowledgment Statement reads

> sits on the traditional homelands of the Pueblo of Sandia; as an institution of learning, UNM has a stated commitment to honoring "the original peoples of New Mexico—Pueblo, Navajo, and Apache" and their "deep connections to the land and . . . significant contributions to the broader community statewide."

Further, as this extends to program development, we know many of our students speak Spanish, Navajo, and languages in the Keres language family (plus many others); thus, we believe it is important to honor the linguistic diversity that shapes New Mexican culture. For us, then, as administrators of a technical and professional communication program, we knew that our SLOs needed to reflect the diverse reality our students know well already.

After much discussion with faculty (both full- and part-time), students, and other stakeholders, in 2017, we ran a pilot of our foundation-level technical writing course by taking inspiration from the "nine areas of competency" articulated by the Society for Technical Communication, as noted above. As Craig Baehr explains in a 2016 *Intercom* article, these competencies reflect "key terminology, facts, concepts, and techniques"; "These areas encompass a broad range of processes, practices, strategies, and roles that comprise the work of technical communicators and teams they serve on and manage" (p. 10). In idea-gathering workshops with our colleagues in the lead-up to the pilot, we did not suggest that our course could do the work of preparing students effectively for the STC Foundation certification. However, we did argue that the taxonomy of the "nine competencies" could offer a compelling language through which we could articulate the professional nature of our curricular goals. In redesigning our student learning outcomes (SLOs) to be more responsive to the STC's language while dedicatedly attending to our own students' needs at the University of New Mexico (UNM), we crafted more precession into our descriptions and a stronger effective connection between both the theory and practice of technical communication. The nine competencies afforded us the "through-line" we believed our curriculum needed.

This chapter describes several parts of our journey from one set of SLOs to the revised, industry-informed set we use now. While what resulted from this process for us will not fit every program's needs, we believe we provide here an example of an effective administrative structure for undertaking an SLO revision project when there are several important and diverse stakeholders to respond to. Below, we will describe some of the theory that informed the earlier manifestations of our SLOs, and we offer a brief survey of how scholars have recently sought to more effectively link industry standards to curricular decision-making. Next, we explain the exigence of our task in revising our existing SLOs, including describing how we navigated stakeholders as we revised and solicited (and achieved) buy-in from our faculty. We then reflect on the sorts of challenges posed by incorporating industry standards into undergraduate academic contexts, as well as what sorts of approaches we may have taken differently in hindsight. We close by examining how we marked the experience as "effective," relative to our annual assessment of the 200-level course, suggesting what may be helpful as we continue an interactive process of assessing and evaluating the SLOs.

While we are generally satisfied with what we have achieved thus far in the SLO redesign and deployment process, this chapter argues that, certainly, our SLO redesign project's quantitative and qualitative "effectiveness" was only *partially* a result of our commitment to creating an academy-to-profession through-line. What our data (discussed in what follows) and anecdotal feedback from colleagues and students—and impressions of the experience identified by us, as project leads—suggests is something *more important* than what we did by refashioning our curriculum via industry-informed SLOs and curricular infrastructure suited to them. The "something" that is more important from our perspective is the commitment to an inclusive, iterative, and cautious approach, which lead up to the implementation of the changes we made. This kind of approach is what we advise other programs foreground in SLO and program revisions, as many already do. We hope that this chapter will serve as a model of effective administration in the midst of competing stakeholders and exigencies that swirled around our growing technical communication program.

■ "Technical Writing" SLOs, Reimagined

In 2016, the authors of this article began exploring a redesign of the TPC curriculum at the University of New Mexico. At that point, the SLOs used for this MMPW ("multimajor professional writing") course, ENGL 219, had been in use long enough that no faculty member could remember when, exactly, they had come about. Without a sense of what exigencies compelled these "legacy" SLOs into being, we could nevertheless observe the appeal of their simplicity. They were composed of four well-articulated outcomes that served us well in our university assessment protocols. These legacy SLOs "aligned" with the State of New Mexico's Higher Education Department (HED) Area 1 "Communications" Competencies, though they were not verbatim the same outcomes, in the same language (UNM Office of Assessment, 2015). We will bracket out this larger discussion concerning "alignment" with State HED Competencies because it exists out of the scope of our present discussion of our own programmatic curriculum revision and its particular emergence in new SLOs for our foundation-level technical writing ENGL 219 course.

Table 8.1. University of New Mexico ENGL 219 student learning outcomes, circa AY 2015–2016

Student Learning Outcome (SLO) Number	SLO Abbreviated Description	SLO Full Description
1	Analyze Rhetorical Situation	Students will analyze the subject, purpose, audience, and constraints that influence the documents they write to ensure they achieve specific and useful results.
2	Find and Evaluate Information	Students will gather information from professional, academic, and government sources, evaluating the information they find for quality, validity, and usefulness.
3	Compose Information	Students will develop strategies for generating content and organizing it into a logical structure that is appropriate for their intended users; they will consider ethical influences for the documents they compose; they will work effectively with others to create documents.
4	Present Information	Students will edit and revise their writing to provide unambiguous meaning and coherent structure; they will incorporate visual elements to improve the reader's understanding; they will create an overall design that enhances readability and shows professionalism.

Table 8.1 shows the ENGL 219 SLOs circa Academic Year (AY) 2015–16. These SLOs were organized around four capacious concept areas: analysis, research, composition, and presentation. Particularly attached to face-to-face (F2F) sections of ENGL 219, these flexible SLOs afforded our diversely-skilled (and here we specifically use "skills" to connote instructors' own histories as TPC practitioners or researchers) ENGL 219 faculty to "teach to" these SLOs in a wide range of ways, with a wide range of assignments and a wide range of final course projects (ranging from professional portfolios to recommendations reports to proposal presentations). Yet, at this same point, our robust online ver-

sion of ENGL 219, eTC (electronic Technical Communication), was using its own SLOs, which aligned with these four SLOs, which themselves aligned with the State of New Mexico HED Communication Competencies. Already, readers of this chapter can surely understand the opportunities for better refinement and synthesis of SLOs across all ENGL 219 sections, as all of the threads of connection just mentioned caused us to be somewhat confusingly organized. The online ENGL 219 (eTC) SLOs circa 2015–16 reflected the research of our colleagues Andrew Bourelle, Tiffany Bourelle, and Natasha N. Jones (2015). They had specifically modified the legacy face-to-face SLOs (see Table 8.1) to make them uniquely appropriate for the online, multimodal curriculum they lead. These scholars drew upon the five rhetorical canons to explore the applicability of the ancient tradition to a modern and multimodal context; thus, the eTC SLOs were, in effect, the classical rhetorical canons adapted for twenty-first-century application and specifically attuned to the multimodal mandate of the eTC curriculum. The exigence for the eTC SLOs, then, was that the legacy F2F ENLG 219 SLOs did not address multimodality at all.

So, one clear goal for creating new SLOs for the entire ENGL 219 course array—face-to-face, online, and hybrid—was to affect curricular consistency. The director of eTC, Tiffany Boruelle, welcomed the opportunity to holistically revise *all* ENGL 219 SLOs so that every section, across modes of delivery, featured SLOs that involved twenty-first-century communication principles, specifically concerning multimodal communication. Collaboration amongst program directors, then, was vital to ensuring that the newly selected SLOs could be modified to support curricular nuances in all modes of delivery of the course so that all 219 students, regardless of their section, would be ensured a certain degree of curricular uniformity. An additional benefit was that the annual assessment of ENGL 219 would then be able to capture programmatic efficacy (or lack thereof) across the entire spectrum of course sections.

Upon the launch of the Society for Technical Communication (STC) Foundation Exam competencies in 2016, which Craig Baehr carefully described in his *Intercom* article that January, the authors of this chapter began discussing the adaptation of these competency areas, and the skills that lie within them, to fit within the framework of a university-sited TPC education at the lower-division level. Moving from four course-wide SLOs to nine, we worried, might concern our instructors, so we quickly moved to planning a series of "listening sessions" with all ENGL 219 instructors. Two such sessions were held in Fall 2016. At the first of these sessions, we circulated the STC Foundation Exam competencies as originally written. We asked our instructors the following questions about those competencies:

- How (or how well or how poorly) might this industry-level certification-exam framework function as a set of learning outcomes in our course? Why?
- What are the limitations of this framework?

- How would the adoption of this framework impact the assignments we teach in ENGL 219?
- Could this framework better support student education in
  - emerging technologies?
  - the needs of diverse users of communication?
  - workplace realities for technical communicators, post-graduation?
- How would we need to revise the STC's language to make the SLOs more focused on our students' needs and discourses?

We collected instructor feedback during this session and then, for the second session, we initiated a conversation around a revised version of the competencies, with language better attuned to our students' needs. Table 8.2 shows the original competencies and our revised versions. Additionally, we shared what we called a "menu" of assignment options that would scaffold appropriately relative to our larger curricular mandates/strategies for ENGL 219 and that would still leave instructors opportunities to employ their own teaching innovations in their online and F2F classrooms.

Table 8.2. STC Foundation Exam competencies (Baehr, 2016) and UNM's ENGL 219 competencies that synthesize these skills

Competency/ SLO Number and Title	UNM SLO Number	STC Description of Competency (Baehr, 2016, pp. 10–11)	ENGL 219 Revised SLOs
Project Planning	1	Project planning focuses on the work involved in planning and managing technical communication work teams and documents through a lifecycle process. It includes process planning, goal setting, progress tracking, and strategic planning activities.	Planning, researching, and composing technical documents (as a lifecycle process) in teams and individually.
Project Analysis	2	Project analysis involves the work of identifying readers and document contexts, including the development of reader profiles. This includes identifying types of audiences, users, readers, and their preferences regarding document use and readability. It also focuses on the analysis of document contexts, including working in global contexts and rhetorical situations.	Identifying a document's readers and a document's context relative to practices of composing for specific global, diverse, and multicultural audiences. Understanding how technical documents occupy and respond to social justice and community service contexts.

Creating the "Through-Line" by Engaging Industry Certification Standards   155

Competency/ SLO Number and Title	UNM SLO Number	STC Description of Competency (Baehr, 2016, pp. 10–11)	ENGL 219 Revised SLOs
Content Development	3	This category focuses on the development of content and technical information products. It addresses technical genres, their content, and use, including: memos, technical descriptions and specifications, instructional content, proposals, activity or status reports, and analytical reports. It also focuses on researching, including finding source materials, defining the scope of research questions and methods, and documenting sources and intellectual property concerns.	Understanding how genre conventions impact writing. Using contextual information to place specialized information into the appropriate genre.
Organizational Design	4	Organizational design focuses on guidelines and techniques for organizing and drafting technical documents. It covers organizational patterns and rhetorical moves for introductions and conclusions to technical reports, as well as patterns for specific technical genres including memos, technical descriptions and specifications, instructional content, proposals, activity or status reports, and analytical reports.	Practicing strong research skills with primary and secondary sources to generate appropriate content. Generating strong research questions and developing clear research practices.
Written Communication	5	Written communication covers general guidelines for composing content and communicating in written and electronic forms. It covers writing style, persuasion, tone, and general readability. It includes techniques for writing sentences and paragraphs for both print and electronic documents, and in global contexts.	Composing clear, stylistically responsible prose that avoids errors and pays attention to audience needs.
Visual Communication	6	This area focuses on general visual communication principles and practices, including using graphics, data displays and other kinds of information graphics, such as bar charts, line graphics, tables, pie charts, flow charts, etc. It covers the use of design principles, such as balance, alignment, grouping, consistency, and contrast. It also addresses the use of visual information and related technologies when giving presentations.	Using visual design principles to develop audience-friendly data displays, including charts, tables, infographics, line graphics, and presentations.

Competency/ SLO Number and Title	UNM SLO Number	STC Description of Competency (Baehr, 2016, pp. 10–11)	ENGL 219 Revised SLOs
Reviewing and Editing	7	This category addresses reviewing and editing processes and guidelines, and general usability. It encompasses the various levels of editing, including revising, substantive editing, copyediting, and proofreading. Additionally, it covers common grammatical and mechanical errors.	Across media and contexts, ensuring final clear style, user-centered writing, and error-free spelling and mechanics.
Content Management	8	This area focuses on managing content of information products, as well as the management of information development teams. It addresses Web content development, including the basic features of Web sites and general guidelines for developing Web-based content. It also covers the uses of social networks, wikis, blogs, microblogs, videos, and podcasts in working settings. From a teaming standpoint, it covers the roles and practices for managing content and roles across a work team.	Gaining knowledge of the organization and management of digital and textual information and receiving an introduction to information architecture, web content management, and social networking.
Production and Delivery	9	This category focuses on the production and delivery of information products, specifically how project outcomes relate to and inform the development of final production deliverables. It also addresses the importance of setting objectives for final deliverables and using them to measure effectiveness and outcomes of technical information products.	Developing skills in presenting information in multiple modes and in various media: web, paper, oral, and video. Applying delivery skills to emerging technologies.

The language shared in Table 8.2's column four was collectively composed by the ENGL 219 directors and the ENGL 219 instructors who attended the "listening session" workshops used to develop and refine the SLOs. The goal was to simplify the language in the STC's "nine areas" to make them reader-friendly to our audience: ENGL 219 students. In addition, we enhanced some of the categories to comply with our programmatic goals, such as to ensure that our beginning-level technical communicators are always mindful of, as we wrote in SLO 2 (Project Analysis), "a document's readers and a document's context relative to practices of composing for specific global, diverse, and multicultural audiences." The language for this SLO was considerably revised from the STC's original, and this change conveys our programmatic—and eventual assessment-level—interest in a targeted cultivation of ethical and culturally situated understandings of audience needs among our students.

We also became concerned that our students and instructors might not be comfortable with a "large" number of SLOs. Of course, we did not have to map the STC's nine competencies one-for-one onto competencies in a redesigned curriculum for our MMPW course. But we did. Our rationale for using nine competencies was that a) our first-year composition (FYC) courses (which are prerequisites for our 200-level MMPW course) have six SLOs (one of which has four subcomponents), and thus students and instructors are familiar with both the scope and the approach of working toward competency in a broad range of skill areas, and b) for students using our MMPW course as an entrée into the TPC field, familiarization with the Foundation Exam's skill areas as written, we believed, was an advantage to them. While we do not have data on the number of students who come through our MMPW course who ultimately pursue this field or take the Foundation Exam years down the line, what we liked about the number of SLOs was its breadth and its flexibility and potential for subdivision across major projects in the course. We received no specific feedback from students (as recorded in their end-of-semester course reflections) about the number of SLOs being unwieldy or challenging to pace through. Our colleagues, whose expertise on SLOs would obviously exceed that of our students, felt that using the nine made sense relative to our desire to create an academy-to-industry through-line.

As Table 8.2 indicates, while the number of SLOs is substantial, we worked to streamline within that number to attend to the fact that our instructors believe that students find SLOs useful—to the degree that students ever find the articulated SLOs in their syllabi and throughout their courses useful—when the language of these SLOs is simple and concise. With simplicity and concision as our watchwords, we aimed to extract the key concepts from the STC's original articulation and re-create them in a user-friendly way for our students, adding, where necessary, concepts that resonated with the unique mission and principles of our program. SLO 2 explicitly reflects our desire to integrate social justice concerns into our learning objectives, while SLOs 7–9 add forthright language about communicating across different media and modes.

One final feature of our development of these SLOs was a Fall 2016 workshop with Rick Johnson-Sheehan, the author of our textbook, *Technical Communication Today*. At the workshop session, all instructors who were past, present, or future ENGL 219 instructors (we determine "future" status by including teaching assistants who were enrolled in our "Teaching Technical Communication" graduate-level practicum) worked with Johnson-Sheehan to connect concepts from the course textbook to the analytic and applied framework established by the new SLOs. Johnson-Sheehan also encouraged attendees to think of model assignments, which fit within the established "menu," that would support the new edition of the textbook's focus on entrepreneurial thinking and related communication-skill development. Johnson-Sheehan expressed to workshop members that the new edition of *Technical Communication Today* already aligned with the STC's conceptual schema, and thus, our goal of creating a "through-line" was supported by our textbook's terminologies and overall perspective, vis-a-vis TPC workplace standards.

In Spring 2017, we launched the pilot semester of the new curriculum. At our mid-August mandatory Convocation for all teachers of ENGL 219, we shared a model syllabus, the assignment "menu," and one fully developed (scaffolded) sequence, which we called a "Job Materials Dossier." In the session, we described the two versions of our SLOs: those that were instructor- or profession-facing (in the original STC language) and those that were precisely designed for our students and our program (as shown in Table 8.2). We then worked through our sample sequence to explore how it attended to the skills highlighted by the news SLOs (the sample assignment sequence practiced three SLOs in particular) and how it could be deployed using specific textbook chapters. Since many of the instructors in attendance had already participated in the listening session in Fall 2016 and the workshop with Rick Johnson-Sheehan, they also manifested ownership over the whole curricular objective and were eager to share their own ideas regarding ways to integrate the new SLOs into adapted versions of their previous tried-and-true assignments. All new ENGL 219 sections would culminate in a multimodal electronic portfolio, featuring a comprehensive two-page reflective memo in which students would discuss their learning, with examples and evidence, of the course SLOs. For assessment purposes for AY 2017 (and for AY 2018), we chose to examine student engagement with SLOs 2 and 9.

We want to add one final note in this section regarding assessment (though we will turn to our assessment results below). We knew that we would need to determine how well, and if, our new SLOs were impacting our students' learning, through the measurement tool of their reflective writing across their entire portfolios and in their final memoranda, but we also knew we were required to assess students' engagement with university-level SLOs. Adding further complexity to the assessment framework already introduced in Table 8.1, UNM adopted all-campus (meaning the main campus and all regional branches, of which there are five) learning objectives for ENGL 219. We were to assess for these university-wide three SLOs and were "welcome" to assess for our own programmatic SLOs, which were the two we named above (2 and 9). In the following section concerning assessment, we will briefly discuss ENGL 219 program performance relative to both the university-wide, "all-campus," set of three SLOs and performance relative to our specifically designed new SLOs. Incidentally, as of Spring 2019 (when we are writing this chapter), we are no longer obligated to assess for the university-wide SLOs and thus, we only assess for our programmatically determined ones, which are the subject of this chapter.

■ Evidence from Assessment

Determining the "effectiveness" of our new SLOs is a complex process. The concept of "effectiveness" operates at the intersection of multiple valences. Regarding our new SLOs, the first and most important question to us is this: does the new SLO structure function in our classrooms in a way that improves student learning? Without meeting a standard of usability and improvement (as in, our instructors find the SLOs easy to use/"teach to" and students engage with them "more meaningfully" than previous iterations), we cannot assert that any other measures of "effectiveness" matter. However, the measure of efficacy of student learning is not a zero-sum game. We may find that some measures are more effective than others, and we readily acknowledge that what we have proposed here probably fits somewhere on a continuum alongside other iterations of SLOs devised by other scholars and programs. Time will tell where our exercise truly fits in the company of others. Still, we believe what we measured here was specific enough to give us tangible (and in some ways, hereto unseen) ideas about the health of our courses that were instructive toward strengthening our program.

What we have identified as one "take-away" at the outset of this chapter concerns our evidence, thus far collected, that there is quantitative and qualitative data suggesting that our new SLOs are an improvement on our old SLOs in the realms of students' own articulations of learning relative to our two assessed SLOs (2 and 9; more on this below). Because our assessment hinges on a reading and scoring of a final course reflection memo as a leading feature of an online, multimodal portfolio, our assessment attempts to honor the portfolio and its reflection as a social ecology, to paraphrase Yancey (2018). We can definitely improve in this vein so that the portfolios we assess are not just a "set of finished projects fronted by a mandated argumentative text [the reflection] in which a student is required to claim in terms of outcomes that he or she has met those outcomes—even if she or he hasn't" (Yancey, 2018, p. 259). We believe that linking our assessment to industry practices adds a level of urgency for students as they realize they are not responding to arbitrary learning outcomes crafted by academics, but to skills that have salience and applicability in the next steps of their writing lives.

While it is beyond the scope of this article to deeply discuss the nuances of our assessment practice for our MMPW course, we do want to explain the rationale behind, and the limitations of, attempting to detect declines, stasis, or improvements in student learning by using an end-of-term, reflection-fronted, ePortfolio for assessment. In our study of our SLO revision and its potential effectiveness relative to detecting and measuring student learning, we did use the final reflections by students, as we agree that they make a "distinctive contribution" to the "learning showcased in, and the assessment of, electronic portfolios" (Yancey, 2018, p. 269). In short, we believe that students' reflections gave us a clue into their metacognitive engagement with the principles that underlie the new SLOs.

Yet, we were not "just" assessing end-of-term, reflection-led portfolios as a measure of student learning; we were also interested in learning about how well, and whether, we were achieving more conspicuous alignment between our new SLOs and TPC scholarship that is interested in responding to industry (or larger "tech comm community") concerns. To this end, we think the redrawn SLOs are particularly suggestive of effectiveness. Students taking our courses are not only exposed to terms and values that are consistent with workplace expectations, but those terms become a part of how students understand the field. Supportive of this is the quantitative data we share below, which reveals interesting "learning" pathways drawn from the assessment data we collected over the past two years and suggests students' enmeshments with industry vocabularies and their deepened knowledge about the field.

The assessments from Fall 2017 and Fall 2018 each examined 22 portfolios that were collected across 22 different sections in each year.[1] All portfolios were reviewed by two readers and scored on a scale from 0–3 (with 0 being the lowest rating). Because our methodology for assessing ENGL 219 was inherited from practices used in first-year composition (FYC) assessment, the Likert scale in use (0, 1, 2, and 3) was a writing program fixture, in effect. As in the assessment of our ENGL 110 and 120 courses, our assessors (all of whom were experienced assessors who had participated in previous 110 and 120 assessment teams) used the scores "0" and "1" to indicated two varying degrees of "Needing Improvement" and the scores "2" and "3" to indicated two varying degrees of "Meeting Requirement."

Of course, ample literature exists to guide writing program and technical communication program administrators on shaping assessment practices, determining numerical scales, and designing reflection prompts for portfolio usage in a particular university context. The practices used in the assessment discussed here for the MMPW course in question descended from an FYC assessment overhaul in our program during 2011–2012, wherein the assessment protocol was informed by the work of Linda Adler-Kassner and Peggy O'Neill (2010), Bob Broad (2003), and Brian Huot (2002). Similar to strategies used at many other institutions, we elected to use rubrics for assessment that were table-driven for numerical value entry but were appended with "Summary Comments" sections. As Jane Detweiler and Maureen McBride (2009) have written, the numerical aspects allow "us to take our outcomes to administrators in terms they can identify with . . . and translate to their audiences as well" (pp. 64–65). Echoing Detweiler and McBride again, we found that numerical data allowed/allows us to show that our "program appear[s] to be succeeding . . . so our external audiences could also point to how the program, a key part of the Core curriculum, appeared to be accomplishing some of the university's 'stated goals'" (p. 65). So, while numerical

---

1. There were in fact 26 sections of the course that ran in Fall 2017 and 31 that ran in Fall 2018. We were unable to perform an assessment of some of these sections due to issues such as file corruption or no portfolio being submitted.

data delivered via a limited Likert scale, accompanied by narrative comments, allowed us to comply with our university's requirements, it also allowed us to detect measurable changes from year to year, which in this case we are correlating to student-learning changes resulting from a thoughtful SLO revision.

In terms of the "nuts and bolts" of our findings, then, we proceeded in the ENGL 219 assessment with a scheme wherein when readers were more than one point away from each other (say, one reader gives a score of 1 and another a score of 3 to the same portfolio), the portfolio would be reviewed by a third reader. All scores were then averaged to give us a picture of how students were generally achieving the SLOs of the course, curriculum wide. Tables 8.3 and 8.4 report on the revised SLOs 2 and 9; the former table defines the SLOs and the latter offers the year-to-year averaged results of this assessment.

Table 8.3. The SLOs, in our program-specific verbiage, assessed in the Fall 2017 and Fall 2018 sections of the MMPW course

SLO	Description
2	**Project Analysis.** Identifying a document's readers and a document's context relative to practices of composing for specific global, diverse, and multicultural audiences. Understanding how technical documents occupy and respond to social justice and community service contexts.
9	**Production and Delivery.** Developing skills in presenting information in multiple modes and in various media: web, paper, oral, and video. Applying delivery skills to emerging technologies.

Table 8.4. Average scores of rated portfolios for the redesigned SLOs

Year	SLO 2	SLO 9
2017	2.09	2.02
2018	2.28	2.45

These results appear to us quite dramatic. They represent a substantial improvement in both SLOs over the course of one year (an improvement of 9 percent for SLO 2 and 21 percent for SLO 9). While we cannot make generalized claims based off of just two years of data, this sort of data provides an interesting baseline from which we can assess in future years. Since the SLOs were new, we are not testing them against a similar SLO that could offer a figure from which we could draw a null hypothesis, so it would be inappropriate for us to offer a sense that the year-to-year change is statistically significant. This, it may seem, is one of the challenges of drastic innovation: data that reflects effectiveness needs to be collected over a long period of time to truly measure the impact of our curricular interventions. This is perhaps doubly true when programs decide that learning-outcome infrastructure that was important in the past no longer serves pressing needs.

We will not include a discussion here of how the scores (and change in scores) in the redesigned STC-influenced SLOs compare to the UNM Core SLOs, as such a discussion would take us too far afield from the scope of this study. However, considering how widely general those SLOs were, it may be no surprise that the redesigned SLOs offered more precision through which we analyzed the health of the course. The emphases designed in both revised SLO 2 and 9 do not have easy corollaries to the SLOs described in Table 8.1. While one Core SLO discusses the importance of presenting material, it does not consider the importance of production across media, including using emerging technologies. None of the Core SLOs pay attention to the role of analyzing the ecology of a technical writing project, particularly in such a way that highlights social justice and community engagement contexts. We believe these newly revised SLOs do well to align with the emerging values and practices within both TPC scholarship and industry. Part of the rise in scores, we believe, is due to the fact that for the first time in the program these professional objectives were instantiated and given a position of privilege in the curriculum.

Naturally, one might wonder if what we are reading as "success" is due to the fact that we developed the new SLOs from the STC competencies, or if simply making more specific SLOs in the first place would have sufficed in creating a more effective curriculum. From our position, we cannot separate the specificity of the SLO from its professional inspiration. If nothing else, the professional currency offered by the STC competencies offered us a new prism through which we could reevaluate the SLOs. What is important to us now is that we were able to find a systematic way to integrate profession-connected expectations into the SLOs, thus making the work that students perform in class more responsive to TPC scholarship, writing studies research, and industry concerns. If effectiveness can be measured in terms of joining these three areas while keeping student learning at the center, our early research results suggest that we may have good reason to be optimistic about this approach to designing student learning outcomes for TPC courses.

Future work will discuss qualitative data (drawn from students' final portfolio reflections, which were the foundation for the rubric-driven scoring described above) that will add nuance to what we have briefly shared here, which figures as interesting statistical evidence of a program, via its foundation-level course moving, changing, and refining its curricular vision.

Conclusion: Pedagogical Challenges of and Justifications for Incorporating Industry Standards Into Curriculum Design

Developing SLOs for a TPC course is tricky. While we concur with Read and Michaud's assertion that a professional writing course studies "professional writing" in terms of the "literacy practices of professionals-who-write in any of the diverse professional contexts of business, industry, government, and the nonprofit sector" (2015, p. 430), we also believe that some of the disciplinary concerns of TPC pedagogies lend themselves well to a more precise set of challenges than most other written composition courses. First, TPC has a robust theoretical heritage that often takes theories of rhetoric and compositions and rearticulates them with more precision toward professional contexts. One example of this is in the way the literature of TPC engages more robustly with user-centered design and usability—two concepts that relate well to rhetorical concerns about audience—than composition has (although both concepts are picking up more momentum in rhetoric and composition scholarship).

Second, students taking TPC courses are immersed in a more focused curricular environment than FYC students. It has been our experience that we rarely encounter TPC students who have not declared a major once they take one of the MMPW courses. Indeed, we know many students enroll in our MMPW courses to complete curricular requirements related to writing or some other general education outcome. While the same case is generally true for FYC courses, TPC courses tout themselves as relating to writing beyond the university. Our course bulletin description describes as much: "Practice in writing and editing of workplace documents, including correspondence, reports and proposals." TPC courses do not abandon important concepts in rhetoric and composition but should represent a rearticulation of them to a new and increasingly profession-oriented context. We wonder if the language of rhetoric and composition is truly the most effective way of imparting values about writing to students taking TPC courses in our SLOs.

The scholarship of outcomes statements that derive from the Council of Writing Program Administrators (CWPA) includes robust discussion within the FYC community for which it was designed and has been extended into TPC curricular design. Barry M. Maid (2004) takes care to adapt the CWPA outcomes into his technical communication course, taking advantage of the invitation of the Outcomes to adapt and adjust the document to meet new disciplinary needs. Maid remarks that he was surprised by how little he had to adjust the outcomes to address his own context. Indeed, he finds that the emphasis on having writers learn how to write from general principles about writing is instructive to TPC students, rather than something that is more focused. Rather than learning to write for a more precise genre or audience, students are instead guided to think of the general principles that underscore all effective writing, and, the thought is, that such awareness will transfer to new writing situations, even ones in professional contexts.

We share this enthusiasm and optimism for the effectiveness of the CWPA outcomes, and we use our own locally adapted approach to these outcomes in our FYC courses. Moreover, the challenges we faced were similar to Maid's, as we too were working to kick off a new (perhaps better classified as "returning") minor and certificate in TPC. Like Maid, we also had to be cognizant of the ways our language would help define our discipline to potential certificate-pursuers and minors. For us, relying upon the language of compositionists might reflect a greater disposition toward the theory of writing rather than its practice. While we know that the two are both sides of the same coin, we did not know if we could articulate such nuance to students, especially the demographics we were targeting who were already majors in a different area with clearer professional trajectories. Likewise, we found ourselves in a similar position to K. Alex Ilyasova and Tracy Bridgeford (2016), who argue that drawing too much from composition "hinders efforts in the field to define technical communication, its theories, its practices, and its identity" (p. 54). As Bourelle, Bourelle, and Jones (2015) similarly adapted the SLOs for our companion courses' online sister from the classical rhetorical tradition, we wanted to experiment with another direction that would draw from the best theories in TPC. To do this, we moved away from the spaces where theory is most prominent and sought to orient ourselves toward practice.

Of course, our openness toward the practice(s) of technical and professional communication and the industries in which these outputs are represented meant we also had to resist the urge to mimic and rearticulate the blind spots in industry that may be driven by financial expedience. Our approach is fundamentally one rooted in humanism, respect for all persons, and the environments in which they dwell. Our approach intends to resist any racism, sexism, discrimination, or bigotry that has been codified in professional practice. We have found, through our program-specific adaptation of the STC's Foundational Exam competencies, that we are in the process of curricularly creating the through-line that we value and that benefits our students, a through-line that begins in first-year writing, gains potency and precision in 200-level MMPWs, and prepares students for their future profession. We hope that our SLOs' increased attention (as our new versions of SLOs 2 and 9 reveal) to social justice issues and ethical obligations in communication-creation allows us to better prepare our students to best serve diverse users and audiences in their university educations and in industry scenarios in future years.

References

Adler-Kassner, L. & O'Neill, P. (2010). *Reframing writing assessment.* Utah State University Press.
Baehr, C. (2016, January). Certified Professional Technical Communicator: The foundation exam and its nine areas of competency. *Intercom.* 10–11.
Blakeslee, A. M. (2001). Bridging the workplace and the academy: Teaching professional genres through classroom-workplace collaborations. *Technical Communication Quarterly, 10*(2), 169–192.
Bourelle, A., Bourelle, T. & Jones, N. (2015). Multimodality in the technical communication classroom: Viewing classical rhetoric through a 21st century lens. *Technical Communication Quarterly, 24*(4), 306–327.
Broad, B. (2003). *What we really value: Beyond rubrics in teaching and assessing writing.* Utah State University Press.

Brumberger, E. & Lauer, C. (2015). The evolution of technical communication: An analysis of industry job postings. *Technical Communication, 62*(4), 224–243.

Detweiler, J. & McBride, M. (2009) Designs on assessment: University of Nevada, Reno. In B. Broad, L. Adler-Kassner, B. Alford & J. Detweiler (Eds.), *Organic writing assessment: Dynamic criteria mapping in action* (pp. 52–72). Utah State University Press.

Huot, B. (2002). *(Re)articulating writing assessment for teaching and learning.* Utah State University Press.

Ilyasova, K. A. & Bridgeford, T. (2016). Establishing an outcomes statement for technical communication. In T. Bridgeford, K.S. Kitalong & B. Williamson (Eds.), *Sharing our intellectual traces: Narrative reflections from administrators of professional, technical, and scientific programs* (pp. 53–80). Baywood Publishing.

Jones, N. N. (2016). The technical communicator as advocate: Integrating a social justice approach in technical communication. *Journal of Technical Writing and Communication, 46*(3), 342–361.

Maid, B. M. (2004). Using the outcomes statement for technical communication. In S. Harrington, K. Rhodes, R. Fischer & R. Malenczyk (Eds.), *The outcomes book: Debate and consensus after the WPA Outcomes Statement* (pp. 139–149). Utah State University Press.

Melonçon, L. & Henschel, S. (2013). Current state of US undergraduate degree programs in technical and professional communication. *Technical Communication, 60*(1), 45–64.

Read, S. & Michaud, M. J. (2015). Writing about writing and the multimajor professional writing course. *College Composition and Communication, 66*(3), 427–457.

University of New Mexico Division for Equity and Inclusion. (2019). *University of New Mexico land acknowledgement.* https://diverse.unm.edu/about/land-acknowledgement.html.

University of New Mexico Office of Assessment. (2015). *NM HED Area I: Communications Competencies; UNM Core Area 1: Writing and Speaking.* https://valencia.unm.edu/academics/faculty-resources/assessment/core-assessment/area-1-core-comps.pdf.

Yancey, K. B. (2018). It's Tagmemics *and* the Sex Pistols: Current issues in individual and programmatic writing assessment. In S.W. Logan & W. H. Slater (Eds.), *Academic and professional writing in an age of accountability* (pp. 257–275). Southern Illinois University Press.

Part Three: Incorporating Technology

9. The Rhetoric, Science, and Technology of Twenty-First Century Collaboration

Ann Hill Duin
UNIVERSITY OF MINNESOTA

Jason Tham
TEXAS TECH UNIVERSITY

Isabel Pedersen
ONTARIO TECH UNIVERSITY

Abstract: We contend that collaboration is an imperative disciplinary assumption in technical and professional communication (TPC). Theorists, researchers, and practitioners grapple with ever-changing modes and models for collaborative work in academia, industry, and with communities. Technical and professional communicators today must be prepared to collaborate with engineers, subject matter experts, and programmers; they must be adept at using collaborative software and working with global virtual teams. The purpose of this chapter is to synthesize the rhetoric, science, and technology of collaboration to consolidate a guiding framework for understanding, teaching, and practicing TPC collaboration in the twenty-first century and beyond. This unified framework provides guidance from which to structure one's own collaboration and the collaborative projects we assign throughout our curriculum. We discuss collaborative software and team communication platforms and share example projects for preparing students for collaborative and global workplaces.

Keywords: collaboration, rhetoric, technology, platforms, global virtual teams

Key Takeaways:

- Collaboration across local and global contexts is an imperative disciplinary assumption in technical and professional communication (TPC).
- TPC instructors must prepare students for the collaborative frameworks and tools that practitioners use, including team management platforms, online repositories, and social media in support of local and global virtual teamwork.
- TPC students need experience in collaborating with clients, gathering customer feedback, and working as part of content development teams.

As an ongoing topic in our field, collaboration is multifaceted. We are invested in studying the rhetoric of collaboration, exploring the socio-cultural and social scientific factors influencing collaboration practices, and keeping our collective fingers on the pulse of collaboration technologies. However, we have yet to create a guiding framework for collaboration specific to technical and professional communication (TPC) that integrates these multiple dimensions of collaboration. Given the criticality of collaboration, the purpose of this chapter is to synthesize the rhetoric, science, and technology of collaboration to consolidate a guiding framework for understanding, teaching, and practicing TPC collaboration in the twenty-first century and beyond (see Figure 9.1). This unified framework provides guidance from which to structure one's own collaboration and the collaborative projects we assign throughout our curriculum.[1]

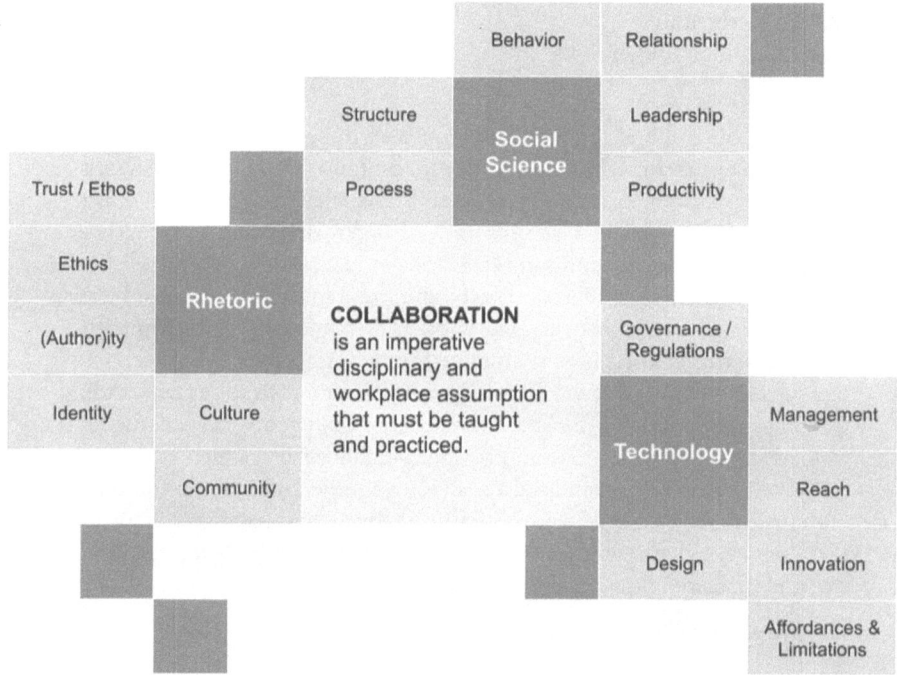

Figure 9.1. Twenty-first-century collaboration framework.

Our method in this chapter is focused literature review and constructivist theory building (Mills et al., 2006). We subscribe to constructivism as the epistemology for our study as it is congruent with our values and purpose. Knowing that no one project can truly encompass the magnitude of collaboration as a theory as well as practice, we do not attempt to infer a singular definition for

---

1. Duin and Pedersen, in *Writing Futures: Collaborative, Algorithmic, Autonomous* (forthcoming), also discuss the socio-rhetorical roots in collaboration theories and the science of collaboration as evidenced in NSF and NIH publications.

collaboration. However, given the aforementioned exigence, we are motivated to construct a consolidated framework based on existing threads of scholarly discussions and exemplary cases. A constructivist theory building methodology allows for co-construction of meaning between us as authors, the participants in studies we feature, and the literature we reference. Our co-constructed framework does not assume objectivity but instead acknowledges the social, cultural, and structural contexts within which our findings emerge.

Considering these contexts, in co-authoring this chapter, we demonstrate cross-generational, cross-disciplinary, and cross-cultural collaboration. Ann Hill Duin is a U.S. writing studies professor from the University of Minnesota with 30+ years of research on collaboration and shared leadership. Originating from Malaysia, Jason Tham is an assistant professor at Texas Tech, with his research positioned at the intersections of rhetoric, communication design, and emerging technologies. Isabel Pedersen, Canada Research Chair in Digital Life, Media, and Culture, is a professor at The University of Ontario Institute of Technology, where she studies digital life and transmedia cultures and leads the internationally known archive, the Fabric of Digital Life, discussed later in this chapter.

We begin with definitions of collaboration, exploring the explicit and implicit messages about collaboration and critiquing the romantic notion of "sole" authorship. We discuss how "successful" collaboration is defined in TPC and the models of collaboration that have most influenced our field. We then explore the effect of collaboration on the traditional rhetor-audience relationship, with emphasis on dialectic as invention, discovery at the intersection of collaborative work, and the ethics and ethos of co-authorship and collaboration. We move next to science, highlighting the increased focus on team science and what makes a scientific team effective. We conclude by discussing technology, emphasizing our need to understand and deploy collaborative software and team communication platforms with our students, sharing example projects for preparing students for collaborative and global workplaces.

Tracing the Socio-Rhetorical Roots in Collaboration Theories

It is well established that technical communicators are expected to work in coordination, cooperation, and collaboration with content experts, designers, and developers to build products and test processes. Isabelle Thompson (2001) observes that "collaboration as a research issue and as practice seems firmly rooted in technical communication as a discipline" (p. 167). Over the last three decades, research on collaboration has generated a body of scholarship with broad conceptions of collaborative writing, group interactions, and team-based learning (e.g., Bruffee, 1984, 1998; Ede & Lunsford, 1990, 2001; Jones, 2007). Technical communication scholars borrow collaboration theories from rhetoric and composition scholars who have studied collaboration at the intersection of collaborative writing and learning. Kenneth Bruffee's (1984) influential scholarship emphasizes the usefulness of conversation and collaborative learning in the classroom. William Duffy (2014), in his review of the decades of scholarship on collaboration, notes that Bruffee's "conversational imperative" sets the stage for what is known largely as the social constructivist epistemology, or the "social turn" in our larger discipline (p. 417).

The social turn has served as a lasting lens within which rhetoricians theorize collaborative efforts. Some challenged the rigidity of style and value that views scholarly work (Sullivan, 1994), while others began to pay attention to the influence of cultural, emotional, and gender factors on rhetoric (Bleich, 1995). Kathleen Blake Yancey and Michael Spooner (1998), in echoing Charlotte Thrall's (1992) argument that "all writing is inherently collaborative" (p. 79), reflected on the impact of collaboration on the writer's sense of self. These pioneering works show that collaboration changes the traditional rhetor-audience relationship. David Frank and Michelle Bolduc (2010) demonstrate this notion through the examination of Lucie Olbrechts-Tyteca's collaboration with Chaim Perelman in their field-defining magnum opus, *The New Rhetoric: A Treatise on Argumentation* (1958/1969). The Perelman/Olbrechts-Tyteca partnership not only produced a groundbreaking audience theory, but also revealed the complexity (or blurred lines) in scholarship collaboration in terms of the author/rhetor's agency and relationships that "defy rigid classifications and proscribed roles" from the perspective of the audience (Frank & Bolduc, 2010, p. 160).

Of note is the emphasis on dialectic as invention in the body of scholarship that Lisa Ede and Andrea Lunsford (1984, 1985, 1990, 2001, 2009; Lunsford & Ede, 2011) have co-created. While their early research showed the focus on organizational patterns (hierarchical structure) in collaborative writing, Ede and Lunsford (1990) invoked a "dialogic" collaboration, which focuses on the dialectical tensions in the collaboration process. The dialogic approach is concerned with roles and process rather than the end product (Qualley & Chiseri-Strater, 1994). Likely building on Bruffee's conversation paradigm, Lunsford and Ede (1990) continued to explore how collaboration opens our disciplinary hearing of a "new key" that has been struck "clearly and repeatedly by many of the women and a few men [they] have mentioned, but which has not often been heard—by our professional organizations, by our institutions, by the culture within which we are all so deeply inscribed" (p. 240). Lunsford and Ede's work has inspired scholars to focus on gender and gender-related conflict or differences in collaboration (Blair & Nickoson, 2018; Burnett & Ewald, 1994; Fredlund, 2016; Karach & Roach, 1992; Lay, 1989; Monk et al., 2003; Morgan, 1994). Moreover, Bruce McComiskey (2015) offers a historic overview of the function of dialectic in its relationship to invention as a means to engage writing students who are learning about argumentation. In combination with invention, dialectic becomes the basis for a heuristic approach to teaching that helps avoid predetermined outcomes for writing.

Another important rhetorical investigation into the impact of collaboration on invention and discovery is made at the intersection of ethics and ethos. Mary Lay (Schuster) and William Karis' (1991) early agenda in cultivating collaboration between academia and industry has generated a huge following among technical communication scholars. At a time when "micro" computers are entering mainstream workplaces and homes, technical communicators are in high demand. The work of technical communication became more hybrid to accommodate the needs of content producers and consumers alike. At this point, discussions of ethics emerged. Steve Katz (1992) blazed the trails by leading an important conversation on the ethics of technical communication through the examination of the so-called "productivity" or expediency that's afforded by communication technology. In the next two decades, scholars have continued to challenge, critique, and propose strategic frameworks for collaboration within technologically enhanced environments. For instance, Heidi McKee and James Porter (2017) in their recent examination of networked interactions urge technical communicators to be aware of the rhetorical situation in professional communication practices mediated by social networks and "smart" or assistive technologies such as artificial intelligence (AI) agents. In *Rhetoric as a Posthuman Practice* (2018), Casey Boyle wants us to reflect on how modern information technology practices, including technical communication, can be "transindividual" practices (p. 187) that require our attention to embodiment, nonhuman agents, and ethical consequences.

Pedagogical Implications

The rhetorical perspectives on collaboration translate into practical implications in technical communication and writing pedagogy. To simulate collaboration, instructors usually assign group projects in technical communication courses so students can gain such experience. Typically, students are asked to collectively brainstorm ideas, draft outlines, conduct research, write sections of a paper, and present findings as a group. Coherence in the work students produce as well as the team working process are normally expectations from instructors. However, instructors often face challenges in motivating students to strive while completing group work and in finding systematic ways to evaluate progress and the quality of collaborative projects.

Early research has revealed some issues dealing with collaborative writing in the technical communication and writing classroom, including resistance from students, students' lack of experience in working together, group conflict and friction, and the instructor's evaluation of group work (Chisholm, 1990). In *Foundations for Teaching Technical Communication*, Rebecca Burnett, Christianna White, and Ann Hill Duin (1997) argued that the nature of collaboration is revealed through exploration of culture, authority, conflict, and gender. More recently, Laurie Cella and Jessica Restaino (2014) remind us that many instructors and students still struggle with practicing team projects. In stories we have heard about team projects, students often describe negative experiences working with others they have just met during the semester, while instructors battle with the stigma about the slackers and sluggards in student teams—and together they paint an unattractive picture for team work. To that end, Elizabeth Adams St. Pierre (2014) invites us to consider the ontology in posthumanism and how such perspective may shift our perspectives on collaborative writing. St. Pierre argues that collaborators may not always be "present" in collaborative projects, but collaboration is always already enabled through an assemblage view of reading, writing, and the world.

Following the proliferation of new theoretical perspectives and advancement of collaboration technologies, evidence-based guides to creating group projects, such as Joanna Wolfe's (2010) *Team Writing*, as well as innovative approaches, such as Agile project management (Moses, 2015; Pope-Ruark, 2012, 2015), design thinking powered collaboration models (Duin et al., 2017), and makerspaces (Gierdowski & Reis, 2015; Tham, 2019b) are making their way into the classroom as potential remedies for negative student learning experience. These approaches focus on a flexible "openness" that supports individual team members as they move from "peripheral participants" to potentially "longstanding members engaged in ongoing projects" (Gierdowski & Reis, 2015, p. 17). Social constructivists in writing studies believe "individual writers compose not in isolation but as members of communities whose discursive practices constrain the ways they structure meaning" (Nystrand et al., 1993, p. 289). The primary assumption behind this learning theory is that social interaction and participation, particularly with instructors, peers, and other members of the knowledge community, have a significant impact on learning (Chism, 2006; Lave & Wenger, 1991; Wenger, 1998). Jean Lave (1991) has contended, "learning, thinking, and knowing are relations among people engaged in activity in, with, and arising from the socially and culturally structured world" (p. 67).

When students work in cross-functional teams to support others through cross pollination of knowledge and skills, they offer different perspectives to spur innovation and challenge conventional practices (i.e., "we have always done it that way"). Peer collaboration also levels the "playing field" for learning—students at any level or with any amount of content knowledge can participate in innovation and execution of ideas, which may increase overall engagement. The role of the instructor is to facilitate a learning atmosphere that encourages students to claim shared ownership of their project. Kenneth Rainey, Roy Turner, and David Dayton's (2005) research on technical communication core competencies is notable for its emphasis on collaboration and collaborative knowledge. Their work has been tested (Hart & Conklin, 2006) and continues to influence competency-based education for both scholarly and professional organizations, such as the Association for Teachers of Technical Writing (ATTW) and the Society for Technical Communication (STC) as well as for technical communication program administrators, even impacting the collaborative design of international curricula and the development of competency statements and learning objectives (Paretti et al., 2007).

Contemporary technical communication scholars follow Rainey et al.'s direction, voicing parallel calls to better understand and meet the needs of students demanding a learning-focused education. For example, Annie Mendenhall (2013) argues that "we need to think vertically, horizontally, and institutionally about how to create courses and curricula. In other words, minors, majors, and graduate programs increase the field's legitimacy by shaping it into a model discipline, but our work might also operate outside the vertical model to engage other disciplines and communities in writing instruction or interdisciplinary programs of study" (p. 97). Sally Henschel and Lisa Melonçon (2014) push for a similar collaborative shift. They state that "even though technical communication programs maintain specific strengths tied to faculty expertise and to local situations, programs should be embracing common conceptual and practical skill sets that will prepare students to become successful professionals" (p. 22). Henschel and Melonçon's (2014) suggestion to make comparisons and to discover commonalities within and outside of a TPC program demonstrates the valuable connections and information that can be gained by joining the learning paradigm shift, which means choosing to collaborate and share conceptual and practical skill sets across writing programs, departments, and an institution.

The programmatic and practical emphasis on collaboration is justifiably ever-present. Thomas Kent (1993) argues that "without collaboration . . . no communicative interaction is possible. . . . If we are communicating, we are collaborating" (cited in Burnett et al., 1997, pp. 136–137). This point resonates in both academic and workplace settings, yet collaboration in academic contexts varies from collaboration in the workplace. For instance, Thompson (2001) has found evidence of these differences in how collaboration is considered in the academy and in industry after conducting a qualitative content analysis of articles on collaboration in technical communication. In workplace terms, Rebecca Burnett, Andrew Cooper, and Candice Welhausen (2013) assert that "[c]ollaboration is important because virtually all workplaces rely on group-based decision making and projects, often increasing creativity, productivity, and the quality of both process and product" (p. 454). Empirical studies of writing in workplace settings (Allen et al., 1987; Cross, 2001; Jones, 2007; Winsor, 2003) have helped to clarify the nature of workplace writing collaboration as well.

■ Scientific and Workplace Collaboration

What is clear is that scientific and workplace collaboration is common, that it is necessary, and that it has become a critical competency for practicing technical communicators. As teachers, we have a pedagogical imperative to learn about and practice collaboration so that we can instruct our students proficiently in its practice. As scholars, we have an experiential imperative to collaborate; this collaboration, as we argue below, reinforces our pedagogy.

A search on "the science of collaboration" results in a plethora of articles emphasizing the importance of collaboration across the academy and industry, the increase in demand for those with collaboration skills, and the exponential increase in tools that support collaboration. Blog postings, webinars, and "top ten" lists populate these search findings; numerous collaborative visualizations allow for articulation of scientific collaborations and future research questions (Isenberg et al., 2011); and publishers such as Elsevier encourage and possibly mandate researchers to visualize their data and scientific research networks (Elsevier, 2019).

Adjacent to rhetoric and writing studies, researchers in speech and organizational communication studies have examined group dynamics and collaborative interactions through functional and interpretive perspectives. In the twentieth century, scholars like George Herbert Mead (1934) and Herbert Blumer (1986) applied philosophical methodologies to theorize group communication as symbolic interactionism. Ernest Bormann's (1972) symbolic convergence theory pulls from the rhetorical and socio-psychological traditions, arguing that sharing of common "fantasies" can transform collaborative groups. Symbolic convergence occurs when group members spontaneously create fantasy chains that display an energized, unified response to common goals. The analysis of these themes may reveal a rhetorical vision that contains vision to enact the joint objectives of the group. Paul Watzlawick (1978) followed a cybernetic tradition and theorized collaborative dynamic as merely the interaction of content (what) and relationship (how). These central realms of group interactions have influenced early theories of collaboration. Bruce Tuckman (1965) hypothesized a four-stage model—what's well known today as the Tuckman Model—in which each stage needed to be navigated sequentially in order to reach effective group functioning. The four stages are forming, storming, norming, and performing. Tuckman and Mary Ann Conover Jensen (1977) later revised this model to include adjourning as the final stage of group interactions.

Perhaps the most respected group interaction theorists, Randy Hirokawa (1994) and Dennis Gouran (1988, 2003) are known for their functional perspective on group decision making. They dismissed pessimistic views about collaboration as unwarranted by actual group processes. Gouran's early writing on group decision making laid the groundwork for Hirokawa's later functional roles in collaborative groups. Their collective work theorized that groups make high-quality decisions when members fulfill four requisite functions: 1) problem analysis, 2) goal setting, 3) identification of alternatives, and 4) evaluation of positive and negative consequences. Erring on the interpretive end, Marshall Scott Poole's (1997, 2003; Poole & Doelger, 1986) adaptive structuration theory uses a "phase" model to complicate Anthony Giddens's (1984) structuration theory, which refers to "the production and reproduction of [sic] social systems through members' use of rules and resources in interaction" (Poole, 2003, p. 50). Poole's phases concern the production of change and reproduction and stability through a duality structure—

what is affected by the group and its effect upon rules and resources. More recent, Barnett Pearce (2004, 2008) and Vernon Cronen (2001) used co-constructionism to understand collaboration as coordinated management of meaning.

What also is clear across the literature is that workplace and scientific collaboration is imperative. As research questions increase in complexity and science struggles "to swim through big data, major funders, including the National Science Foundation (NSF) and the National Institutes of Health (NIH), are pushing scientists to collaborate more across disciplines, institutions, and even nations under the banner of *team science*" (Baker, 2015, p. 639). In the past decade, a new field—the science of team science (SciTS)—has emerged, with its aim "to better understand the circumstances that facilitate or hinder effective team-based research and practice and to identify the unique outcomes of these approaches in the areas of productivity, innovation, and translation [of science]" (Stokols et al., 2013, p. 4). The Team Science Toolkit (2019) states:

> Over the past two decades, there has been an emerging emphasis on scientifically addressing multi-factorial problems, such as climate change, the rise of chronic disease, and the health impacts of social stratification. This has contributed to a surge of interest and investment in team science. Increasingly, scientists across many disciplines and settings are engaging in team-based research initiatives. These include small and large teams, uni- and multi-disciplinary groups, and efforts that engage multiple stakeholders such as scientists, community members, and policy makers. Academic institutions, industry, national governments, and other funders are also investing in team science initiatives.

According to Nancy Cooke and Margaret Hilton (2015), team science "focuses on science teams and groups and their individual members as the principal units of study" (p. 49). Most recently, as part of a review of 109 empirical articles on collaboration in science, Kara Hall and colleagues (2018) define team science as "the approach of conducting research in teams within complex social, organizational, political, and technological milieu that heavily influence how that work occurs" (p. 533), ultimately finding that "the degree to which researchers achieve team-based and integrative science is driven by a complex mix of attitudes, behaviors, and cognition, which, in turn, may be influenced by features of the team, organization, and broader context" (p. 544).

Also key to team science is the specific study of what makes a scientific team effective. Dimensions of team science under study include diversity of team or group membership, disciplinary integration, team or group size, goal alignment across teams, permeable team and organizational boundaries, proximity of team or group members, and task interdependence (Cooke & Hilton, 2015). In their study of key elements critical to the success of collaboration and team science, L. Michelle Bennett and Howard Gadlin (2012) found the most important element to be that of trust: "without trust, the team dynamic runs the risk of deteriorating over time" (p. 768). Other key elements included "developing a shared vision, strategically identifying team members and purposefully building the team, promoting disagreement while containing conflict, and setting clear expectations for sharing credit and authorship" (p. 768).

Margaret Hinrichs and colleagues (2017), in their review of the 2015 National Academies report, address the need to attend to the relational side of collaboration. Their recommendations include "a renewed focus on the process of organizing through communication rather than focusing on organization as an outcome or consequence of teamwork" (p. 144). We use technology as the ultimate means to organize communication.

∎ The Technology of Collaboration

Advances in writing technology bolster collaboration. As James Porter (2009) notes, "The computer plus the internet and the World Wide Web provide publishing capacity to the individual writer" (p. 219). The individual writer's capacity is motivated by social impulses: "people write because they want to interact, to share, to learn, to play, to feel valued, and to help others. And that drive to interact socially is a key feature of the new digital era" (Porter, 2009, p. 219). Laura Gurak and Ann Hill Duin (2004) contend that emerging digital technologies foster collaboration in technical communication pedagogy and research. Powered by open access and open collaborative tools, many modern classrooms are reimagined as hubs of learning where individuals come to share ideas and work on projects together. These spaces invite students to come out from their silo workspaces and combine resources to tackle complex communicative issues. Such tendency is deemed favorable by public and private sectors today where collective intelligence (Levy, 2000) is considered valuable in social capital. Thus, to integrate such learning in technical communication education is to acculturate learners into their future work environments, where collaboration and cross-functional teams are already commonplace.

Jessica Behles, in her 2013 survey of the use of collaborative writing technologies by technical communication practitioners and students, identified wikis, online word processors, learning management systems, SharePoint, and Google Docs as tools used daily by practitioners, but at that time, only weekly by students. She found that students were "features driven" while practitioners primarily used tools chosen by their companies (p. 28). More recently, Stephanie Vie (2017), based on her national survey of 30 TPC programs' use of social media, identified as "crucial that online TPC courses consider moving past the familiar 'big three' of Facebook, Twitter, and YouTube and examine other social media tools of interest to the field" (p. 353), calling for "pedagogical artifacts and reflections that specifically respond to the exigencies of increased social media use" (p. 354). Jason Tham (2017) edited a collection on collaboration technologies, with technical communication instructors sharing such a suite of artifacts and collections in their discussion of how Join.me, Facebook Messenger, Scalar, and WebEx technologies transform our collaborative work and pedagogy.

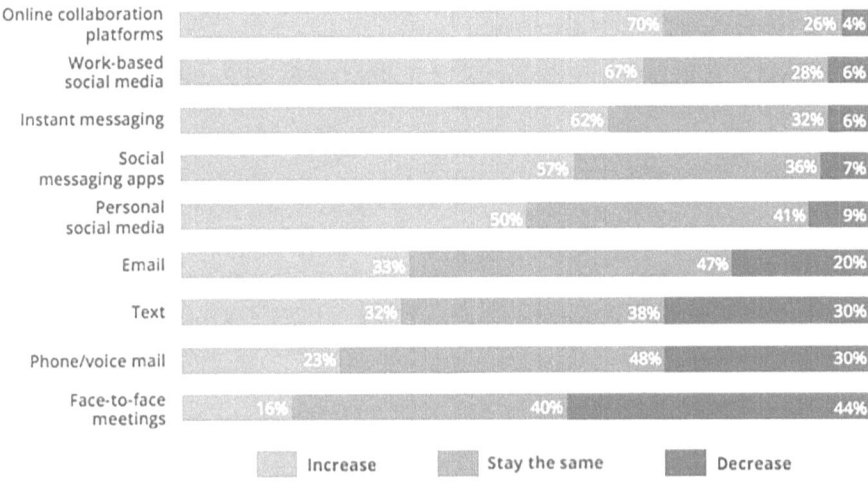

Figure 9.2. Expected use of communications channels in the next three to five years. (Deloitte, 2018, p. 82).

Professionals indeed get things done through the use of social, collaborative, and virtual tools, and a myriad of such tools now crowds the marketspace (Capterra, 2019). The most recent Deloitte *Global Human Capital Trends* survey examined this "flood of new tools" in support of "the hyper-connected workplace," finding online collaboration platforms, work-based social media, instant messaging, and social messaging apps to be increasing, while face-to-face meetings, the use of the phone and voice mail, text, and email are decreasing (see Figure 9.2). In comparison to Behles' 2013 study, it's commonplace for technical communication students to regularly use email, learning management systems, and collaboration systems such as shared Google Drive files and folders as part of their coursework. However, they are likely to have less knowledge of specific collaboration software and management directions designed for industry use.

Industry reviews of collaboration software or groupware note that use of these tools "allows the managing, sharing and processing of files, documents and other types of data among several users and systems anytime and anywhere" (Seymour, 2019). Given that industry markets the benefits of these tools as saving time, enhancing project management, strengthening team relationships, and improving overall organization, as instructors, we should expand our pedagogy well beyond the "group project," framing, discussing, and studying our use of technologies in these expanded terms as a means to prepare students for industry.

In terms of project management as collaboration, Nancy Allen and Steven Benninghoff's (2004) survey of TPC programs in the US found 30 of 42 programs to include project management courses, and more recently, Lisa Melonçon and Sally Henschel (2013) found that 18 percent of 65 TPC programs included emphasis on project management. This comes with increased discussion and use of Agile project management strategies across TPC. As Rebecca Pope-Ruark (2015) notes, when working in Agile software development environments, "writers have much more opportunity to advocate for users, express concerns and insights, and create more lightweight external documentation throughout short, iterative development cycles rather than focus on heavyweight internal documentation, ensuring better products and better supporting documents" (p. 113). Pope-Ruark used Scrum, "the most popular Agile framework," to structure her course on Grant Writing for Nonprofit Organizations. Use of Scrum resulted in breaking down the main collaborative project into small slices where teams meet regularly to share what has been done, what each member will do next, and the challenges or issues needing team input. She emphasizes that Scrum "was designed for complex, multifaceted projects that require close collaboration" (p. 129), recommending the use of Agile practices for more complex collaborative projects. Joseph Moses, Trey Conner, and Jason Tham (2019) agree with Pope-Ruark, using Agile-informed strategies to inspire team-based learning through collaborative projects in the classroom. Their framework is focused on using design thinking as guiding principles for making team commitment, adaption, and evaluation visible components of the collaborative process (Tham & Moses, 2019).

In short, we must prepare our students for the collaborative frameworks and tools that technical communication practitioners now are called upon to use. These include team management platforms, online repositories, and any number of platforms and social media in support of global virtual teamwork. Therefore, we turn next to more detailed discussion of a team communication platform (Slack), an online repository (GitHub), and the importance of practice in collaborating and leading global virtual teams.

Online Collaboration Platforms

In his opening to a special issue examining the potential of online collaborative platforms, Peter Cardon (2016) notes that most organizations adopt these platforms based on the promise for more open, transparent, and collaborative communication; however, most have experienced "little or no change," leading to research showing "that the transformative potential of these platforms depends on a communication [vs. a technological] perspective" (p. 141). As a means to emphasize the need to expose TPC students to the use of collaborative platforms, we highlight

Abram Anders' (2016) study of team communication (collaboration) platforms and emergent social collaboration practices that concludes Cardon's special issue.

Specifically, Anders examined a prominent team communication platform (TCP), Slack (https://slack.com/), used by one million people at the time of his study, and now (in 2019) used by ten million people a day across all types of industries and organizations. TCPs integrate multiple media in support of collaborative work, and conversations are organized into groups for specific teams and projects and channels for knowledge sharing and topic-based communication. TCPs also include notifications (alerts) managed by the team member as well as mentions or alerts that team members can send to others. Users can integrate services like Google Drive and Dropbox or various video-conferencing services. The overall design makes communication and collaboration visible, searchable, and available across organizational boundaries. In Anders' analysis of 100 self-published blog posts by Slack users, he found this TCP to support knowledge sharing and collaborative workflows: "The communication visibility afforded by TCPs . . . had direct impacts on collaboration processes. Users noted that communication visibility—especially when supported by compartmentalization of groups, projects, and topics—enabled more distributed and self-organized styles of collaboration" (p. 247). The use of Slack also resulted in greater engagement and presence, context awareness, generative role taking, leadership awareness, and synchronicity. As Anders quotes a user, "'It [Slack] compresses a lot of the stuff you might otherwise do in meetings into a Slack channel, so that information is visible to everyone it should be visible to, and it saves people time: They don't necessarily have to meet but can stay updated on a project's status'" (p. 252).

Today's technical communication students need to be prepared for a workplace that deploys collaborative software or team communication platforms. Rich McCue (2015), Systems Administrator for the University of Victoria Libraries, provides access to a vast set of research and collaboration tools for use by students, staff, and faculty as a means to create a "modern memex." As part of our courses, we should integrate media capabilities that make routine communication and collaborative workflows visible and shareable. At the University of Minnesota, we are a Google campus, so students become increasingly adept in their use of Google Drive applications along with sharing of files and folders throughout their collegiate work. However, we could certainly adopt and integrate a TCP such as Slack as part of a group assignment. In doing so, we should focus on the communicative affordances of the collaborative software platform along with how each platform supports community and team development.

Collaborative Use of Online Repositories

The Center for Information Design Management (CIDM) conducts a yearly survey of trends in the development and delivery of content. Based on this ongoing analysis, JoAnn Hackos (2015) emphasizes that technical communication students need experience in collaborating with clients and with gathering customer feedback as well as requisite knowledge and practice with content management systems as preparation for collaborating in content development teams and challenging the assumed authority of product developers. In response to this need, we offer two project directions for providing TPC students with exposure to and practice with collaborative use of online repositories.

As one example, Duin and Tham (2018) used Hackos' work as a springboard for their redesign of a course titled Writing with Digital Technologies. As part of the redesigned course, students are instructed in the use of an open-source online repository, GitHub (https://github.com/), for recording, editing, and sharing of their HTML and XML work. GitHub is an online repository used by over 28 million developers and 1.5 million companies across the world (Wan, 2018). While most popular for use by programmers and software companies, GitHub also serves as a portfolio where technical communication students can showcase their individual projects and contributions to others. While development and use of an online repository such as GitHub is a stretch for most technical communication students, knowledge and use of such a collaborative online repository provides them with greater understanding of how technical communicators and software engineers host, review code, and manage projects; it also results in a student's competitive advantage when entering the workplace. As one student shared in a response to this instruction,

> Once I figured it out, GitHub is one of the greatest tools I have encountered. I am fortunate to have been introduced to this and to better understand its collaborative functions. Since the beginning of the semester, I have used GitHub as a temporary hosting source for a total of five web projects and have also used it to download other resources.

As another example, we encourage instructors and students to identify and use online repositories such as the Fabric of Digital Life ("Fabric") research archive (https://fabricofdigitallife.com/) to examine and/or curate emerging technologies and their impact on technical communication. Fabric monitors the emergence of digital technology prototypes, inventions, news, and research by archiving representations in several categories and media types (text, images, video, etc.), concentrating on platforms of human-computer interaction to reveal the multiple ways that embodied technologies emerge in society.

As students examine and/or curate artifacts, they use instructions that guide them in learning a common language of classification to ground their understanding of technical emergence. Here, a broader goal is to reveal rhetorical motivations across interdisciplinary discourses in order to study sociotechnical tradeoffs among technical innovations (Iliadis & Pedersen, 2018). With specific focus on wearables, carriables, implantables, ingestibles, embeddables, and roboticals, students can use this customized metadata system to archive representations and facilitate simultaneous content collaboration on the database. One component is a keyword schema that helps students to standardize the constantly evolving language used to describe emerging technology. Three categories distinguish between *technology keywords* (e.g., fitness monitor, smartwatch, accelerometer), *marketing keywords* (e.g., Apple, Forbes, Fitbit), and general *thematic keywords* (e.g., health, education, children, manufacturing, climate change).

This classification standard helps when archiving primary research items. For example, the archiving of a video of engineers demonstrating a humanoid robot communicating with factory workers would use keywords to track the technological ability of the robot (e.g., natural language processing; human-robot communication), the corporate backing designated by marketing keywords (e.g., SoftBank Robotics), and the application in a real-world scenario under general keywords (e.g., manufacturing, factory, work, dialog, polite, posthuman). The standardized keyword system becomes useful over time when different archiving teams build upon previous content. To extend the example, if the same robot appears in a concept video for a childcare fitness scenario at a daycare several years later, it might be able to *play*, extending its general keyword profile to be classified under *children, toy, caregiver, fitness*, etc. One important pedagogical goal might be met if rich sociotechnical themes arise, such as the realization that emergent robots were previously trained in factories before daycares, leading to a nuanced understanding technology emergence.

Most important, Fabric enables team-based collaboration. The keyword system affords students the opportunity to revisit previous items in order to enhance them as vocabularies evolve, leading to collaborative archiving practices for globally-dispersed work groups. In our case, a recent collaboration between the University of Minnesota's Emerging Technology Research Collaboratory (ETRC; https://etrc.umn.edu/) and Fabric is resulting in multiple undergraduate and graduate student collections, the first of which is titled "Emerging Technologies for Technical Communication" found directly on the site (Tham, 2019a).

Global Virtual Teams

Moreover, today's technical communicators increasingly perform their work as part of global virtual teams (GVTs). Technical communication researchers such as Clay Spinuzzi (2007) emphasize the need for adjusting to multiple stakeholders in global virtual environments, stating, "Currently we face work structures that were hardly conceivable a few decades ago, and these work structures again require different rhetorical skills and communication practices" (p. 266); and contributions to Rich Rice and Kirk St.Amant's (2018) edited collection, *Thinking Globally, Composing Locally*, provide direction for rethinking perceptions of global communication and reconsidering approaches to writing online. Organizational researchers such as Scott I. Tannenbaum, John E. Mathieu, Eduardo Salas, and Debra Cohen (2012) agree that we have entered a new era in that teams operate in a more fluid, dynamic, and complex environment than in the past. They change and adapt more frequently, operate with looser boundaries, and are more likely to be geographically dispersed. They experience more competing demands, are likely to be more heterogeneous in composition, and rely more on technology than did teams in prior generations. Moreover, teams have become so ubiquitous that many employees and managers take them for granted and assume that they will be effective (Tannenbaum et al., 2012).

We contend that practice as part of global virtual teams should be a required component of TPC curricula. According to Giuseppe Palumbo and Ann Hill Duin, "GVTs are those teams connected via technology and comprised of people in various locations around the globe.... The primary objective of virtual collaboration is for a technology-mediated globally-dispersed work group to launch, develop, and complete its assigned task" (2018, p. 109). A 2018 survey of 1,620 respondents from 90 countries found 89 percent of respondents to be working on global virtual teams, 88 percent reporting that this virtual teamwork is critical to their productivity, and 84 percent reporting virtual collaboration to be more difficult than in-person collaboration, especially considering that 89 percent of virtual teams include at least two cultures (CultureWizard, 2018). This survey aligns with the earlier Deloitte findings, with 48 percent of respondents reporting that they never meet other virtual team members in person. Unfortunately, but perhaps not surprisingly, 80 percent report that they received no formal training in leading or being part of GVTs, often leading to differing assumptions, misunderstandings, and resentment during collaboration across cultures.

To begin, students can read and provide reflections on GVT research, such as a study on building swift trust in global virtual teams (Crisp & Jarvenpaa, 2013), a study on the effects of a dyad's cultural intelligence on global virtual collaboration (Li et al., 2017), and a study on managing multicultural teams (Behfar et al., 2006). More important, however, is the experiential practice of being part of a "real" global virtual team. To address this need, since 1999, Bruce Maylath and colleagues have supported the Trans-Atlantic and Pacific Project (TAPP), a network of partners that "establishes links between students in different countries so that each learns from the other. In so doing, students become aware of the diversity of the world community in which their documents travel" (2019, para. 1).

In the standard TAPP format, U.S. students prepare a set of instructions, conduct a usability test on the document with the help of students from another country, and then finalize the same document for later translation into another language by students from the partnering country. Instructions for joining and materials for beginning this type of project are provided by current TAPP instructors on the Google site, TAPP2018 (Duin et al., 2018, https://sites.google.com/a/umn.edu/tapp2018/). In one study of a standard TAPP project, Palumbo and Duin (2018) reported on their study of U.S.-Italian student interactions and use of visualizations of their personal learning networks as a means to build cross-cultural competence, trust, and learning strategies and attitudes—aspects that the authors found to be characteristic of the students' collaboration besides the obvious and more immediate focus on questions of language and translation.

Given the need for technical communication students to receive training in the management of GVTs, Duin and Palumbo recently adapted the standard TAPP format with the goal of having the U.S. (UMN) students serve as project managers of teams of Italian (University of Trieste) students practicing translation. In this model, each U.S. student leads the GVT of five members, providing oversight of team building, project preparation, translation and submission of final materials as well as post-mortem work to evaluate the overall project, and the professors represent the clients for each team's project. This adaptation of the TAPP model to support a six-week project includes use of an abridged version of *A Guide to Translation Project Management* (2016) by David Russi and Rebecca Schneider (used with permission) as a student guide for this work, videos on translation workflow, articles on the role of a translation project manager, and use of the above noted readings by the U.S. students. During the recent 2019 deployment, project managers (U.S. students) primarily used Skype for weekly meetings, shared Google Drive folders and files for organizing the work and used WhatsApp for secure messaging as part of daily communication.

Artificial Intelligence

As we consider the future of collaboration, we call on readers to recognize our increased collaboration with artificial intelligence (AI) agents and nonhuman collaborators. The higher education landscape already includes "smart writers" to assist with academic writing and AI-based teachers at universities including Michigan, Miami, Georgia Tech, and others, where students no longer can distinguish between human and AI teaching assistants (Goel & Polepeddi, 2016). In industry, Microsoft, Salesforce, and Oracle have integrated AI into their enterprise collaboration platforms, including Slack, discussed earlier (Fluckinger, 2019). And in a recent *Harvard Business Review* article on collaborative intelligence, H. James Wilson and Paul R. Daugherty (2018) found from their research of 1,500 companies that firms achieve the most significant performance improvements when humans and machines work together. Through such collaborative intelligence, humans and AI actively enhance each other's complementary strengths: the leadership, teamwork, creativity, and social skills of the former, and the speed, scalability, and quantitative capabilities of the latter. (p. 117)

Adapting pedagogical models for collaborative AI may well be our next task as we evolve in designing effective teaching of technical communication.

Guiding Framework for Collaboration in TPC

In this chapter, we have employed focused literature review and constructivist theory building to frame how collaboration is defined and practiced in the twenty-first century and to provide pedagogical direction for active use of online repositories and collaboration platforms as a means to prepare TPC students for their current and future work in industry.

Collaboration is an imperative disciplinary assumption that must be taught and practiced in ways that expand student understanding of the rhetoric, science, and technology of collaboration. Returning to Figure 9.1, we demonstrate that collaboration is informed by socio-rhetorical traditions concerned with shared authority, trust, identity, ethics, community, and culture. As a process, collaboration can be examined through social structures, human behaviors, relationships, leadership, and productivity. And since twenty-first century collaboration is greatly affected by technology, it should include considerations of governance, regulations, and management of collaborative technology, its reach, and the affordances and limitations of various technology design and innovation.

We hope that instructors will use this as a guide in designing assignments for students to practice twenty-first century collaboration. A twenty-first century pedagogical collaboration framework includes multiple assignments, projects, and experiences for students to practice co-authoring and collaboration, with emphasis on dialectic as invention. It includes exposure to the complex contexts of team science and workplace collaboration along with understanding of innovative approaches such as Agile project management and design thinking as they approach their work. In our move from the use of the desktop to mobile technologies to social media to desktop videoconferencing and online collaboration platforms, technical communicators increasingly have worked in collaboration with others and with the evolving technologies supporting such collaboration. Powered by open access and open collaborative tools and repositories, we have the ability to reimagine our on-campus and online courses as hubs of learning where individuals come to share ideas and work on projects both together and with collaborators throughout the world.

■ References

Allen, N., Atkinson, D., Morgan, M., Moore, T. & Snow, C. (1987). What experienced collaborators say about collaborative writing. *Journal of Business and Technical Communication, 1*(2), 70–90.

Allen, N. & Benninghoff, S. T. (2004). TPC program snapshots: Developing curricula and addressing challenges. *Technical Communication Quarterly, 13,* 157–185.

Anders, A. (2016). Team communication platforms and emergent social collaboration practices. *International Journal of Business Communication, 53*(2), 224–261.

Baker, B. (2015). The science of team science: An emerging field delves into the complexities of effective collaboration. *BioScience, 65*(7), 639–644.

Behfar, K., Kern, M. & Brett, J. (2006). Managing challenges in multicultural teams. In Y-R. Chen (Ed.), *Research on managing groups and teams: Vol. 9. National culture and groups* (pp. 233–262). Emerald Group Publishing Limited.

Behles, J. (2013). The use of online collaborative writing tools by technical communication practitioners and students. *Technical Communication*, *50(1)*, 28–44.

Bennett, L. M. & Gadlin, H. (2012). Collaboration and team science: From theory to practice. *Journal of Investigative Medicine*, *60*(5), 768–775. https://www.ncbi.nlm.nih.gov/pmc/articles/PMC3652225/.

Blair, K. & Nickoson, L. (Eds.). (2018). *Composing feminist interventions: Activism, engagement, praxis.* The WAC Clearinghouse; University Press of Colorado.

Bleich, D. (1995). Collaboration and the pedagogy of disclosure. *College English*, *57*(1), 43–61.

Blumer, H. (1986). *Symbolic interactionism: Perspective and method.* University of California Press.

Bormann, E. (1972). Fantasy and rhetorical vision: The rhetorical criticism of social reality. *Quarterly Journal of Speech*, *58*, 396–407.

Boyle, C. (2018). *Rhetoric as a posthuman practice.* The Ohio State University Press.

Bruffee, K. A. (1984). Collaborative learning and the conversation of mankind. *College English*, *46*(7), 635–652.

Bruffee, K. A. (1998). *Collaborative learning: Higher education, interdependence, and the authority of knowledge.* Johns Hopkins University Press.

Burnett, R. E., Cooper, A. & Welhausen, C. A. (2013). What do technical communicators need to know about collaboration? In J. Johnson-Eilola & S.A. Selber (Eds.), *Solving problems in technical communication* (pp. 454–478). University of Chicago Press.

Burnett, R. E. & Ewald, H. R. (1994). Rabbit trails, ephemera, and other stories: Feminist methodology and collaborative research. *JAC: Journal of Advanced Composition*, *14*(1), 21–51.

Burnett, R. E., White, C. I. & Duin, A. H. (1997). Locating collaboration: Reflections, features, and influences. In K. Staples & C. Ornatowski (Eds.), *Foundations for teaching technical communication: Theory, practice, and program design* (ATTW contemporary studies in technical communication; Vol.1, pp. 133–160). Ablex Publishing.

Capterra. (2019). *Collaboration software.* https://www.capterra.com/collaboration-software/.

Cardon, P. W. (2016). Community, culture, and affordances in social collaboration and communication. *International Journal of Business Communication*, *53*(2), 141–147.

Cella, L. & Restaino, J. (2014). Lean on: Collaboration and struggle in writing and editing. *Literacy in Composition Studies*, *2*(2), 66–76.

Chisholm, R.M. (1990). Coping with the problems of collaborative writing. *Writing Across the Curriculum*, *2*, 90–108.

Chism, N. (2006). Challenging traditional assumptions and rethinking learning spaces. In D. G. Oblinger (Ed.), *Learning spaces* (pp. 2.1–2.12). Educause.

Cooke, N. & Hilton, M. (Eds.). (2015). *Enhancing the effectiveness of team science.* National Research Council; The National Academies Press. https://doi.org/10.17226/19007.

Crisp, C. B. & Jarvenpaa, S. L. (2013). Swift trust in global virtual teams: Trusting beliefs and normative actions. *Journal of Personnel Psychology*, *12*(1), 45–56.

Cronen, V. (2001). Practical theory, practical art, and the pragmatic-systemic account of inquiry. *Communication Theory*, *11*, 14–35.

Cross, G. A. (2001). *Forming the collective mind: A contextual exploration of large-scale collaborative writing in industry.* Hampton Press.

CultureWizard. (2018). *2018 Trends in High-Performing Global Virtual Teams*. https://content.ebulletins.com/hubfs/C1/Culture%20Wizard/LL-2018%20Trends%20in%20Global%20VTs%20Draft%2012%20and%20a%20half.pdf.

Deloitte. (2018). *The rise of the social enterprise: 2018 Deloitte Global Human Capital Trends*. https://www2.deloitte.com/content/dam/Deloitte/at/Documents/human-capital/at-2018-deloitte-human-capital-trends.pdf.

Duffy, W. (2014). Collaboration (in) theory: Reworking the social turn's conversational imperative. *College English, 76*(5), 416–435.

Duin, A. H., Moses, J., McGrath, M., Tham, J. & Ernst, N. (2017). Design thinking methodology: A case study of "radical collaboration" in the Wearables Research Collaboratory. *Connexions: International Professional Communication Journal, 5*(1), 45–74.

Duin, A. H., Palumbo, G., Arno, E., Goetz, G., Maylath, B., Mousten, B. & Vandepitte, S. (2018). *The Trans-Atlantic and Pacific Project (TAPP) model. Instructions for designing, deploying, and studying internationally networked collaboration*. https://sites.google.com/a/umn.edu/tapp2018/.

Duin, A. H. & Tham, J. (2018). Cultivating code literacy: A case study of course redesign through advisory board engagement. *Communication Design Quarterly, 6*(3), 44–58.

Ede, L. & Lunsford, A. A. (1984). Audience addressed/audience invoked: The role of audience in composition theory and pedagogy. *College Composition and Communication, 35*(2), 155–171.

Ede, L. & Lunsford, A. A. (1985). Let them write—together. *English Quarterly, 18*, 119–127.

Ede, L. & Lunsford, A. A. (1990). *Singular texts/plural authors*. Southern Illinois University Press.

Ede, L. & Lunsford, A. A. (2001). Collaboration and concepts of authorship. *PMLA, 116*(2), 354–369.

Ede, L. & Lunsford, A. A. (2009). Among the audience: On audience in an age of new literacies. In M. E. Weiser, B. M. Fehler & A. M. Gonzalez (Eds.), *Engaging audience: Writing in an age of new literacies* (pp. 42–72). NCTE.

Elsevier. (2019). *Data visualization*. https://www.elsevier.com/authors/author-resources/data-visualization.

Fluckinger, D. (2019). *AI in enterprise collaboration platforms: A comparison*. https://searchcontentmanagement.techtarget.com/ehandbook/Artificial-intelligence-meets-enterprise-collaboration-systems.

Frank, D. A. & Bolduc, M. (2010). Lucie Olbrechts-Tyteca's New Rhetoric. *Quarterly Journal of Speech, 96*(2), 141–163.

Fredlund, K. (2016). Feminist CHAT: Collaboration, nineteenth-century women's clubs, and activity theory. *College English, 78*(5), 470–495).

Giddens, A. (1984). *The constitution of society: Outline of the theory of structuration*. University of California Press.

Gierdowski, D. & Reis, D. (2015). The MobileMaker: An experiment with a mobile makerspace. *Library Hi Tech, 33*(4), 480–496.

Goel, A. K. & Polepeddi, L. (2016). *Jill Watson: A virtual teaching assistant for online education*. Design & Intelligence Laboratory, School of Interactive Computing, Georgia Institute of Technology. https://smartech.gatech.edu/bitstream/handle/1853/59104/goelpolepeddi-harvardvolume-v7.1.pdf.

Gouran, D. (1988). Group decision making: An integrative research. In C. Tardy (Ed.), *A handbook for the study of human communication* (pp. 247–267). Ablex.

Gouran, D. (2003). Reflections on the type of question as a determinant of the form of interaction in decision-making and problem-solving discussions. *Communication Quarterly*, *51*(2), 111–125.

Gurak, L. & Duin, A. H. (2004). The impact of the Internet and digital technologies on teaching and research in technical communication. *Technical Communication Quarterly*, *13*(2), 187–198.

Hackos, J. (2015). Changing times—Changing skills. *Communication Design Quarterly*, *3*(2), 7–12.

Hall, K., Vogel, A. L., Huang, G. C., Serrano, K. J., Rice, E. L., Tsakraklides, S. P. & Fiore, S. M. (2018). The science of team science: A review of the empirical evidence and research gaps on collaboration in science. *American Psychologist*, *73*(4), 532–548.

Hart, H. & Conklin, J. (2006). Toward a meaningful model for technical communication. *Technical Communication*, *53*(4), 395–415.

Henschel, S. & Melonçon, L. (2014). Of horsemen and layered literacies: Assessment instruments for aligning technical and professional communication undergraduate curricula with professional expectations. *Programmatic Perspectives*, *6*(1), 3–26.

Hinrichs, M. M., Seager, T. P., Tracy, S. J. & Hannah, M. A. (2017). Innovation in the knowledge age: Implications for collaborative science. *Environment Systems and Decisions*, *37*(2), 144–155.

Hirokawa, R. (1994). Functional approaches to the study of group decision. *Small Group Research*, *25*(4), 542–550.

Iliadis, A. & Pedersen, I. (2018). The fabric of digital life: Uncovering sociotechnical tradeoffs in embodied computing through metadata. *Journal of Information, Communication and Ethics in Society*, *16*(3), 1–19.

Isenberg, P., Elmqvist, N., Scholtz, J., Cernea, D., Ma, K-L & Hagen, H. (2011). Collaborative visualization: Definition, challenges, and research agenda. *Information Visualization*, *10*(4), 310–326.

Jones, S. L. (2007). How we collaborate: Reported frequency of technical communicators' collaborative writing activities. *Technical Communication*, *54*(3), 283–294.

Karach, A. & Roach, D. (1992). Collaborative writing, consciousness raising, and practical feminist ethics. *Women's Studies International Forum*, *15*(2), 303–308.

Katz, S. (1992). The ethic of expediency: Classical rhetoric, technology, and the Holocaust. *College English*, *54*(3), 255–275.

Kent, T. (1993). *Paralogic rhetoric: A theory of communicative interaction*. Bucknell University Press.

Lave, J. (1991). Situated learning in communities of practice. In L. Resnick, J. M. Levine & S. D. Teasley (Eds.), *Perspectives on socially shared cognition* (pp. 63–82). American Psychological Association.

Lave, J. & Wenger, E. (1991). *Situated learning: Legitimate peripheral participation*. Cambridge University Press.

Lay, M. M. (1989). Interpersonal conflict in collaborative writing: What we can learn from gender studies. *Journal of Business and Technical Communication*, *3*(2), 5–28.

Lay, M. M. & Karis, W. M. (Eds.). (1991). *Collaborative writing in industry: Investigations in theory and practice*. Baywood.

Levy, P. (2000). *Collective intelligence: Mankind's emerging world of cyberspace*. Perseus Book Group.

Li, Y., Rau, P. P., Li, H. & Maedche, A. (2017). Effects of a dyad's cultural intelligence on global virtual collaboration. *IEEE Transactions in Professional Communication, 60*(1), 56–75.

Lunsford, A. A. & Ede, L. (1990). Rhetoric in a new key: Women and collaboration. *Rhetoric Review, 8*(2), 234–241.

Lunsford, A. A. & Ede, L. (2011). *Writing together: Collaboration in theory and practice: A critical sourcebook*. Bedford; St. Martin's.

Maylath, B. (2019). *The Trans-Atlantic & Pacific Project*. North Dakota State University. https://www.ndsu.edu/english/transatlantic_and_pacific_translations/.

McComiskey, B. (2015). *Dialectical rhetoric*. University Press of Colorado.

McCue, R. (2015). *Research & collaboration tools for students, staff & faculty: Creating a modern memex*. University of Victoria Libraries. https://docs.google.com/document/d/1mAzY0aRrWZr16TjOxCxq5i6NVnfAnknMjPM17bC1cvE/edit.

McKee, H. & Porter, J. (2017). *Professional communication and network interaction: A rhetorical and ethical approach*. Routledge.

Mead, G. H. (1934). *Mind, self, and society*. University of Chicago Press.

Melonçon, L. & Henschel, S. (2013). Current state of U.S. undergraduate degree programs in technical and professional communication. *Technical Communication, 60*(1), 45–64.

Mendenhall, A. (2013). The historical problem of vertical coherence: Writing, research, and legitimacy in early 20th century rhetoric and composition. *Composition Studies, 41*(1), 84–100.

Mills, J., Bonner, A. & Francis, K. (2006). The development of constructivist grounded theory. *International Journal of Qualitative Methods, 5*(1), 25–35.

Monk, J., Manning, P. & Denman, C. (2003). Working together: Feminist perspectives on collaborative research and action. *ACME: An International Journal for Critical Geographies, 2*(1), 91–106. https://acme-journal.org/index.php/acme/article/view/710.

Morgan, M. (1994). Women as emergent leaders in student collaborative writing groups. *JAC: Journal of Advanced Composition, 14*(1), 203–219.

Moses, J. (2015). Agile writing: A project management approach to learning. *International Journal of Sociotechnology and Knowledge Development, 7*(2), 1–13.

Moses, J. & Tham, J. (2017, March). *Agile writing: Design thinking approaches to writing instruction*. [Paper presentation]. Conference on College Composition and Communication (CCCC), Portland, OR, United States.

Moses, J. & Tham, J. (2019, February 25). *Four ideas for creating a collaborative writing environment that works*. Medium. http://bit.ly/2Uk8X7w.

Nystrand, M., Greene, S. & Wiemelt, J. (1993). Where did composition studies come from?: An intellectual history. *Written Communication, 10*, 267–333.

Palumbo, G. & Duin, A. H. (2018). Making sense of virtual collaboration through personal learning networks. In B. Mousten, S. Vandepitte, E. Arno & B. Maylath (Eds.), *Multilingual writing and pedagogical cooperation in virtual learning environments* (pp. 109–136). IGI Global.

Paretti, M. C., McNair, L. D. & Holloway-Attaway, L. (2007). Teaching technical communication in an era of distributed work: A case study of collaboration between U.S. and Swedish students. *Technical Communication Quarterly, 16*(3), 327–352.

Pearce, B. (2004). The coordinated management of meaning. In W. B. Gudykunst (Ed.), *Theorizing about intercultural communication* (pp. 35–54). Sage Publications.

Pearce, B. (2008). *Making social worlds: A communication perspective*. Blackwell.
Perelman, C. & Olbrechts-Tyteca, L. (1969). *The new rhetoric: A treatise on argumentation*. University of Notre Dame Press. (Original work published 1958.)
Poole, M. S. (1997). The small group should be the fundamental unit of communication research. In J. Trent (Ed.), *Communication: Views from the helm for the 21st century* (pp. 94–97). Allyn & Bacon.
Poole, M. S. (2003). Group communication and the structuring process. In R. Cathcart, L. Samovar & L. Henman (Eds.), *Small group communication* (7th ed., pp. 48–56). Roxbury Publishing Co.
Poole, M. S. & Doelger, J. (1986). Developmental processes in group decision-making. In R. Y. Hirokawa & M. S. Poole (Eds.), *Communication and group decision-making* (pp. 35–62). Sage.
Pope-Ruark, R. (2012). We Scrum every day: Using scrum project management framework for group project. *College Teaching, 60*(4), 164–169.
Pope-Ruark, R. (2015). Introducing Agile project management strategies in technical and professional communication courses. *Journal of Business and Technical Communication, 29*(1), 112–133.
Pope-Ruark, R., Moses, J., Conner, T. & Tham, J. (2019). Special issue of *Journal of Business and Technical Communication*, July 2019: Design-Thinking Approaches in Technical and Professional Communication. *Journal of Business and Technical Communication, 31*(4), 520–522.
Porter, J. (2009). Recovering delivery for digital rhetoric. *Computers & Composition, 26*, 207–224.
Qualley, D. & Chiseri-Strater, E. (1994). Collaboration as reflexive dialogue: A knowing "deeper than reason." *JAC: Journal of Advanced Composition, 14*(1), 111–130.
Rainey, K., Turner, R. & Dayton, D. (2005). Do curricula correspond to managerial expectations? Core competencies for technical communicators. *Technical Communication, 52*(3), 323–353.
Rice, R. & St.Amant, K. (Eds.). (2018). *Thinking globally, composing locally: Rethinking online writing in the age of the global internet*. Utah State University Press; University Press of Colorado.
Russi, D. & Schneider, R. (2016). *A guide to translation project management*. The COMET® Program with support from NOAA's National Weather Service International Activities Office and the Meteorological Service of Canada. https://courses.comet.ucar.edu/pluginfile.php/27060/mod_resource/content/12/GuideToTranslation Management_V1a_02102017_final.pdf.
Seymour, S. (2019). *What is collaboration software? Analysis of features, types, benefits and pricing*. FinancesOnline. https://financesonline.com/collaboration-software-analysis-features-types-benefits-pricing.
Spinuzzi, C. (Ed.). (2007). Technical communication in the age of distributed work [Special issue]. *Technical Communication Quarterly, 16*(3), 265–277.
St. Pierre, E. A. (2014). An always already absent collaboration. *Cultural Studies <> Critical Methodologies, 14*(4), 374–379.
Stokols, D., Hall, K.L. & Vogel, A.L. (2013). Transdisciplinary public health: Research, methods, and practice. In D. Haire-Joshu and T.D. McBride (Eds), *Transdisciplinary public health: Core characteristics, definitions, and strategies for success* (pp. 3–30). Jossey-Bass.

Sullivan, P. (1994). Revising the myth of the independent scholar. In S. B. Reagan, T. Fox & D. Bleich (Eds.), *Writing with: New directions in collaborative teaching, learning, and research* (pp. 11–30). SUNY Press.

Tannenbaum, S. I., Mathieu, J. E., Salas, E. & Cohen, D. (2012). Teams are changing: Are research and practice evolving fast enough? *Industrial and Organizational Psychology, 5*(1), 2–24.

Team science toolkit. (2019). National Cancer Institute. https://teamsciencetoolkit.cancer.gov/Public/WhatIsTS.aspx.

Tham, J. (Ed.). (2017). *The rhetoric and technology of collaboration: How digital technologies transform our collaborative work and pedagogy.* Digital Rhetoric Collaborative. http://www.digitalrhetoriccollaborative.org/2017/11/27/the-rhetoric-and-technology-of-collaboration-how-digital-technologies-transform-our-collaborative-work-and-pedagogy/.

Tham, J. (Curator). (2019a). Emerging technologies for technical communication. *Fabric of Digital Life.* https://fabricofdigitallife.com/index.php/Browse/objects/facet/collection_facet/id/29.

Tham, J. (2019b). *Multimodality, makerspaces, and the making of a maker pedagogy for Technical Communication and Rhetoric* [Doctoral dissertation, University of Minnesota]. University of Minnesota Digital Conservancy. https://conservancy.umn.edu/handle/11299/206361.

Tham, J. & Moses, J. (2019). *How we use design thinking to support collaborative writing.* Medium. http://bit.ly/2KEhFoN.

Thompson, I. (2001). Collaboration in technical communication: A qualitative content analysis of journal articles, 1990–1999. *IEEE Transactions on Professional Communication, 44*(3), 161–173.

Thralls, C. (1992). Bakhtin, collaborative partners, and published discourse: A collaborative view of composing. In J. Forman (Ed.), *New visions of collaborative writing* (pp. 63–82). Heinemann.

Tuckman, B. W. (1965). Developmental sequence in small groups. *Psychological Bulletin, 65*(6), 384–399.

Tuckman, B. W. & Jensen, M. A. C. (1977). Stages of small-group development revisited. *Group and Organization Studies, 2*(4), 419–427.

Vie, S. (2017). Training online technical communication educators to teach with social media: Best practices and professional recommendations. *Technical Communication Quarterly, 26*(3), 344–359.

Wan, T. (2018). *GitHub's new education bundle equips students with industry-standard coding tools.* EdSurge. https://www.edsurge.com/news/2018-06-19-github-s-new-education-bundle-equips-students-with-industry-standard-coding-tools.

Watzlawick, P. (1978). *The language of change.* W.W. Norton.

Wenger, E. (1998). *Communities of practice: Learning, meaning and identity.* Cambridge University Press.

Wilson, H. J. & Daugherty, P. R. (2018). Collaborative intelligence: Humans and AI are joining forces. *Harvard Business Review, 96(4),* 114–123.

Winsor, D. A. (2003). *Writing power: Communication in an engineering center.* State University of New York Press.

Wolfe, J. (2010). *Team writing: A guide to working in groups.* Bedford; St. Martin's.

Yancey, K. B. & Spooner, M. (1998). A single good mind: Collaboration, cooperation, and the writing self. *College Composition and Communication, 49*(1), 45–62.

10. Using the Community of Inquiry Theory to Assess Online Programs and Help Students to Analyze Their Learning

Julie Watts
UNIVERSITY OF WISCONSIN-STOUT

Abstract: Some program assessment strategies may fall short for assessing online technical and professional communication (TPC) programs. The community of inquiry (COI) theory, when paired with an outcomes approach, provides a well-rounded assessment of both the online learning environment and what outcomes students achieve. COI theory also can impact students by framing a course-embedded, learning-focused online student orientation (OSO) meant to give students a vocabulary with which to analyze the online learning environment and to recognize what behaviors and skills they, their peers, and instructors should exhibit to improve learning. COI is a theoretical framework designed for online learning, showing that instructors and students need to be "present" in different ways to cultivate a learning community conducive to deep learning. COI helps participants to determine what student and instructor behaviors and activities best contribute to student success online and why. Three presences—cognitive, social, and teaching—are used as lenses with which to identify and assess these behaviors and activities. This chapter introduces the content and delivery methods used to develop a COI-framed OSO for online TPC master's students, the methods used to pair COI with an outcomes-based program assessment, and strategies for communicating results and recommendations to program stakeholders.

Keywords: program assessment, outcomes, community of inquiry, online student orientation

Key Takeaways:

- The community of inquiry (COI) theory, a framework designed for online learning, demonstrates that instructors and students need to be "present" in different ways to cultivate a learning community conducive to deep learning.
- COI, when paired with an outcomes approach, provides a well-rounded assessment of both the online learning environment and what outcomes students achieve.
- COI is useful for helping online program students identify and reflect on the behaviors and skills they need to succeed in an online graduate program.

Administrators of academic programs in technical and professional communication (TPC) select from a variety of theories to situate program assessment, including social construction (Coppola, 1999), "layered literacies" (Carpenter, 2011; Cargile Cook, 2002), or theories positioning the "thinking, doing, teaching" in which TPC professionals engage (Johnson-Eilola & Selber, 2001). Program administrators often use assessment strategies that are outcomes based (Coppola & Elliot, 2007; Say, 2015; Williams, 2010), helping to show the "evidence of impact" (Allen, 2004, p. 94) that curricula have on learning (Coppola et al., 2016).

While TPC program assessment practices have developed throughout the last decades (see special issue on programmatic research in *Programmatic Perspectives*, 2016), many fall short for assessing online TPC programs. Because of lingering doubts concerning the efficacy and rigor of online learning (Allen et al., 2016), directors of online TPC programs need to coordinate assessment that evaluates outcomes and the learning environment.

As director for an online TPC program for over a decade, I have found that only assessing student learning outcomes for program assessment is not sufficient to properly communicate the value of my program to its stakeholders: I also have to assess the online learning environment (Watts, 2017). Why? Because online learning (despite its growth) is still regarded as less valuable than face-to-face learning. Research indicates that employers (Fogle & Elliott, 2013; Linardopoulos, 2012) and students (Chant, 2013; Parker et al., 2011) hold reservations about the value of online degrees and online learning in general. Thus, if program directors can provide evidence of how students and instructors are participating in a robust community of learners, they can better communicate the value of their online program.

Most important, though, this focus on the learning environment not only assists with assessment, it also helps students in online programs to take control of their learning. Faculty and program directors understand all too well the digital native myth: students—even those who have experience playing and working online—often are not prepared for online learning (Brumberger, 2011; Kennedy et al., 2008). Students need help "learning how to learn" online, and the theoretical tool I discuss below can give students a vocabulary with which to analyze their online learning experiences and to reflect on and improve them (Watts, 2019).

The community of inquiry (COI) theory—developed to characterize how students and instructors ideally need to be present as participants in online learning (Garrison et al., 2000)—can serve both a program assessment and a student reflection purpose. When paired with an outcomes approach, COI provides a comprehensive assessment of the online environment and its impact on student learning. Additionally, COI gives students a language to recognize what behaviors and skills they, their peers, and instructors should exhibit to promote "deep learning" and improve their learning experiences (Phillips & Graeff, 2014).

■ Community of Inquiry Theory

COI is a theory analyzing the online learning environment (Garrison et al., 2000), showing how instructors and students need to be "present" to cultivate a learning community conducive to "deep learning" (Rourke & Kanuka, 2009, p. 23). With deep learning as the desired outcome, the bar is set high for student performance. To achieve deep learning, students must move beyond surface learning and instead "utilize critical thinking skills by looking for meaning in the course content and trying to relate it to personal experiences and ideas" (Phillips & Graeff, 2014, p. 242).

COI examines the online learning environment as a key facet of program assessment, addressing stakeholder concerns that often question online programs' rigor: Do students fully contribute to an engaging learning experience (Lear et al., 2009)? Do instructors foster learning experiences meaningfully (Cameron et al., 2009; Jones, 2013)? Do students feel connected with a community or isolated without peer and instructor support (Shackelford & Maxwell, 2012)? Because COI examines the learning environment and the presence of students and instructors in it, questions such as these can be investigated and addressed. Specifically, COI helps participants determine what student and instructor behaviors and activities contribute to student success (deep learning) in an online course and why (Shearer et al., 2015). Three presences—cognitive, social, and teaching—are used as lenses with which to identify and assess these behaviors and activities.

Cognitive presence is characterized by students' sustained interaction with and reflection about course material: students "question their existing assumptions" and need to "construct" and apply "new knowledge" (Stewart, 2017, p. 71). Students create meaning and reflect on their learning to confirm their understanding of complex processes (Garrison & Cleveland-Innes, 2005). Instructors assist by scaffolding the "process of critical inquiry": setting up a complex problem and helping students to research, apply, and test their recommendations (p. 134). The goal is for students to acquire a set of behaviors and actions constituting cognitive presence, with the other presences supporting this.

Social presence recognizes that interacting with peers and the instructor fosters an individual's cognitive presence and cultivates deep learning (Oztok & Brett, 2011; Wang & Wang, 2012). A critique of online learning environments is that they lack the traditional structures of support and community often taken for granted in face-to-face classes (Bejerano, 2008). Thus, a common misconception about online learning is that those who succeed do so without support (Wooten & Hancock, 2009). COI does not support the myth of the isolated learner. Instead, instructors need to cultivate social presence by creating a trusting learning environment and facilitating student collaboration around a common set of intellectual tasks (Swan et al., 2009).

Teaching presence is achieved through thoughtfully designing the course, facilitating discourse among participants, providing direct instruction, and offering timely feedback about student work. To be present, instructors should not be omnipresent: "Too much instructor presence can actually impede students from taking more responsibility for their learning, prevent critical thinking, and downplay the value of student-to-student interaction" (Peery & Veneruso, 2011). Teaching presence changes over time, with different strategies (e.g., direct instruction, facilitating discourse) more frequent at different times (Akyol & Garrison, 2008). Specific practices such as audiovisual commentary on student work can play a role in establishing teaching presence (Grigoryan, 2017). A positive correlation exists between teaching presence and student motivation (Baker, 2010) and between effective teaching presence and healthy social presence (Shea et al., 2006).

Using COI to Help Students Reflect on Their Online Learning

Before I examine how COI complements an outcomes-based program assessment, I first consider the uses of COI as a reflective tool for students to use to improve their own online learning experiences. To introduce students to this use of COI, I designed a course-embedded online student orientation (OSO) using COI as its theoretical framework to help students articulate and nurture the behaviors and skills that they and others (peers, instructors) need to enact that most help them achieve deep learning (Watts, 2019). Below, I discuss the rationale for creating the OSO, the OSO's content and delivery strategy, and some benefits students achieved.

The COI-framed OSO that I devised gives students a vocabulary with which to identify those factors cultivating teaching and social presence and how these impact their cognitive presence. Importantly, the OSO involves all participants in cultivating cognitive presence and deep learning. Working adult students (the majority of whom enroll in my program) are a particularly vulnerable population, experiencing family issues, gaps in previous education experiences (Brewer & Yucedag-Ozcan, 2012–2013), and employment responsibilities (Ashby, 2004), often leading to "limited persistence" in which their enrollment demonstrates frequent starts and stops that delay degree benefits and increase costs (Hutchens, 2014). Unfortunately, while OSOs often are available to online students, OSOs for this group tend to focus on orienting students to technologies used in the learning environment (Taylor et al., 2015) or introducing them to university resources (Jones, 2013): learning-focused OSOs are less common (Wozniak et al., 2012).

While OSOs are typically perceived as a precursor to coursework, studies suggest that course-embedded OSOs are preferable and help to increase completion rates (Taylor et al., 2015; Wozniak et al., 2012). I embed my OSO into the first and last week of one of my program's 15-week courses, a class introducing students to TPC theory and research and one students are advised to take early in the master's program.

The week 1 OSO consists of several activities:

- Students view a 10-minute video of me explaining COI and read an accompanying blog article, "Five-Step Strategy for Student Success with Online Learning" (Morrison, 2015).
- Students participate in a discussion board to deliberate what strategies or behaviors could be used to inculcate cognitive, social, or teaching presence in the seminar.
- Students individually write a two-page response paper about a study (Lambert & Fisher, 2013) that examines COI in an online graduate seminar.

At the end of the semester, students respond individually to a final exam prompt asking them to analyze the ways they have "learned how to learn online" during their time in the seminar, using the OSO COI concepts. Students are given the prompt during the last day of classes, and they have the remainder of the one-week evaluation period to submit their responses.

During week 1 of the OSO, I introduce students to the concepts of COI; I have facilitated this OSO several times, and generally, students are not familiar with COI prior to this. Students examine the video of me explaining COI concepts; the video is accompanied by a slideshow, listing cogent COI definitions. While the Morrison (2015) blog article does not invoke COI, I ask students to compare and contrast ideas between the article and the video. On a discussion board, I ask students to respond to these questions: (a) How do the concepts discussed in the COI video align with the five-step strategy proposed in the blog post? (b) Name and define one "strategy" (it doesn't necessarily need to be one mentioned in the blog post) that could be used to inculcate cognitive, social, or instructor presence in this class. I encourage students to read the discussion board and reply to at least one student's post. Doing so affords students the opportunity to see how others understand and apply COI. My presence on this board is limited to identifying errors or misjudgments in the definition or application of COI. I provide students with participation points for this activity. Students also read the Lambert and Fisher (2013) article, which uses COI as the framework for a study examining an online graduate program. In a two-page paper, students individually identify the main ideas of the study and discuss how the findings relate to them. I provide feedback and a grade on this activity.

The final exam is a useful way for students to assess and reflect on the semester and the progress they and their peers and instructor have made in cultivating a COI. For the exam, students individually revisit the three presences and analyze how they "learned how to learn" online in this course, pointing to particular instances concerning themselves, their peers, and/or the instructor while identifying specific learning activities. Students receive feedback and a grade for completing the exam.

Results of the OSO study showed that the majority of students used the COI language as a vocabulary to analyze their learning, and even students who were more experienced online learners still found COI to be useful for this task (Watts, 2019). Additionally, students perceived the benefit of social presence to their learning, even though research shows that adult working professional learners tend to characterize such relationships "as a bonus" or "not something . . . expected" (Ke, 2010, p. 816). Importantly, OSO's notion of cognitive presence seemed to bolster students' awareness of their responsibility for their learning; in particular, students who connected their academic and industry lives practiced a central concept of cognitive presence.

Using COI With Outcomes Program Assessment

COI is a useful framework with which to assess online courses (Hilliard & Stewart, 2019; Lear et al., 2009; Stewart, 2017) and programs (Lee et al., 2006). My outcomes-based assessment (IRB approved) identifies which program learning outcomes students achieve, how well they achieve them, and which assignments and activities students perceived helped them to practice the outcomes. I generally focus on the program's 12–credit core curriculum (four three-credit required classes taught each year), collecting data annually and communicating results biennially.

To conduct outcomes assessment, I use course-embedded assessment (CEA) and a three-question student survey (see Table 10.1). These direct and indirect measures allow me to continually examine and refine program curriculum (Say, 2015). For the CEA, instructors select a major assignment and reflect on each student's performance, identifying which outcomes each student achieved by completing it. Appendix A shows an example spreadsheet given to instructors: the outcomes are listed and the instructor rates student performance for each as below expectations, acceptable, or exceeds expectations, or by indicating that an outcome is not applicable.

Table 10.1. COI and outcomes program assessment (core courses)

		Measure	
Schedule	Outcomes Assessment	Direct	Indirect
Fall, Spring	Course-embedded assessment (CEA) completed by instructors to ascertain which program outcomes students practiced and how effectively	X	
Fall, Spring	Survey to students: 3 questions about how well students were provided opportunities to practice program outcomes		X
	COI Assessment		
Fall, Spring	Survey to instructors: 10 questions about COI presences		X
Every 3rd Year	Focus Group with all program faculty: 10 questions asking about COI presences (*Note*: Replaces instructor survey)		X

Students receive a three-question survey, asking them to identify which outcomes they practiced in the class and which assignments or other tasks helped them to practice the outcomes (Appendix B). Used together, the CEA and survey help me to map which outcomes students practiced and perceived they practiced and to identify gaps between curriculum and outcomes.

This outcomes-based assessment has been useful for refining program curriculum. For example, program assessment showed students did not have adequate opportunities to practice "evaluating and executing team-building and interpersonal communication strategies." We collaborated with communication studies colleagues to include an interpersonal communication seminar and a speech communication-for-industry course as new program offerings. We also use the assessment to refine what assignments and activities we ask students to complete, focusing on aligning these to outcomes.

With the outcomes approach focusing on student learning, pairing it with a COI assessment can help demonstrate the cognitive, social, and teaching presences constructing the program's online communities of learners. I continue to refine my approach using COI for program assessment below, I discuss my pilot results and the changes to methodology that I plan to make COI assessment (see Table 10.1).

Student Survey

To assess student perceptions of the COI teaching, social, and cognitive presences, I have streamlined a popular survey tool used by J. B. Arbaugh and his colleagues (2008). The instrument—a 34-question, multiple-choice survey—has been validated by Swan and her colleagues (2008). Arbaugh's survey has been used in hundreds of studies and lends consistency to the COI literature, especially in terms of assessing COI at a large scale, across institutions and courses (Stenbom, 2018). Designed as a statistical tool, survey results can show the relationship among the presences (Akyol & Garrison, 2008; Stewart, 2019): how do student perceptions of teaching presence affect cognitive presence (Garrison et al., 2010), or how do student perceptions of social presence impact cognitive presence (Shea & Bidjerano, 2009)?

When I initially piloted Arbaugh's lengthy survey for program assessment, my students experienced survey fatigue (especially when students were enrolled in multiple core courses during a given semester), and my participant numbers seemed to reflect that. I also found that student-participant numbers did not lend themselves to the statistical analysis for which the instrument was designed. Thus, I revised the survey from 30+ questions to ten (Appendix C) and plan to pilot it during Fall 2020. When devising the survey, Arbaugh and his colleagues relied on COI literature (Garrison et al., 2000) and used it to determine the questions. The majority of COI studies have retained these definitions (Stenbom, 2018), and I keep these intact in my adapted survey, despite literature arguing for other iterations (Kozan & Caskurlu, 2018).

Students who are cognitively present are more likely to retain course concepts and apply them in other settings. Given this, I asked two questions related to course-concept retention and a third related to course-concept application (Appendix C). These help ascertain whether students have opportunities to take advantage of a "triggering event" (i.e., ill-defined problem), which they then have opportunities to "explore" and investigate (Garrison et al., 2000). Next, students synthesize existing and new concepts to apply content to contexts beyond the course, another important feature of cognitive presence (Rourke & Kanuka, 2009).

Social presence is constituted by interaction among students and instructors, which fosters cognitive presence and enables students to retain and apply course concepts. A key feature of social presence is that students have opportunities for "open communication," and I asked whether students had opportunities to interact and discuss course concepts (Arbaugh et al., 2008). Having students collaborate on common intellectual tasks also cultivates social presence (Oztok & Brett, 2011), so I included two questions related to this, one querying about small-group interaction and the other concerning collaboration and new knowledge.

Teaching presence is achieved through proper course design, discourse facilitation, and direct instruction and feedback (Shea et al., 2010). Thus, I devised three questions, one related to opportunities for idea exchange, another to course design, and a final concerning feedback about student work.

Instructor Survey and Focus Group

Rather than conduct individual interviews to compile instructor perceptions of the COI presences as I did in my pilot study, I now plan to conduct a Qualtrics survey with faculty, which contains ten open-ended questions: three teaching presence questions, four social presence, and three cognitive presence (Appendix D). These questions allow me to track instructor perceptions over time and across courses, identifying themes and patterns, and comparing these to student perceptions. I also plan to collect focus group data from program instructors every three years in lieu of surveys. Focus groups can be enriching, valuable experiences—a useful way to enable "enhanced data quality" in that participants hear other responses and contribute to a conversation, rather than simply responding singularly (Patton, 2015, p. 478).

My pilot study COI program assessment showed that a handful of activities helped to cultivate cognitive, social, and teaching presence: conducting more frequent formative assessment, encouraging student reflection, and facilitating team-based active learning (Watts, 2017). These findings prompted faculty to discuss how such activities are happening in our seminars and how they could be further promoted. We have found that COI assessment enables a better understanding of the practices and behaviors that help students (and help instructors to help students) achieve cognitive, teaching, and social presence.

Articulating and analyzing these presences allows faculty to continue discovering how what they are doing works and what can be done to help students achieve success.

■ Communicating COI Program Assessment

A key feature of any program assessment design is communicating results to stakeholders and collaboratively deciding ways to implement recommendations, always striving for continual improvement (Walvoord, 2010). Below, I suggest ways to communicate results with five groups and the benefits of doing so: faculty, advisory board members, university administrators, students, and prospective students.

Faculty, Advisory Board Members, Administrators

Faculty are valuable stakeholders with which to discuss assessment results because of their power to effect change pedagogically and curricularly. Those of us lucky enough to work with enthusiastic, collaborative peers find these assessment conversations energizing. Often, we do not find time to discuss pedagogy, curriculum, and student challenges and successes with our colleagues, but conversations about assessment afford us this opportunity. Faculty also want to hear input about these results from advisory board members, who hold points of view from industry and beyond the department and institution. Thus, I adopted a strategy for assessment dissemination probably used by many programs: write one report (with faculty and board members' input) that is distributed to university administrators, containing the blueprint for proposed change over the long- (5+ years) and short-term.

As director, my institution requires that I submit a biennial "assessment in the major" (AIM) report, listing my program's outcomes, assessment and results, and recommendations. To draft AIM, I discuss results first with faculty, conferring about what to implement and how. Then faculty and I seek feedback from board members at our advisory meeting about these proposed plans for action. The finalized report—a result of conversations with faculty and board members—is submitted to administrators.

Using COI to frame program assessment encourages stakeholders to better understand the practices and behaviors that help students achieve key learning outcomes. Stakeholders seem to appreciate this approach. Board members and administrators want to hear how effectively students achieve outcomes but also how aware students are of cultivating social and cognitive presences and the importance for doing so. They are interested in what constitutes teaching presence and the value of social presence in cultivating deep learning. Faculty appreciate understanding what practices help them facilitate a positive online learning environment that helps set up students for success.

Students and Prospective Students

Prospective and current students do not receive the AIM report, but rather it is communicated to them through informational and promotional web content. Results are woven into the degree's promotional materials and sharing these with prospective students helps to showcase the program's value, combatting perceptions of online learning as isolating and without depth or engagement. Students receive results through program informational content. For example, one AIM report revealed that changes needed to occur with the program's core to align it more effectively with the outcomes. Online advisement materials were updated to reflect the change. Thus, students benefit from the COI theory through its impact on program curriculum and through students' OSO participation. In particular, students seem to welcome the OSO and the opportunities it gives them to reflect on and hopefully improve their online learning experience (Watts, 2019).

The OSO continues, and the cycle of data collection, dissemination and discussion, action-planning, and revision of informational and promotional program content is ongoing. With the merit of online programs still scrutinized, framing outcomes-based assessment using COI and incorporating a learning-focused OSO into the curriculum not only shows what students accomplish and the skills and activities used to achieve outcomes but also encourages students to learn a vocabulary to help them reflect on those behaviors and skills they can cultivate in themselves (and request in others) to help manage and deepen their learning.

References

Akyol, Z. & Garrison, D. R. (2008). The development of a community of inquiry over time in an online course: Understanding the progression and integration of social, cognitive and teaching presence. *Journal of Asynchronous Learning Networks, 12*(3–4), 3–22.

Allen, I. E., Seaman, J. E., Poulin, R. & Straut, T. T. (2016). *Online report card: Tracking online education in the United States.* Babson Survey Research Group and Quahog Research Group. http://onlinelearningsurvey.com/reports/onlinereportcard.pdf.

Allen, J. (2004). The impact of student learning outcomes assessment on technical and professional communication programs. *Technical Communication Quarterly, 13*(1), 93–108.

Arbaugh, J. B., Cleveland-Innes, M., Diaz, S. R., Garrison, D. R., Ice, P., Richardson, J. C. & Swan, K. (2008). Developing a community of inquiry instrument: Testing a measure of the community of inquiry framework using a multi-institutional sample. *Internet and Higher Education, 11*, 133–136.

Ashby, A. (2004). Monitoring student retention in the Open University: Definition, measurement, interpretation and action. *Open Learning, 19*(1), 65–77.

Baker, C. (2010). The impact of instructor immediacy and presence for online student affective learning, cognition, and motivation. *Journal of Educators Online, 7*(1), 1–30.

Bejerano, A. R. (2008). The genesis and evolution of online degree programs: Who are they for and what have we lost along the way? *Communication Education, 57,* 408–414. https://doi.org/10.1080/03634520801993697.

Brewer, S. A. & Yucedag-Ozcan, A. (2012–2013). Educational persistence: Self-efficacy and topics in a college orientation course. *Journal of College Student Retention, 14*(4), 451–465.

Brumberger, E. (2011). Visual literacy and the digital native: An examination of the millennial learner. *Journal of Visual Literacy, 30*(1), 19–46.

Cameron, B. A., Morgan, K., Williams, K. C., Kostelecky, K. L. (2009). Group projects: Student perceptions of the relationship between social tasks and a sense of community in online group work. *American Journal of Distance Education, 23*(1), 20–33.

Cargile Cook, K. (2002). Layered literacies: A theoretical frame for technical communication pedagogy. *Technical Communication Quarterly, 11*(1), 5–29. https://doi.org/10.1207/s15427625tcq1101_1.

Carpenter, J. H. (2011). A 'layered literacies' framework for scientific pedagogy. *Currents in Teaching and Learning, 4*(1), 17–33.

Chant, I. (2013). Research: As online degrees become more prevalent, questions linger. *Library Journal, 138*(18), 23.

Coppola, N. W. (1999). Setting the discourse community: Tasks and assessment for the new technical communication service course. *Technical Communication Quarterly, 8*(3), 249–267.

Coppola, N. & Elliot, N. (2007). Technology transfer model for program assessment in technical communication. *Technical Communication, 54,* 459–474.

Coppola, N., Elliot, N., Newsham, F. (2016). Programmatic research in technical communication: An interpretive framework for writing program assessment. *Programmatic Perspectives, 8*(2), 5–45.

Fogle, C. D. & Elliott, D. (2013). The market value of online degrees as a credible credential. *Global Education Journal, 3.* http://ssrn.com/abstract=2326295.

Hutchens, M. K. (2014). Nontraditional students and student persistence. In D. Hossler & B. Bontrager (Eds.), *Handbook of strategic enrollment management* (pp. 333–350). Jossey-Bass.

Garrison, D. R., Anderson, T. & Archer, W. (2000). Critical inquiry in a text-based environment: Computer conferencing in higher education. *The Internet and Higher Education, 2*(2–3), 87–105.

Garrison, D. R. & Cleveland-Innes, M. (2005). Facilitating cognitive presence in online learning: Interaction is not enough. *American Journal of Distance Education, 19,* 133–148. https://doi.org/10.1207/s15389286ajde1903_2.

Garrison, D. R., Cleveland-Innes, M. & Fung, T. S. (2010). Exploring causal relationships among teaching, cognitive and social presence: Student perceptions of the community of inquiry framework. *The Internet and Higher Education, 13,* 31–36.

Grigoryan, A. (2017). Audiovisual commentary as a way to reduce transactional distance and increase teaching presence in online writing instruction: Student perceptions and preferences. *Journal of Response to Writing, 3*(1), 83–128.

Hilliard, L. P. & Stewart, M. K. (2019). Time well spent: Creating a community of inquiry in blended first-year writing courses. *The Internet and Higher Education, 41,* 11–24.

Johnson-Eilola, J. & Selber, S. (2001). Sketching a framework for graduate education in technical communication. *Technical Communication Quarterly, 10*(4), 403–437.

Jones, K. R. (2013). Developing and implementing a mandatory online student orientation. *Journal of Asynchronous Learning Networks, 17*(1), 43–45.

Ke, F. (2010). Examining online teaching, cognitive, and social presence for adult students. *Computers and Education, 55*, 808–820. https://doi.org/10.1016/j.compedu.2010.03.013.

Kennedy, G. E., Judd, T. S., Churchward, A. & Gray, K. (2008). First year students' experiences with technology: Are they really digital natives? *Australasian Journal of Education Technology, 24*(1), 108–122.

Kozan, K. & Caskurlu, S. (2018). On the Nth presence for the community of inquiry framework. *Computers & Education, 122*, 104–118.

Lambert, J. L. & Fisher, J. L. (2013). Community of inquiry framework: Establishing community in an online course. *Journal of Interactive Online Learning, 12*, 1–16.

Lear, J. L., Isernhagen, J. C., LaCost, B. A. & King, J. W. (2009). Instructor presence for Web-based classes. *Delta Pi Epsilon Journal, LI*(2), 86–98.

Lee, J., Carter-Wells, J., Glaeser, B., Ivers, K. & Street, C. (2006). Facilitating the development of a learning community in an online graduate program. *Quarterly Review of Distance Education, 7*(1), 13–33.

Linardopoulos, N. (2012). Employers' perspectives of online education. *Campus-Wide Information Systems, 29*, 189–194.

Morrison, D. (2015). Five-step strategy for student success in online learning. *Online Learning Insights*. https://onlinelearninginsights.wordpress.com/2012/09/28/five-step-strategy-for-student-success-with-online-learning/.

Oztok, M. & Brett, C. (2011). Social presence and online learning: A review of the research. *Journal of Distance Education, 25*(3). http://www.ijede.ca/index.php/jde/article/view/758/1299.

Parker, K., Lenhart, A. & Moore, K. (2011). *The digital revolution and higher education: College presidents, public differ on value of online learning*. Pew Research Center. https://www.pewresearch.org/internet/2011/08/28/the-digital-revolution-and-higher-education-2/.

Patton, M. Q. (2015). *Qualitative research and evaluation methods: Integrating theory and practice* (4th ed.). SAGE Publications.

Peery, T. S. & Veneruso, S. S. (2011). Balancing act: Managing instructor presence and workload when creating an interactive community of learners. *Online Classroom, 3*(8).

Phillips, M. E. & Graeff, T. R. (2014). Using an in-class simulation in the first accounting class: Moving from surface to deep learning. *Journal of Education for Business, 89*, 241–247. https://doi.org/10.1080/08832323.2013.863751.

Rourke, L. & Kanuka, H. (2009). Learning in communities of inquiry: A review of the literature. *Journal of Distance Education, 23*(1), 19–48.

Say, B. H. (2015). Developing learning outcomes in professional writing and technical communication programs: Obstacles, benefits, and potential for graduate program improvement. *Programmatic Perspectives, 7*(2), 25–49.

Shackelford, J. L. & Maxwell, M. (2012). Sense of community in graduate online education: Contribution of learner to learner interaction. *The International Review of Research in Open and Distance Learning, 13*, 228–249.

Shea, P. & Bidjerano, T. (2009). Community of inquiry as a theoretical framework to foster 'epistemic engagement' and 'cognitive presence' in online education. *Computers & Education, 52*(3), 543–553.

Shea, P., Li, C. S. & Pickett, A. (2006). A study of teaching presence and student sense of learning community in fully online and Web-enhanced college courses. *Internet and Higher Education, 9*, 175–190.

Shea, P., Vickers, J. & Hayes, S. (2010). Online instructional effort measured through the lens of teaching presence in the community of inquiry framework: A re-examination of measures and approaches. *International Review of Research in Open and Distance Learning, 11*, 127–154.

Shearer, R. L., Gregg, A. & Joo, K. P. (2015). Deep learning in distance education: Are we achieving the goal? *American Journal of Distance Education, 29*(2), 126–134.

Stenbom, S. (2018). A systematic review of the community of inquiry survey. *The Internet and Higher Education, 39*, 22–32.

Stewart, M. K. (2017). Communities of inquiry: A heuristic for designing and assessing interactive learning activities in technology-mediated FYC. *Computers and Composition, 45*, 67–84.

Stewart, M. K. (2019). The community of inquiry survey: An assessment instrument for online writing courses. *Computers and Composition, 52*, 37–52.

Swan, K., Garrison, D. R. & Richardson, J. C. (2009). A constructivist approach to online learning: The community of inquiry framework. In C. R. Payne (Ed.), *Information technology and constructivism in higher education: Progressive learning frameworks* (pp. 43–57). IGI Global.

Swan, K. P., Richardson, J. C., Ice, P., Garrison, D. R., Cleveland-Innes, M. & Arbaugh, J. B. (2008). Validating a measurement tool of presence in online communities of inquiry. *E-Mentor, 2*(24), 1–12.

Taylor, J. M., Dunn, M. & Winn, S. K. (2015). Innovative orientation leads to improved success in online courses. *Online Learning 19*, 112–120.

Walvoord, B. E. (2010). *Assessment clear and simple: A practical guide for institutions, departments, and general education* (2nd ed.). John Wiley & Sons.

Wang, J. & Wang, H. (2012). Place existing online business communication classes into the international context: Social presence from potential learners' perspectives. *Journal of Technical Writing and Communication, 42*(4), 431–451.

Watts, J. (2017). Beyond flexibility and convenience: Using the community of inquiry framework to assess the value of online graduate education in technical and professional communication. *Journal of Business and Technical Communication, 31*(4), 481–519. https://doi.org/10.1177/1050651917713251.

Watts, J. (2019). Assessing an online student orientation: Impacts on retention, satisfaction, and student learning. *Technical Communication Quarterly, 28*(3), 254–270.

Williams, J. M. (2010). Evaluating what students know: Using the RosE portfolio system for institutional and program outcomes assessment. *IEEE Transactions on Professional Communication, 53*(1), 46–57.

Wooten, B. & Hancock, T. (2009, February–March). Online learning offers flexibility and convenience for teacher education. *Momentum*, 28–31.

Wozniak, H., Pizzica, J. & Mahony, M. J. (2012). Design-based research principles for student orientation to online study: Capturing the lessons learnt. *Australasian Journal of Educational Technology, 28*, 896–911.

Appendix A: Course-Embedded Assessment Sample Spreadsheet

Course Number_____ Course Name_____
Semester and Year _____

Directions. Select one assignment from your course and type in the name of the assignment below. Submit a copy of the assignment sheet with this completed table. Use the Rating Key to assess how well each student's assignment achieved the ten program outcomes.

Rating Key. 1) Below Expectations, 2) Meets Criteria at Acceptable Level, 3) Exceeds Expectations, 9) Not Applicable

	Outcome 1	Outcome 2	Outcome 3	Outcome 4	Outcome 5
Student 1					
Student 2					
Student 3					
Student 4					
Student 5					
Student 6					
Student 7					
Student 8					
Student 9					
Assignment Used in this Rating:					

Appendix B: Outcomes Student Survey

1. Which of the following program learning outcomes do you believe ENGL-XXX helped you to practice? (Indicate all those that apply.)
   - Survey and synthesize theoretical concepts and principles about major TPC issues.
   - Select and apply theoretical concepts and principles to the interpretation of technical and professional communication phenomenon.
   - Evaluate relevant scholarship as a means of informing inquiry in technical and professional communication.
   - Select, design and conduct research, using proper methods and methodology, making sound recommendations and drawing logical conclusions.
   - Compose texts, designs and other deliverables, demonstrating ethical, rhetorical, and user-centered strategies.

- Assess documentation for accuracy, adequacy, correctness, accessibility and usability.
- Appraise international and intercultural issues in technical and professional communication, recommending strategies for addressing these issues.
- Evaluate the ways emerging media and digital technologies impact technical and professional communication.
- Plan a documentation schedule and monitor project progress against that schedule.
- Evaluate and execute team-building and interpersonal communication strategies.

2. Rate the usefulness of the following parts of ENGL-XXX in helping you practice these outcomes:
    - Learning Activity 1: Very Helpful / Helpful / Not Helpful / Not Applicable
    - Learning Activity 2: Very Helpful / Helpful / Not Helpful / Not Applicable
    - Learning Activity 3: Very Helpful / Helpful / Not Helpful / Not Applicable
3. What other comments do you have concerning the ways you were encouraged to practice these program outcomes in ENGL-XXX?

Appendix C: Community of Inquiry Student Survey

The community of inquiry (COI) theory was developed to identify what behaviors and practices students and instructors could engage in to help students learn best in online classes. Three presences (cognitive, social, and teaching) are used as lenses with which to identify and assess these behaviors and activities. Please respond to the following questions about your _____ class.

Cognitive Presence: Students who are cognitively present are more likely to retain course concepts and be able to apply them in other settings.

1. This course set up an ill-structured problem for me to research. (Ill-structured problems are multifaceted: they may not have clear solution paths or expected solutions.)
2. This course asked me to discover new ways to address or solve problems.
3. I can see ways to apply aspects of this course's content to other areas of my life (i.e., to other courses, my work).

Social Presence: COI argues that interaction among students and the instructor fosters cognitive presence and enables students to retain and apply course concepts.

1. Interacting with my peers enabled me to construct new knowledge that I would not have been able to construct otherwise.
2. This course gave me opportunities to interact with my peers to discuss problems or concepts.
3. This course asked me to work together in pairs or on a team to collaborate on some aspect of a course assignment or activity.

Teaching Presence: Teaching presence is achieved through properly designing the course, facilitating discourse, and offering direct instruction and feedback about student work.

1. I was given opportunities to exchange ideas about course topics with my peers.
2. The course was designed such that I could identify the activities/assignments I needed to complete and when.
3. I was provided with sufficient and timely feedback about my work.

Final Thoughts

1. If you wish, please comment on any aspects of cognitive, social, and/or teaching presence as they relate to this class.

Appendix D: Community of Inquiry Faculty Survey

1. What are the top three strategies (pedagogies, assignments) that you believe helped to characterize your teaching presence in the online course you are teaching this semester?
2. In what ways do you set boundaries about the limits/scope of your teaching presence to students in the online course that you are teaching this semester? Overall, how are students responding to these strategies?
3. How pleased are you in the ways that you crafted your teaching presence in the online course that you are teaching this semester: what is working well and what do you think needs improvement?
4. Did you actively attempt to cultivate a sense of community (social presence) among the students in the course that you are teaching this semester? If so, how? If not, why?
5. Do you believe that students perceive a sense of community in the course that you are teaching this semester? What led you to draw this conclusion?
6. In general, do you believe that students learn more or learn more deeply when they have an active social presence in online courses? What led you to draw this conclusion?
7. Researchers argue that each student needs to be cognitively engaged in the material, assignments, discussions, etc. in order for deep learning to occur. In what ways do you help prompt students to engage cognitively during the course that you are teaching this semester?
8. What helps you to recognize when students have engaged in deep learning?
9. What do you believe are the roadblocks to students' ability to engage in deep learning in the course that you are teaching this semester?
10. Do you have anything else you'd like to say about teaching presence, social presence or cognitive presence?

11. Designing a Team-Based Online Technical Communication Course

Luke Thominet
Florida International University

Abstract: This chapter describes how team-based learning (TBL), a pedagogical strategy used in high-enrollment in-person business and science classes, can foster effective collaborative writing practices in online technical communication service courses. While collaborative writing projects reflect common workplace communication practice and can help to lessen students' perceptions of isolation in online courses, they often come into conflict with online students' needs for flexible schedules and with the difficulty of establishing interpersonal trust in online environments. TBL offers a conceptual structure for designing effective collaborative learning experiences by organizing courses into units with repeated stages for preparation, content application, and team accountability. The course design presented in this chapter also used the conceptual frame of multimodal editing, where professional writers start from preexisting documents rather than blank pages to create cases conducive to repeated, rapid units that helped students learn to work together over time. The units moved through cycles of collaborative analysis and evaluation of sample documents to a scaffolded, divided, and layered approach to collaborative writing. This course design offers a starting point for considering the strategic integration of collaborative writing processes throughout an online technical communication course.

Keywords: pedagogy, online course design, collaboration, team-based learning, multimodal editing

Key Takeaways:

- Collaborative writing assignments reflect common workplace practice and can help to reduce students' perceptions of isolation in online courses.
- Team-based learning (TBL) offers a strategic approach for creating a replicable group project structure that helps students learn to work together over time.
- Adapting TBL to the exigencies of an online technical communication course increases student communication, engagement, and retention.

My first online technical communication course was based on a flawed design. I copied much of the existing assignment sequence from the institution's face-to-face course while trimming or altering elements that seemed less suited to the online environment. For example, I eliminated many of the smaller in-class exercises to simplify due dates. I replaced real conversations and discussions with extended forum posts. And, based on advice from experienced colleagues, I also removed all collaborative writing assignments.

In retrospect, the primary flaw of the course design was the focus on perceived anti-affordances. When encountering objects or technologies, users perceive the potential for certain kinds of interactions or affordances. Anti-affordances are perceptions of "the prevention of interaction" (Norman, 2013, p. 11).[1] They are the interactions that users think are difficult or impossible with respect to a specific technology, but, as perceptions, anti-affordances need not reflect the actual capabilities of a technology. My initial course design was driven by my perceived anti-affordances of online learning environments: asynchronous student communication occurred slowly; disembodied, online communication curtailed students' mutual trust; and the lack of active conversation stymied collaboration.

This critique is not offered as a strawman representation of all online technical communication courses. The field's literature has provided ample evidence of robust, engaging, and varied approaches to teaching technical communication online: over the past 15 years, there have been two edited collections and two special issues devoted to the topic (Cargile Cook & Davie, 2013; Cargile Cook & Grant-Davis, 2005; Hewett & Bourelle, 2017; Hewett & Powers, 2007). The most recent special issue focused on training online technical communication instructors (Bartolotta et al., 2017; Bay, 2017; Grover et al., 2017; Vie, 2017), including for cross-cultural and global communication courses (Gonzales & Baca, 2017; St.Amant, 2017; Thrush & Popham, 2013). Other literature has discussed online program administration by examining the balance between instructor autonomy and curricular consistency (Maid & D'Angelo, 2013; Rodrigo & Ramírez, 2017; Tillery & Nagelhout, 2013) and assessing the effectiveness of online program orientations for students (Watts, 2019). Other authors have also adapted popular pedagogical practices such as service-learning (Bourelle, 2014; Nielsen, 2016; Soria & Weiner, 2013) and multimodal writing (Bourelle et al., 2017) to online environments. Finally, there have been discussions of best practices for organizing and scaffolding work in online technical communication courses (Grant-Davie & Hailey, 2014; Jones & Jenkins, 2013). So rather than acting as a generalization, the critique of my previous course design is only intended to depict the context for the subsequent redesign that re-centered student-to-student interaction in the form of team-based learning.

This emphasis on collaboration in the redesign sought to improve students' social learning experiences and to increase retention rates. Research has shown that students often feel isolated in online courses, which leads to lower retention rates (Bolliger & Inan, 2012; Bowers & Kumar, 2015). Collaborative projects have been shown to address this isolation and improve retention (Bergin, 2015; Bolliger

---

1. Anti-affordances are slightly different from constraints, which are features of a technology that guide interactions in specific ways (Norman, 2013).

& Inan, 2012; Hazari & Thompson, 2015), while also providing a real social context (Bruffee, 1984) and audience for the assignments (Blair, 2005). Additionally, collaborative writing is an important and challenging workplace communication practice, making it a core element of many technical communication service courses (Bremner, 2010; Burnett et al., 1997, 2013; Hewett & Robidoux, 2010; Johnson-Eilola, 1996; Lunsford & Ede, 2011; Stratton, 2015).

Despite these many benefits, there is evidence that online students often dislike and resist collaborative work. Students primarily choose online courses because of the flexibility and convenience they offer (Benbunan-Fich & Hiltz, 2003; Clark et al., 2018; Eaton, 2013; Jaggars, 2014; Jaggars & Bailey, 2010; Kariya, 2003; Mahoney, 2009; Smart & Cappel, 2006). Therefore, intensive group projects, which make students rely on each other's schedules, can conflict with this flexibility. Additionally, collaborative writing can be prone to unequal contributions (Hewett, 2015; Wolfe, 2010), thus increasing the potential for interpersonal discord. Furthermore, online students may not trust their group members, which can make it difficult to coordinate group work (Burton & Goldsmith, 2002). Even students in highly collaborative online writing courses have reported that "they did not benefit from interacting with their peers" and that they "could have gotten just as much out of it if it were individual work" (Stewart, 2018, "Findings," sec. 3.2.3).

However, there has also been an increase in research on online collaborative writing instruction. Scott Warnock (2009) briefly described a range of group projects online, including a collaborative argument website, peer review, and group message boards. Jeffrey Bergin (2015) adapted Karen B. LeFevre's (1987) classification of social learning approaches to describe online group writing as ranging from projects where students interact but submit separate deliverables (e.g., discussion boards or peer review) to projects where all students work together on a single deliverable (e.g., a wiki page).[2] Beth L. Hewett (2015) recommended small, permanent teams and pointed, low-stakes assignments—an approach that this chapter largely adopts and expands. And Teresa Mauri and Javier Onrubia (2015) and Carola Strobl (2015) described how providing students with a script of the recommended work process could help them to collaborate effectively online.

This chapter contributes to the ongoing discussion by presenting an additional pedagogical approach for building collective writing projects online. It

---

2. Bergin described the former as "collaborative writing" and the latter as "collective writing," but other authors have offered alternative ways of differentiating between similar terminology. For example, Pope-Ruark (2017) described "collaboration" as an intensive process with shared goals where the result is greater than the sum of its parts. Conversely, she described "cooperation" as focused on coordinating and combining individual efforts, thus aligning it more directly with Bergin's definition of "collaboration." Given the disagreement over definitions, this chapter will generally use "collective writing" and "collaboration" interchangeably to mean a writing process with shared goals and a single, shared product. Students might still divide up work during this process, but they must also do substantial work together to complete the final product.

argues that we can and should design more effective contexts for online student collaboration. It employs and adapts team-based learning (TBL), a popular pedagogical approach in high-enrollment science and business courses, to structure an online technical communication service course. And it provides examples of how a case-based, multimodal editing perspective can structure writing tasks according to the needs of online student writing teams.

In the following sections, I briefly review the literature describing the design of TBL classes and discuss the primary limitations of TBL for online writing courses. Then I present the adaptation of TBL for my technical communication course. Finally, I reflect on how the core elements of this course design can be expanded and adapted for the needs of other online technical communication courses.

Literature Review: Team-Based Learning and Its Limitations

Team-based learning is a teaching strategy for systematically integrating teamwork throughout a course. It was developed by Larry Michaelsen in the 1970s and has grown into a significant body of pedagogical literature. TBL has been used in a range of disciplines, including health, business, and science (Emke et al., 2016; Huang & Lin, 2017; Ratta, 2015; Sharma et al., 2017; Stepanova, 2018), and, more recently, in humanities courses as well (Harde, 2012; Restad, 2012; Roberson & Reimers, 2012). Despite this broad usage, there is no literature on implementing TBL in writing courses and only limited work on adapting TBL to online courses (Clark et al., 2018; Freeman, 2002; Hosier, 2013; Palsolé & Awalt, 2008).

TBL is best understood as a prescribed set of course design elements rather than a fully-fledged pedagogical theory. While TBL has a theoretical basis in social-constructivism and cognitive apprenticeship (Fink, 2002; Sweet & Michaelsen, 2012), most TBL literature has prioritized observed practical benefits over theoretical foundations. For example, Michaelsen invented TBL as a way to help instructors manage increasing course enrollments (Michaelsen et al., 2002). Likewise, other authors have highlighted benefits such as maintaining instructors' enthusiasm for teaching (Knight, 2002) and supporting nontraditional students (Goodson, 2002), diverse students (Croyle & Alfaro, 2012), and students with disabilities (Nakaji, 2002).

Michaelsen (2002) defined TBL against generalized student group work through four principles: 1) intentionally-designed student teams, 2) strong student-accountability measures, 3) assignments designed for active collaboration, and 4) immediate and regular feedback. These principles are built into the recommended TBL course and unit structures.

First, effective TBL teams are large, diverse, and permanent. Large teams of 5–7 students ensure "that the vast majority of groups will have ample resources" (Michaelsen, 2002, p. 40). Likewise, diverse teams fairly distribute student knowledge and perspectives across the course. Finally, permanent teams help students to learn to work together over time (Michaelsen, 2002).

After teams are formed, TBL courses proceed through a series of units with three phases each: preparation, application, and evaluation (Fink, 2002; Sweet & Michaelsen, 2012). The preparation phase has three components. First, students read the assigned texts. Then they complete a Readiness Assurance Process (RAP), which includes both an individual and a team version of the same test. The individual test is meant to foster students' accountability to the content, while the team test is intended to encourage students to teach each other (Michaelsen, 2002). Finally, the instructor gives a corrective lecture focused on the most commonly missed questions in the tests.

During the subsequent application phase, student teams apply "course concepts to make and justify discipline-based decisions" (Sweet & Michaelsen, 2012, p. 10). To encourage deep learning and team cohesion, the application phase is structured as a series of increasingly difficult "4–S problems," where all teams work on the *same, significant* problem, answer with a *specific* choice, and report answers *simultaneously* (Sweet & Michaelsen, 2012, pp. 24–26). While the original version of TBL depicted 4–S problems primarily as challenging, case-based, multiple-choice questions, more recent work has described a range of deliverables, including posters, Excel charts, and overheads (Sibley, 2012).

The final phase of TBL units involves evaluating each student on their content knowledge and their contributions to the team. The evaluation of content knowledge has not been given much attention in TBL literature. For example, Fink (2002) alternately described it as an exam or as solving a 4–S problem individually. But team evaluations have been discussed in more detail. They are meant to build accountability among team members and to address issues of unequal contribution (Fink, 2002). They also encourage constructive feedback and improve team cohesion in future units (Lane, 2012). Overall, this unit structure of preparation, application, and evaluation is intended to create engaging and active collaboration.

However, there are two significant issues for adapting TBL to an online writing course. First, core TBL literature has explicitly rejected collaborative writing as an appropriate team activity:

> It is our experience that the worst assignment when trying to build group cohesiveness is to ask students to write a term paper as a group. Group papers seldom provide any support for building group cohesiveness and almost universally result in social loafing, or at least what is perceived by other students as social loafing. Writing is inherently an individual activity; therefore, the rational way to accomplish the overall task is to divide up the work so that each member independently completes part of the assignment. . . . As a result, there is seldom any significant discussion after

the initial division of labor, and feedback is generally unavailable until after the project is handed in.... In fact, high-achieving students often express the feeling that getting an acceptable grade on a group term paper feels like having crossed a freeway during rush hour without being run over. (Michaelsen & Knight, 2002, p. 61)

By stating that writing is inherently individual, Michaelsen and Knight diverged from decades of research on how writing functions (e.g., Cooper & Holzman, 1989; Ede & Lunsford, 1990; Flower, 1994; Kroll, 1984; LeFevre, 1987; McComiskey, 2000; Swales, 2017). They also overlooked valid reasons for teaching team-based writing even if it is difficult, including the continued importance of collaborative writing practices in workplace environments (Blythe et al., 2014; Brumberger & Lauer, 2017). Finally, by arguing that collaborative writing projects lack support for building group cohesiveness, they ignored composition studies literature that has introduced numerous successful strategies for such projects (e.g., Beard et al., 1989; Bilansky, 2016; Conklin, 2017; Kittle & Hicks, 2009). Joanna Wolfe (2010), in particular, has offered an invaluable guide on effective team-based writing practices, including supporting diverse teams, managing projects, creating constructive conflict in discussions, and developing effective revision and feedback processes. She even described specific strategies to address the problem of an unequal division of labor, such as the development of task schedules around layered collaboration, where each student adopts a specific role within the project, such as researcher, writer, or editor (Wolfe, 2010).

However, if Michaelsen and Knight's criticism is limited solely to collaborative "term papers," it might merit further exploration. Some research has shown that extended report projects are effective collaborative writing assignments because the complexity of the genre requires meaningful contributions from multiple people (Rentz et al., 2009). And many instructors scaffold these projects through several phases and deliverables to create accountability and encourage discussion (Wolfe, 2010). Still, collaborative report assignments have sometimes been appended to courses where all other writing assignments are completed individually, and research has shown that students benefit from consistent online course design and structures (Dhilla, 2017; Swan, 2001). This shift from largely individual work to a high-stakes collective project at the end of the semester might not give students sufficient time to build mutual trust, leading to increased anxiety and group dysfunction (Allan & Lawless, 2003). In other words, while extended reports can be effective collaborative writing assignments, online courses likely need to build strong networks between students first.

The second significant issue for adapting TBL to an online writing course is that the major unit structures, including the RAP and 4–S application, assume in-class time when students can engage in regular synchronous communication. In fact, L. Dee Fink (2002) argued that teamwork should occur exclusively during class sessions to encourage students to work together rather than splitting up work. While synchronous online meetings might address this issue, asynchronous modalities are often a better fit for online students' needs for flexible schedules (Mick et al., 2015). Consequently, instructors have sought to create best practices for adapting TBL to asynchronous online environments. The most comprehensive advice on this can be found in Michelle Clark et al.'s (2018) white paper, which built on previous articles describing online TBL practices in individual courses (Hosier, 2013; Palsolé & Awalt, 2008). The white paper described principles for aligning each of the main TBL phases with Quality Matters standards, a set of widely used principles for online education. Their primary advice for the RAP was to slow down the process so it takes several days, to use timed quizzes for the individual test, and to write questions that move beyond memorization (Clark et al., 2018). For the application phase, they discussed the difficulty of adapting the 4–S aspect of simultaneous reporting to online settings and thus recommended a two-step process where teams submit answers whenever they are ready and then gain access to other teams' responses at a predetermined time. They also suggested using the learning management system's tools to support collaboration and analytics to measure each student's contributions. For peer evaluation, they recommended using multiple formative and summative evaluations, being transparent about the impact of peer evaluations, and using analytics to support evaluation. Finally, in contrast to traditional TBL practices, they recommended assigning students the roles of team leaders and reporters to help facilitate the teamwork. Clark et al. offered useful advice, but they also had to generalize this information for a broad audience, and they often focused on technological solutions (e.g., learning management system tools) for fixing potential issues with asynchronous, online collaboration. In short, there remains room for further exploration of how technical writing courses specifically might adopt the TBL structure in an online environment.

Adapted-TBL Online Technical Communication Course Design

I adapted the online TBL model for ENC 3213: Professional and Technical Writing, an upper-division undergraduate course that introduces students to the expectations of writing in the workplace. It functions primarily as a multi-major technical communication service course that draws students from a range of disciplines, including engineering, computer science, nursing, healthcare administration, business, international relations, and English. Students also enter the course with a range of professional experience: some are already working professionals, while others have recently completed high school or community college and have little experience in writing for non-academic audiences. Other sections of the course at my institution have typically begun with two to three brief units on professional correspondence, job application documents, marketing materials, or instruction sets. Then they transitioned to an extended project that included a research proposal, an analytic report, and a project presentation.

My adapted-TBL course situated students in the fictional Writing@FIU team, which provided freelance writing services for local organizations through a series of brief, rapid, low-stakes units. Each unit was focused on a multi-stage, complex, realistic case that asked students to make specific decisions and to craft documents within messy problem spaces. This case-based approach has been shown to help students develop teamwork skills (Thondhlana & Smith, 2013) and audience awareness (Robles & Baker, 2019). While technical communication scholarship has identified numerous benefits of online service-learning projects, including increasing students' self-accountability and engagement (Nielsen, 2016), the speed and structure of the adapted-TBL units largely precluded working with real community partners. As described below, the unit structure included elements of individual and team-based work, and it used the same structure across units to allow students to cohere and grow as teams throughout the duration of the course. Matching this structure with real partners' needs and schedules would have been difficult, though there is certainly room to explore this approach in the future. Still, by using messy, document-based cases as the foundation for each assignment, the course was able to keep some of the benefits found in work with real partners.

Each case was also intentionally designed as a multimodal editing process, which Claire Lauer and Eva Brumberger (2019) described as an essential practice in contemporary professional writing:

> Many writers actually act as multimodal editors—people who work with myriad modes of content—often encountered in medias res after the content has been originated by coworkers or consultants. Multimodal editors are responsible for modifying, adapting, designing, editing, selecting, and constructing content in ways that are dispersed, non-linear, collaborative, and responsive. (p. 637)

Throughout the article, Lauer and Brumberger (2019) gave numerous examples of multimodal editing, including revising rough content from a legal/compliance team, reworking and repackaging clients' video content, and translating technical content for lay audiences on social media. They also recommended adopting similar practices in technical communication courses:

> Setting up situations in which students start not with their own blank page, but with textual or visual material developed by others . . . can help situate them in a professional situation that might lead to more authentic, transactional writing experiences. (p. 657)

Within the adapted-TBL course, this meant that cases were built around existing, flawed documents: rough drafts of correspondence, a brief usability report, an email with ideas and notes for a proposal, etc. These documents grounded students in the case and reduced the time needed for initial ideation phases. They helped students produce long and complex documents quickly while also prompting difficult decisions on content.

The overall unit sequence introduced progressively more complex genres and situations, but the later assignments were still structured to support rapid production cycles. While the exact assignments have changed throughout the iterations of the course design, the most recent unit sequence was:

- Unit 1: Students individually create functional résumés to apply to the Writing@FIU team.
- Unit 2: Student teams evaluate informal team charters and draft correspondence related to realistic group problems.
- Unit 3: Student teams evaluate an instructional video and remediate it as written instructions to help faculty update their bios on a local college's website.
- Unit 4: Student teams evaluate past grant proposals for a local fund and then produce a brief grant proposal to create a community garden.
- Unit 5: Student teams evaluate presentation graphics and speaker's notes and then produce and record a PowerPoint presentation for a local initiative to support bicycle safety.
- Unit 6: Students individually research the writing practices of professionals in their field and produce a memo connecting course topics to their profession.

The following three subsections will break down the key TBL concerns of team formation, unit phases, and student evaluation. Then I will provide some basic information on the results of the course design thus far.

Team Formation

The course was designed to have the first and last units completed individually by students in order to minimize anxiety at the beginning and end of the semester and to create more positive and productive team environments. The first unit covered the common principles of technical communication and document design to provide students with a shared knowledge set and language for the remaining assignments (though we also open room for problematizing and revising these principles throughout the course). The unit also gave the class an opportunity to build a positive social environment: we started by posting introductions and had additional channels for casual off-topic discussions. Finally, the first unit created a time buffer so course enrollment could stabilize while I intentionally constructed student teams.

During those two weeks, I gathered information through a survey and an assignment. I then constructed teams of four to five students based on three factors: 1) typical weekly availability (so teams could collaborate synchronously if they wished to), 2) professional writing experience, and 3) performance on the functional résumé assignment. When the course was taught as multiple combined sections (allowing for a greater student population), factors such as students' majors and gender and cultural identities were also intentionally distributed amongst the teams. After the teams were formed, they were effectively permanent for the duration of the course. Over more than two years of using this course design, only one team has had to be reorganized. Two other teams that lost a member were offered the option to dissolve their membership into other teams, but they both chose to remain in a smaller team rather than divide up.

Finally, while official TBL approaches reject the practice of giving team members specific roles (Fink, 2002), the adapted-TBL course had a student assigned as the project manager for each unit. This role largely mirrored Wolfe's (2010) description of a project manager: they began conversations, scheduled teamwork, and produced meeting minutes. This role helped to improve overall team coordination while also offering project management experience to each student during the semester.

Unit Structure

While the team-based units have gone through several iterations, the general structure always followed the TBL phases of preparation, application, and evaluation. The first version mirrored Allison Hosier (2013) and Sunay Palsolé and Carolyn Awalt's (2008) course structures with interwoven RAP and 4–S processes, which each included individual and team-based elements. Recent iterations simplified this structure to create a more predictable weekly schedule. The most recent version used three-week team-based units with the following structure:

- Week 1: Preparation and case introduction
- Week 2: Cooperative organization and individual drafting
- Week 3: Collective revision and peer evaluation

Since this structure was identical across all units, I provide examples below from Unit 4, which introduced students to grant proposals.

Students began the first week of Unit 4 by reading excerpts from our textbook on proposal writing as well as a few outside texts on related topics. Then they completed a short reading quiz on grant proposals. This quiz has evolved over the course iterations from a ten-question, multiple-choice test focused on recall to a five-question short answer test with mixed recall and evaluation questions. For example, one recent question asked students to describe the purpose of the introduction section of a grant in their own words. Another question asked them to evaluate a specific example of a grant task description based on the information in our textbook. These quizzes were intended to encourage each student to familiarize themselves with the content of the unit in order to create more productive conversations throughout the rest of the teamwork.

During the second half of the first week, teams discussed and evaluated sample documents related to the unit's case. In Unit 4, teams were provided with a call for proposals and four sample grants for the fictional Keep Miami Beautiful Small Grant Program, which was based on the real Keep Oakland Beautiful Small Grant Program (*KOB Small Grant Program*, n.d.). The Call for Proposals (CFP) requested proposals for small, local projects that create or improve community spaces in Miami. The four sample proposals covered a range of topics, including the creation of a new mural in Wynwood, a beach cleanup in South Miami, and the construction of a pocket park in Sweetwater. Teams subsequently discussed, evaluated, and ranked the proposals. This discussion was designed to encourage constructive conflict, or "the healthy, respectful debate of ideas and competing solutions to a problem" (Wolfe, 2010, p. 51), through the following features:

- The discussion occurred on the team's private Slack channel. The structure of this software as an instant messaging platform encouraged a more fluid and active conversation than learning management system forums.
- Qualitative evaluations were tied to quantifiable ratings (e.g., asking students to rank the proposals), which increased the potential for disagreement and debate.
- The texts being evaluated were of varying quality, but they all included both effective and ineffective features. For example, the beach cleanup proposal had a persuasive problem statement, but it included only a generalized budget with no itemized breakdown. This created room for debate around the relative importance of various features.
- Students controlled their own discussions, with two limitations: each team member needed to contribute actively, and the work could not be subdivided amongst the group (i.e., everyone needed to be able to discuss every grant proposal).
- Finally, most group members were given credit simply for participating actively. The only deliverable for the assignment was a set of meeting minutes created by the project manager. These minutes were expected to summarize the discussion while clearly attributing contributions to individual team members.

During the second week, students were introduced to the team writing task for the unit, which built on the situation introduced in the first week's discussion. For Unit 4, teams were asked to develop a grant proposal for starting a community garden in response to the Keep Miami Beautiful Small Grants Program. The prompt for the assignment was presented as an email from Josiane, a representative of the community garden who asked for help with the grant. The email included both relevant and irrelevant information. For example,

> Our proposed garden is at 58th St. and NE 4th Ave. We have a contract for a 10–year, low-cost lease in hand. The owner of the

plot is a local resident who is very supportive of our project. So, we're pretty secure in the longevity of the garden. We hope to expand to similar plots in the Little Haiti area in the coming years but decided that we want to get this one up and running first. We might also try to expand our goals in future years to support in-home gardens of local residents, but again, we don't have enough resources yet.

The lengthy email went on to list the potential uses for funds (e.g., hedge plants for boundary beautification, lumber and soil for creating raised garden beds, a rototiller, compensation for volunteers, etc.), the potential positive impact of the garden (e.g., improving the local availability of fresh vegetables and herbs, increased physical activity, stress release, etc.), and other thoughts on the project. Eventually, the email asked the Writing@FIU team for their help in developing the proposal. I passed this email on to the team with some additional instructions for the project, including a schedule for initial drafts and an expected final completion date. Students were then prompted to divide the work into four sections that aligned with key pieces of information in the grant proposal: 1) problem statement, 2) benefit statement, 3) methods plan, and 4) itemized budget. They drafted these sections individually but had to coordinate the work, so all the sections contained consistent information. They then submitted the sections both to me on our learning management system and to their teammates in a shared Google Doc. While this initial divided approach did not reflect Wolfe's (2010) recommendation for layered collaboration, it has helped students to establish more individual accountability to initial drafts, which has lessened some of the concern over fully team-based grades in the online class.

At the outset of the final week of the unit, teams were given new correspondence with slight alterations to the existing prompt, such as new length limits, new content expectations or limitations, or new formatting procedures. For Unit 4, these changes included 1) a reduced availability of funds (from $1,200 to $800 per grant), 2) additional requested information (on the community garden organization's ethos for carrying out the project), and 3) an email response to Josiane that explained the team's decisions in crafting the proposal. These changes were designed to prompt alterations to the existing content so that individual drafts could not simply be pasted together. It also allowed teams to focus more on a layered, actively collaborative approach to designing their final drafts.

The final phase of each adapted-TBL unit asked students to complete a 180-degree performance review by evaluating both their own and their teammates' contributions to the teamwork. These evaluations had three parts:

1. A self-reflection that described their contributions and identified their effective and ineffective professional writing and teamwork strategies.
2. Numerical evaluations of each peer's contributions. Based on TBL literature, the evaluation scale was effort above or below the average for the

team. This evaluation system intentionally foregrounded perceived effort over the perceived quality of contributions. This way students who were active throughout the group project were rewarded rather than those who simply wrote effective prose at the last minute. Also, the rating system intentionally limited scale inflation: if one student contributed more than average effort, another had to be rated as contributing less than average.
3. Constructive feedback to peers. Students were provided with models for constructive feedback, and all comments were reviewed before being distributed to other students.

Overall, these regular team evaluations ensured an additional level of accountability while providing channels for discussing and improving teamwork.

Grading

Collaborative learning causes anxiety partially due to shared grades (Allan & Lawless, 2003). For that reason, this course was designed around low-stakes projects and a mix of individual and team grades. For each of the four team-based units, students received grades on five assignments: a reading quiz, a team discussion, an individual draft, a collective draft, and a team evaluation. Only one of these assignments was a fully shared grade (the collective draft). The other grades were either entirely individual or included individualized elements (e.g., the team evaluation grade included credit for completing the evaluation and credit for peers' evaluations of the student). In total, there were 26 graded assignments in the most recent version of the course, making most relatively low stakes (3–4% of the final grade). And many of the projects (e.g., participation in discussion and individual drafts) were graded on a full credit/no credit basis. Collectively, this meant that final grades were primarily based on completing the assigned work and on contributing actively to the team's efforts. If a student received positive peer evaluations and completed all the quizzes, discussions, and individual drafts, they universally earned a passing grade in the class.

This grading strategy was explained to students at the beginning of the semester and was reinforced throughout the course. Reassuring students that they truly did have individual control over their grades helped to ease initial fears about the potential chaos of an online team-based course.

Course Design Results

I can only provide anecdotal evidence of the success of the course design, but by most available measures, every iteration of the adapted-TBL design has been effective:

- Student retention rates were high: Many institutions use a DFW rate, or the percentage of students earning Ds, Fs, and withdrawing from the

course, to identify students at risk of dropping out. The adapted-TBL course had a 6.5 percent DFW rate. During the same time period, other online sections of the same course had a 15.8 percent DFW rate.
- Most students were engaged: They asked and answered questions about course concepts, related content to their own experiences, and talked to each other about topics outside of the course (e.g., they recruited each other for student clubs or online gaming guilds, shared information about events, posted pictures of pets, etc.). Over the 12-week collaboration, team channels had an average of 953 messages (or approximately 20 messages per student per week). Some messages were short affirmations or project management questions, but many messages engaged in substantive conversations on the assignment cases or the rhetorical decisions for documents.
- Anxiety over teamwork appeared to be minimized: After they started working in teams, students generally did not complain about the course structure. Only one student directly requested to complete the coursework on their own (due to personal reasons unrelated to the class). Students also regularly rated their teammates and their teamwork highly in their peer evaluations.
- Student teams grew more effective over time: Their initial projects had some confusion over the best way to schedule and structure the teamwork, but these processes became much smoother by the second or third time through the same unit structure. The teams also continued to submit more and more effective final products throughout the semester.
- Finally, students valued their teams: they recognized and discussed the value of collaborative writing in their end-of-semester reflections and course evaluations, and they regularly reported learning useful strategies for professional writing simply by managing and negotiating shared online writing projects with their peers.

Likewise, my experience of teaching the course also shifted. My first online course design was focused on managing course content and feedback: recording lecture videos, explaining assignments, getting in touch with missing students, giving feedback on drafts, and grading assignments. In the adapted-TBL design, students did some of this work themselves: they explained the core concepts to each other as they worked through the cases, contacted teammates who were falling behind, and offered feedback on each other's drafts. The instructional work shifted more toward the so-called "guide on the side" role, which is characterized by "being a facilitator who orchestrates the context, provides resources, and poses questions to stimulate students to think up their own answers" (King, 1993, p. 30). The instructional role also shifted toward higher-level management of team dynamics and production: I set up teams, handled disagreements, made suggestions, and amplified students' ideas. Even with the rapid unit structures, the grading load decreased significantly so more time could be spent on providing additional resources and engaging students in conversations about professional communication practices.

■ Conclusion

This chapter has argued that TBL structures can create a consistent and productive approach to building collaborative writing assignments in online technical communication courses. Of course, this specific course design is not universally applicable. The collective writing projects were pertinent to a multi-major survey course, but they might not be as effective in advanced courses. Likewise, the rapid, low-stakes units might not be ideal client-based or service-learning projects. Still, there are elements of the adapted-TBL design that can transfer relatively easily across contexts.

First, the creation of permanent teams and of units with repeated, predictable structures allowed students to know what to expect. TBL can help reorient us to seeing teamwork as a practice that grows and improves over time. By providing students with opportunities to practice their teamwork, we can help them develop into high-performing teams.

Second, TBL ensures that we hold students accountable both for their knowledge of the course content and for their contributions to teamwork. The regular assignments and discussions encouraged positive practices of preparation and engagement, which, in turn, helped to build trust among teammates.

Third, changing the grading structure to primarily assess labor and effort gave control over final grades back to individual students. And making this grading philosophy and students' individual control explicit in course documents helped to minimize anxiety over shared grades. At the same time, retaining shared grades for the collective drafts ensured that students were encouraging each other to do their best work.

Finally, by framing professional writing as multimodal editing, the course gave teams concrete starting points, sped up the planning phases of teamwork, and grounded the work in realistic contexts. At the same time, the projects built in multiple decision points to create a range of potential results. This encouraged real discussions about priorities while maintaining student engagement throughout the duration of the assignment.

In closing, we can return to the broader question of collaboration in our online courses. Collaboration is a powerful pedagogical tool. It can combat feelings of isolation, encourage student engagement and persistence, and contribute to deep learning. It can also conflict with students' desire for flexibility and autonomy. But rather than focusing on potential anti-affordances of online communication systems, we can use the affordances of the environment to construct better teamwork online. This project started by centering collaboration at the outset of course design so it was strategically integrated throughout the semester in a consistent and cohesive manner. This process can require a significant re-thinking of existing course structures, but it can also build better learning experiences for students.

References

Allan, J. & Lawless, N. (2003). Stress caused by on-line collaboration in e-learning: A developing model. *Education + Training*, *45*(8/9), 564–572. https://doi.org/10.1108/00400910310508955.

Bartolotta, J., Bourelle, T. & Newmark, J. (2017). Revising the online classroom: Usability testing for training online technical communication instructors. *Technical Communication Quarterly*, *26*(3), 287–299. https://doi.org/10.1080/10572252.2017.1339495.

Bay, J. (2017). Training technical and professional communication educators for online internship courses. *Technical Communication Quarterly*, *26*(3), 329–343. https://doi.org/10.1080/10572252.2017.1339526.

Beard, J. D., Rymer, J. & Williams, D. L. (1989). An assessment system for collaborative-writing groups: Theory and empirical evaluation. *Journal of Business and Technical Communication*, *3*(2), 29–51. https://doi.org/10.1177/105065198900300203.

Benbunan-Fich, R. & Hiltz, S. R. (2003). Mediators of the effectiveness of online courses. *IEEE Transactions on Professional Communication*, *46*(4), 298–312. https://doi.org/10.1109/TPC.2003.819639.

Bergin, J. (2015). Fostering persistence and student connections in online writing course: A social constructivist approach. *Journal of Teaching Writing*, *29*(1), 45–69. http://journals.iupui.edu/index.php/teachingwriting/article/view/20734.

Bilansky, A. (2016). Using Wikipedia to teach audience, genre, and collaboration. *Pedagogy*, *16*(2), 347–355. https://doi.org/10.1215/15314200-3435996.

Blair, L. (2005). Teaching composition online: No longer the second-best choice. *Kairos*, *8*(2). http://kairos.technorhetoric.net/8.2/binder.html?praxis/blair/index.html.

Blythe, S., Lauer, C. & Curran, P. G. (2014). Professional and technical communication in a web 2.0 world. *Technical Communication Quarterly*, *23*(4), 265–287. https://doi.org/10.1080/10572252.2014.941766.

Bolliger, D. U. & Inan, F. A. (2012). Development and validation of the Online Student Connectedness Survey (OSCS). *The International Review of Research in Open and Distributed Learning*, *13*(3), 41–65. https://doi.org/10.19173/irrodl.v13i3.1171.

Bourelle, T. (2014). Adapting service-learning into the online technical communication classroom: A framework and model. *Technical Communication Quarterly*, *23*(4), 247–264. https://doi.org/10.1080/10572252.2014.941782.

Bourelle, T., Bourelle, A., Spong, S. & Hendrickson, B. (2017). Assessing multimodal literacy in the online technical communication classroom. *Journal of Business & Technical Communication*, *31*(2), 222–255. https://doi.org/10.1177/1050651916682288.

Bowers, J. & Kumar, P. (2015). Students' perceptions of teaching and social presence: A comparative analysis of face-to-face and online learning environments. *International Journal of Web-Based Learning and Teaching Technologies*, *10*(1), 27–44. https://doi.org/10.4018/ijwltt.2015010103.

Bremner, S. (2010). Collaborative writing: Bridging the gap between the textbook and the workplace. *English for Specific Purposes*, *29*(2), 121–132. https://doi.org/10.1016/j.esp.2009.11.001.

Bruffee, K. A. (1984). Collaborative learning and the "conversation of mankind." *College English*, *46*(7), 635–652. https://doi.org/10.2307/376924.

Brumberger, E. R. & Lauer, C. (2017). International faces of technical communication: An analysis of job postings in three markets. *Technical Communication*, *64*(4), 310–327.

https://www.stc.org/techcomm/2017/11/01/international-faces-of-technical-communication-an-analysis-of-job-postings-in-three-markets/.

Burnett, R., Cooper, L. A. & Welhausen, C. A. (2013). What do technical communicators need to know about collaboration? In J. Johnson-Eilola & S. A. Selber (Eds.), *Solving problems in technical communication* (pp. 454–478). University of Chicago Press. https://doi.org/10.7208/chicago/9780226924083.001.0001.

Burnett, R., White, C. & Duin, A. H. (1997). Locating collaboration: Reflections, features, and influences. In K. Staples & C. M. Ornatowski (Eds.), *Foundations for teaching technical communication: Theory, practice, and program design* (pp. 133–160). Greenwood Publishing Group.

Burton, L. J. & Goldsmith, D. (2002). *Students' experiences in online courses: A study using asynchronous focus groups*. Connecticut Distance Learning Consortium.

Cargile Cook, K. & Davie, K. G. (2013). *Online education 2.0: Evolving, adapting, and reinventing online technical communication*. Routledge. https://doi.org/10.4324/9781315231471/

Cargile Cook, K. & Grant-Davis, K. (2005). *Online education: Global questions, local answers*. Routledge.

Clark, M., Merrick, L., Styron, J., Dolowitz, A., Dorius, C., Madeka, K., Bender, H., Johnson, J., Chapman, J., Gillette, M., Dorneich, M., O'Dwyer, B., Grogan, J., Brown, T., Leonard, B., Rongerude, J. & Winter, L. (2018). *Off to on: Best practices for online team-based learning* [White paper]. Team-Based Learning Collaborative. http://www.teambasedlearning.org/wp-content/uploads/2018/08/Off-to-On_OnlineTBL_WhitePaper_ClarkEtal2018_V3.pdf.

Conklin, R. C. (2017). Composing at the threshold: Collaborative composition and innovative form. *Pedagogy*, *17*(2), 359–365. https://doi.org/10.1215/15314200–3770261.

Cooper, M. & Holzman, M. (1989). *Writing as social action*. Heinemann.

Croyle, K. L. & Alfaro, E. (2012). Applying team-based learning with Mexican American students in the social science classroom. In M. Sweet & L. Michaelsen (Eds.), *Team-based learning in the social sciences and humanities: Group work that works to generate critical thinking and engagement* (pp. 203–220). Stylus Publishing.

Dhilla, S. (2017). The role of online faculty in supporting successful online learning enterprises: A literature review. *Higher Education Politics & Economics*, *3*(1). https://digitalcommons.odu.edu/aphe/vol3/iss1/3.

Eaton, A. (2013). Students in the online technical communication classroom: The next decade. In K. Cargile Cook & K. Grant-Davie (Eds.), *Online education 2.0: Evolving, adapting, and reinventing online technical communication* (pp. 113–132). Baywood Publishing. https://doi.org/10.4324/9781315231471.

Ede, L. S. & Lunsford, A. A. (1990). *Singular texts/plural authors: Perspectives on collaborative writing*. Southern Illinois University Press.

Emke, A. R., Butler, A. C. & Larsen, D. P. (2016). Effects of team-based learning on short-term and long-term retention of factual knowledge. *Medical Teacher*, *38*(3), 306–311. https://doi.org/10.3109/0142159X.2015.1034663.

Fink, L. D. (2002). Beyond small groups: Harnessing the extraordinary power of learning teams. In L. Michaelsen, A. B. Knight & L. D. Fink (Eds.), *Team-based learning: A transformative use of small groups in college teaching* (pp. 3–25). Stylus Publishing.

Flower, L. (1994). *The construction of negotiated meaning: A social cognitive theory of writing* (1st ed.). Southern Illinois University Press.

Freeman, M. (2002). Team-based learning in a course combining in-class and online interaction. In L. Michaelsen, A. B. Knight & L. D. Fink (Eds.), *Team-based learning: A transformative use of small groups in college teaching* (pp. 213–232). Stylus Publishing.

Gonzales, L. & Baca, I. (2017). Developing culturally and linguistically diverse online technical communication programs: Emerging frameworks at University of Texas at El Paso. *Technical Communication Quarterly, 26*(3), 273–286. https://doi.org/10.1080/10572252.2017.1339488.

Goodson, P. (2002). Working with nontraditional and underprepared students in health education. In L. Michaelsen, A. B. Knight & L. D. Fink (Eds.), *Team-based learning: A transformative use of small groups in college teaching* (pp. 119–129). Stylus Publishing.

Grant-Davie, K. & Hailey, D. (2014). 18 Years of teaching technical communication online: Tricks and traps, Dos and Don'ts, strengths and weaknesses. In *2014 IEEE International Professional Communication Conference (IPCC)* (pp. 1–5). https://doi.org/10.1109/IPCC.2014.7020340.

Grover, S. D., Cargile Cook, K., Harris, H. S. & DePew, K. E. (2017). Immersion, reflection, failure: Teaching graduate students to teach writing online. *Technical Communication Quarterly, 26*(3), 242–255. https://doi.org/10.1080/10572252.2017.1339524.

Harde, R. (2012). Team-based learning in the first-year English classroom. In M. Sweet & L. Michaelsen (Eds.), *Team-based learning in the social sciences and humanities: Group work that works to generate critical thinking and engagement* (pp. 143–158). Stylus Publishing.

Hazari, S. & Thompson, S. (2015). Investigating factors affecting group processes in virtual learning environments. *Business and Professional Communication Quarterly, 78*(1), 33–54. https://doi.org/10.1177/2329490614558920.

Hewett, B. L. (2015). *Reading to learn and writing to teach: Literacy strategies for online writing instruction.* Macmillan Higher Education.

Hewett, B. L. & Bourelle, T. (2017). Online teaching and learning in technical communication: Continuing the conversation. *Technical Communication Quarterly, 26*(3), 217–222. https://doi.org/10.1080/10572252.2017.1339531.

Hewett, B. L. & Powers, C. E. (2007). Online teaching and learning: Preparation, development, and organizational communication. *Technical Communication Quarterly, 16*(1), 1–11. https://doi.org/10.1207/s15427625tcq1601_1.

Hewett, B. L. & Robidoux, C. (Eds.). (2010). *Virtual collaborative writing in the workplace: Computer-mediated communication technologies and processes* (1st ed.). IGI Global. https://doi.org/10.4018/978-1-60566-994-6.

Hosier, A. (2013). When teachers are taught to learn: Using team-based learning in an online, asynchronous information literacy course. *Journal of Library Innovation, 4*(2), 111–121.

Huang, C.-K. & Lin, C.-Y. (2017). Flipping business education: Transformative use of team-based learning in human resource management classrooms. *Educational Technology & Society, 20*(1), 323–336. https://eric.ed.gov/?id=EJ1125969.

Jaggars, S. S. (2014). Choosing between online and face-to-face courses: Community college student voices. *American Journal of Distance Education, 28*(1), 27–38. https://doi.org/10.1080/08923647.2014.867697.

Jaggars, S. S. & Bailey, T. (2010). *Effectiveness of fully online courses for college students: Response to a department of education meta-analysis.* Community College Research Center. https://eric.ed.gov/?id=ED512274.

Johnson-Eilola, J. (1996). Relocating the value of work: Technical communication in a post-industrial age. *Technical Communication Quarterly, 5*(3), 245–270. https://doi.org/10.1207/s15427625tcq0503_1.

Jones, D. & Jenkins, P. (2013). Expanding the scaffolding of the online undergraduate technical communication course. In K. Cargile Cook & K. Grant-Davie (Eds.), *Online education 2.0: Evolving, adapting, and reinventing online technical communication* (pp. 237–256). Baywood Publishing Company, Inc. https://doi.org/10.4324/9781315231471.

Kariya, S. (2003). Online education expands and evolves. *IEEE Spectrum, 40*(5), 49–51. https://doi.org/10.1109/MSPEC.2003.1200179.

King, A. (1993). From sage on the stage to guide on the side. *College Teaching, 41*(1), 30–35. https://doi.org/10.1080/87567555.1993.9926781.

Kittle, P. & Hicks, T. (2009). Transforming the group paper with collaborative online writing. *Pedagogy, 9*(3), 525–538. https://doi.org/10.1215/15314200-2009-012.

Knight, A. B. (2002). Team-based learning: A strategy for transforming the quality of teaching and learning. In L. Michaelsen, A. B. Knight & L. D. Fink (Eds.), *Team-based learning: A transformative use of small groups in college teaching* (pp. 201–211). Stylus Publishing.

KOB Small Grant Program. (n.d.). Keep Oakland Beautiful. Retrieved April 18, 2019, from https://www.keepoaklandbeautiful.org/kob-small-grant-program.html.

Kroll, B. M. (1984). Writing for readers: Three perspectives on audience. *College Composition and Communication, 35*(2), 172–185. https://doi.org/10.2307/358094.

Lane, D. (2012). Peer feedback processes and individual accountability in team-based learning. In M. Sweet & L. Michaelsen (Eds.), *Team-based learning in the social sciences and humanities: Group work that works to generate critical thinking and engagement* (pp. 51–62). Stylus Publishing.

Lauer, C. & Brumberger, E. (2019). Redefining writing for the responsive workplace. *College Composition and Communication, 70*(4), 634–663.

LeFevre, K. B. (1987). *Invention as a social act.* Southern Illinois University Press.

Lunsford, A. A. & Ede, L. (2011). *Writing together: Collaboration in theory and practice.* Bedford; St. Martin's.

Mahoney, S. (2009). Mindset change: Influences on student buy-in to online classes. *Quarterly Review of Distance Education, 10*(1), 75–83.

Maid, B. & D'Angelo, B. J. (2013). What do you do when the ground beneath your feet shifts? In K. Cargile Cook & K. Grant-Davie (Eds.), *Online education 2.0: Evolving, adapting, and reinventing online technical communication* (pp. 11–24). Baywood Publishing. https://doi.org/10.4324/9781315231471.

Mauri, T. & Onrubia, J. (2015). Online collaborative writing as a learning tool in higher education. In M. Deane & T. Guasch (Eds.), *Learning and teaching writing online: Strategies for success* (pp. 94–110). Brill. https://doi.org/10.1163/9789004290846_007.

McComiskey, B. (2000). *Teaching composition as a social process.* Utah State University Press.

Michaelsen, L. (2002). Getting started with team-based learning. In L. Michaelsen, A. B. Knight & L. D. Fink (Eds.), *Team-based learning: A transformative use of small groups* (pp. 213–232). Praeger.

Michaelsen, L. & Knight, A. B. (2002). Creating effective assignments: A key component of team-based learning. In L. Michaelsen, A. B. Knight & L. D. Fink (Eds.), *Team-based learning: A transformative use of small groups* (pp. 53–76). Praeger.

Michaelsen, L., Knight, A. B. & Fink, L. D. (Eds.). (2002). *Team-based learning: A transformative use of small groups*. Praeger.

Mick, C. S., Middlebrook, G., Hewett, B. L. & DePew, K. E. (2015). Asynchronous and synchronous modalities. In B. L. Hewett & K. E. DePew (Eds.), *Foundational practices of online writing instruction* (pp. 129–148). The WAC Clearinghouse; Parlor Press. https://doi.org/10.37514/PER-B.2015.0650.2.03.

Nakaji, M. C. (2002). A dramatic turnaround in a classroom of deaf students. In L. Michaelsen, A. B. Knight & L. D. Fink (Eds.), *Team-based learning: A transformative use of small groups in college teaching* (pp. 129–136). Stylus Publishing.

Nielsen, D. (2016). Facilitating service learning in the online technical communication classroom. *Journal of Technical Writing and Communication*, 46(2), 236–256. https://doi.org/10.1177/0047281616633600.

Norman, D. (2013). *The design of everyday things: Revised and expanded edition*. Basic Books.

Palsolé, S. & Awalt, C. (2008). Team-based learning in asynchronous online settings. *New Directions for Teaching and Learning*, 2008(116), 87–95. https://doi.org/10.1002/tl.336.

Pope-Ruark, R. (2017). *Agile faculty: Practical strategies for managing research, service, and teaching*. University of Chicago Press. https://doi.org/10.7208/chicago/9780226463292.001.0001

Ratta, C. B. D. (2015). Flipping the classroom with team-based learning in undergraduate nursing education. *Nurse Educator*, 40(2), 71. https://doi.org/10.1097/NNE.0000000000000112.

Rentz, K., Arduser, L., Melonçon, L. & Debs, M. B. (2009). Designing a successful group-report experience. *Business Communication Quarterly*, 72(1), 79–84. https://doi.org/10.1177/1080569908330373.

Restad, P. (2012). American history, learned, argued, and agreed upon: Team-based learning in a large lecture class. In M. Sweet & L. Michaelsen (Eds.), *Team-based learning in the social sciences and humanities: Group work that works to generate critical thinking and engagement* (pp. 159–180). Stylus Publishing.

Roberson, B. & Reimers, C. (2012). Team-based learning for critical reading and thinking in literature and great books courses. In M. Sweet & L. Michaelsen (Eds.), *Team-based learning in the social sciences and humanities: Group work that works to generate critical thinking and engagement* (pp. 129–142). Stylus Publishing.

Robles, V. D. & Baker, M. J. (2019). Using case-method pedagogy to facilitate audience awareness. *IEEE Transactions on Professional Communication*, 62(2), 1–16. https://doi.org/10.1109/TPC.2019.2893464.

Rodrigo, R. & Ramírez, C. D. (2017). Balancing institutional demands with effective practice: A lesson in curricular and professional development. *Technical Communication Quarterly*, 26(3), 314–328. https://doi.org/10.1080/10572252.2017.1339529.

Sharma, A., Janke, K. K., Larson, A. & Peter, W. S. (2017). Understanding the early effects of team-based learning on student accountability and engagement using a three session TBL pilot. *Currents in Pharmacy Teaching & Learning*, 9(5), 802–807. https://doi.org/10.1016/j.cptl.2017.05.024.

Sibley, J. (2012). Facilitating application activities. In M. Sweet & L. Michaelsen (Eds.), *Team-based learning in the social sciences and humanities: Group work that works to generate critical thinking and engagement* (pp. 33–50). Stylus Publishing.

Smart, K. L. & Cappel, J. J. (2006). Students' perceptions of online learning: A comparative study. *Journal of Information Technology Education: Research, 5*, 201–219. https://doi.org/10.28945/243.

Soria, K. M. & Weiner, B. (2013). A "virtual fieldtrip": Service learning in distance education technical writing courses. *Journal of Technical Writing and Communication, 43*(2), 181–200. https://doi.org/10.2190/tw.43.2.e.

St.Amant, K. (2017). Of friction points and infrastructures: Rethinking the dynamics of offering online education in technical communication in global contexts. *Technical Communication Quarterly, 26*(3), 223–241. https://doi.org/10.1080/10572252.2017.1339522.

Stepanova, J. (2018). Team-based learning in management. In L. Daniela (Ed.), *Innovations, technologies and research in education* (pp. 78–90). Cambridge Scholars Publishing.

Stewart, M. (2018). Community building and collaborative learning in OWI: A case study of Principle 11. *Research in Online Literacy Education, 1*(1). http://roleolor.weebly.com/stewart-community-building-and-collaborative-learning.html.

Stratton, C. R. (2015). Collaborative writing in the workplace. In D. F. Beer (Ed.), *Writing and speaking in the technology professions* (pp. 260–264). Wiley-Blackwell. https://doi.org/10.1002/9781119134633.ch41.

Strobl, C. (2015). Learning to think and write together: Collaborative synthesis writing supported by a script and a video-based model. In M. Deane & T. Guasch (Eds.), *Learning and teaching writing online: Strategies for success* (Vol. 29, pp. 67–93). BRILL. http://dx.doi.org/10.1163/9789004290846_006.

Swales, J. (2017). The concept of discourse community. *Composition Forum, 37*. http://compositionforum.com/issue/37/swales-retrospective.php.

Swan, K. (2001). Virtual interaction: Design factors affecting student satisfaction and perceived learning in asynchronous online courses. *Distance Education, 22*(2), 306–331. https://doi.org/10.1080/0158791010220208.

Sweet, M. & Michaelsen, L. (2012). Critical thinking and engagement: Creating cognitive apprenticeships with team-based learning. In M. Sweet & L. Michaelsen (Eds.), *Team-based learning in the social sciences and humanities: Group work that works to generate critical thinking and engagement* (pp. 143–158). Stylus Publishing.

Thondhlana, J. & Smith, A. F. V. (2013). Cracking the case: A task-based investigation of a group case-study project at a business school. *Journal of Business and Technical Communication, 27*(1), 32–61. https://doi.org/10.1177/1050651912458922.

Thrush, E. A. & Popham, S. L. (2013). Teaching technical communication to a global online student audience. In K. Cargile Cook & K. Grant-Davie (Eds.), *Online education 2.0: Evolving, adapting, and reinventing online technical communication* (pp. 113–132). Baywood Publishing Company, Inc. https://doi.org/10.4324/9781315231471.

Tillery, D. & Nagelhout, E. (2013). Theoretically grounded, practically enacted, and well behind the cutting edge: Writing course development within the constraints of a campus-wide course management system. In K. Cargile Cook & K. Grant-Davie (Eds.), *Online education 2.0: Evolving, adapting, and reinventing online technical communication* (pp. 25–44). Baywood Publishing Company. https://doi.org/10.4324/9781315231471.

Vie, S. (2017). Training online technical communication educators to teach with social media: Best practices and professional recommendations. *Technical Communication Quarterly, 26*(3), 344–359. https://doi.org/10.1080/10572252.2017.1339487.

Warnock, S. (2009). *Teaching writing online: How and why*. National Council of Teachers of English.

Watts, J. (2019). Assessing an online student orientation: Impacts on retention, satisfaction, and student learning. *Technical Communication Quarterly, 28*(3), 254–270. https://doi.org/10.1080/10572252.2019.1607905.

Wolfe, J. (2010). *Team writing: A guide to working in groups*. Bedford; St. Martin's.

12. Preparing Future Professionals in and for a Global Context: A Case for Telecollaborative Educational Initiatives

Elisabet Arnó-Macià
UNIVERSITAT POLITÈCNICA DE CATALUNYA

Tatjana Schell
INDEPENDENT SCHOLAR

Abstract: This chapter evaluates the educational practices of the Trans-Atlantic and Pacific Project (TAPP), a multinational collaborative network that has connected hundreds of students and their lecturers in the United States, Europe, Asia, and Africa. While these partnerships connect a variety of classes in writing, translation, and English for Specific Purposes for different disciplines through telecollaboration, their main purpose is for college students to hone their professional, technical, and intercultural communication skills in an increasingly globalized professional world. By asking current and former students, educators, and administrators about their experiences with the TAPP, the authors have been able to evaluate how such exchanges provide participants with opportunities to strengthen these skills. The authors argue that including telecollaboration into the technical and professional communication (TPC) education can help educators design more internationalization-focused college curricula and support students in strengthening those skills that extend beyond oral and written communication (project management, intercultural awareness, teamwork, etc.). Furthermore, the results of their analysis point to an overall positive impact of telecollaboration on TPC pedagogy, although greater effort should be made to raise students' awareness of the value of telecollaboration for prospective international employability in the increasingly globalized professional world.

Keywords: education, technical communication, professional communication, telecollaboration, English as a lingua franca (ELF)

Key Takeaways:

- Telecollaboration can be an effective means of teaching technical and professional communication (TPC) within international contexts.
- Including telecollaboration into the TPC curriculum can help instructors design more internationalization-focused assignments and support stu-

dents in strengthening those skills that extend beyond oral and written communication.
- Greater effort should be made to raise students' awareness of the value of telecollaboration for prospective international employability.

Recent college composition scholarship has explored the factors influencing the ways professional writers work. The most obvious one has been the technological development in communication. In their recent article "Redefining Writing for the Responsive Workplace," Claire Lauer and Eva Brumberger (2019) explore various real-life professional writing environments and conclude that technology has redefined the role of professional writers. "Rather than the originator of content, the writer is becoming a sort of multimodal editor who revises, redesigns, remediates, and upcycles content into new forms, for new audiences, purposes, and media," they argue (Lauer & Brumberger, 2019, p. 634). As the technological tools used by professional communicators influence their tasks as writers, so does the cultural and linguistic context in which they perform this work.

Furthermore, because the work of technical and professional communicators often involves speakers of different languages, their ability to communicate effectively becomes even more crucial. With regard to communication in multilingual environments, recent scholarship in our field has focused on the theory and practice of translingualism (Canagarajah, 2009, 2013a, 2013b; Horner et al., 2011). In her in-depth overview on this subject, Joleen Hanson (2015) defines translingual practice as "the strategies, languages, signs, and genres that people can use to communicate effectively in global contact zones" (p. 89). In order to help professionals excel in communicating effectively in their post-graduate workplaces, their learning about such "strategies" should begin while in college. The goal here is to help students hone their skills in effective technical and professional communication (TPC) to various audiences in diverse discourse communities and to manage the complexity of such work in order to adequately prepare them for the demands of their future career (Arnó-Macià et al., 2014; Brewer & St.Amant, 2015; Hanson, 2015).

One way to incorporate this outcome into the teaching process is to encourage intercultural, or international, collaboration within the context of higher education in order to enrich students' knowledge of intercultural communication and work practices in the increasingly globalized professional world. Telecollaboration (Guth & Helm, 2010), or online intercultural exchange (O'Dowd & Lewis, 2016), also recently named virtual exchange (O'Dowd, 2018), has received considerable attention in the past years, as universities strive for greater internationalization (Verzella, 2018). In this chapter, we introduce and evaluate the impact of a specific telecollaborative project that has connected students and instructors at multiple universities in the US and abroad.

This study focuses on the Trans-Atlantic and Pacific Project (TAPP), a collaborative network that has connected hundreds of students and instructors at various institutions of higher education in the United States, Europe, Asia, and Africa, working together to hone students' professional, technical, and intercultural communication skills (Arnó-Macià et al., 2014; Maylath et al., 2008; Sorensen et al., 2015). Through telecollaboration, the TAPP features multiple grassroots partnerships, pairing classes in writing, translation, and English for Specific Purposes for different disciplines. Although such partnerships provide instructors with the flexibility to include any assignments that fit their own course goals, the TAPP usually involves bilateral partnerships (writing-translation, writing/translation-editing, authoring, and peer review) in the technical, scientific, and more recently, the humanities fields (e.g., Humbley et al., 2005; Tzoannopoulou & Maylath, 2018).

In multilateral partnerships, virtual teams of up to six students work on complex projects that involve writing, translating, and usability testing of technical documents (Maylath et al., 2013a). Through such projects, students experience the realities of professional collaboration in international working environments. Yet however realistic these contexts are, they remain a safe learning space for students; "because they are not actual workplaces, the stakes are much lower, allowing [students] to make mistakes and learn from them without incurring losses" (Mousten et al., 2018, p. xx).

As an integral part of regular college-level courses, TAPP partnerships can facilitate the transition between higher education and the workplace. Keeping this idea in mind, this chapter aims to explore the ways telecollaboration can contribute to the development of international professional communication skills through the following research questions:

1. What is the role of a telecollaboration initiative integrated in language, communication, and translation courses in the development of international professional communication?
2. What skills are perceived to be necessary for graduates to participate effectively in international professional communication?
3. How do participants evaluate their own telecollaboration experience and what adjustments do they suggest to better prepare graduates for effective communication in the globalized workplace?

In order to answer these questions, a small-scale exploratory study was conducted to gather stakeholders' perspectives on how telecollaborative initiatives, such as the TAPP, can contribute to preparing students for future careers as professional communicators within the context of a globalized job market.

■ Data Collection and Analysis

Data were collected in the spring term of 2019 through open-ended question surveys about stakeholders' experiences with the TAPP. Taking a qualitative perspective (Croker, 2009; Denzin & Lincoln, 1994), we focused on participants' accounts of their experience and their views on the development of professional communication competencies. An open-response item questionnaire was chosen (Brown, 2009), which consisted mainly of broad questions encouraging lengthy free writing, although a few closed questions were also included to measure participants' evaluation of the project (on a five-point scale) and for comparison across the different categories of participants.

To evaluate the effectiveness of the TAPP from a variety of perspectives, we selected four types of respondents on three different levels: instructional (college instructors and students involved in the TAPP network), workplace (former TAPP students, who have since graduated), and institutional (program and college administrators).

All participants were asked similar questions, although some were adapted to their experiences with or knowledge of the TAPP. Accordingly, the survey included the following parts, following a brief section asking for basic demographic information:

- account and evaluation of own TAPP experience,
- relating the TAPP (and telecollaboration initiatives in general) to effective preparation for future career, and
- general views on (teaching) intercultural TPC.

The students' survey focused especially on their personal experiences as TAPP participants so as to incorporate students' views into the design of future telecollaborative projects. With instructors, survey questions focused on their motivation to join the TAPP network and their opinion about the role of telecollaboration in TPC pedagogy. On the other hand, the graduates' and administrators' surveys focused on the visibility and recognition of telecollaboration programs and on their role in developing professional communication skills, both from the perspective of former students, now graduates in the labor market, and from an institutional perspective in the case of university administrators. In the latter case, respondents were asked whether they were familiar with the TAPP, and about the support it receives at their institution. It was an adaptive survey so that if they were not familiar with the TAPP, they were asked about their general views on telecollaboration and its role in helping students to learn more about TPC within international contexts.

The surveys were sent to prospective participants as Google forms (former graduates were contacted mostly via LinkedIn and Facebook). A total of 44 subjects volunteered to complete the survey; most of them were instructors and students participating in TAPP partnerships (see Table 12.1). There were fewer graduates and administrators, and the latter came from the universities that the authors of this study are or have been affiliated with.

For the closed questions, means and standard deviations (SD) were calculated, while the open-ended questions were analyzed qualitatively, from an exploratory-interpretive perspective (Hobbs et al., 2010), i.e., without predefined assumptions to probe into participants' perceptions and interpret them in the context of each participant's profile. Categories were derived from the themes that emerged in the analysis of the open-ended questions, which were in turn refined through an inductive-deductive process.

Table 12.1. Summary of respondents

Students	14
Instructors	15
Graduates	11
Administrators	4

Findings: Stakeholders' Perspectives on the TAPP

Students

A total of 14 students responded to the survey. Eight were M.A. students of English from Konin, Poland and six were mechanical and computer engineering students in the last year of their bachelor's degree at the Polytechnic University of Catalonia (UPC) in Barcelona, Spain. Both cohorts were involved in bilateral projects with U.S. students. Polish students were engaged in writing instructions, exchanging them with their U.S. partners and giving feedback on and testing the usability of each other's texts. UPC students, on the other hand, participated in the TAPP as part of a project-focused technical communication course. As learners of English for Specific Purposes, they made creative videos on their technical projects, which were reviewed by their U.S. partners. In addition to this collaboration, some of the students had previously participated in the TAPP while in technical English courses focused on writing or speaking, respectively.

Overall, students are highly satisfied with the TAPP, with a rating of 3.93/5 (SD = 0.47), because they improved their written and spoken communication skills, intercultural communication, and had contact with native speakers of English: "TAPP project is an opportunity to learn about the other cultures and also improve your English skills."

On the other hand, students pointed out a few shortcomings connected to project management for assignments, such as unclear instructions and lack of commitment on the part of their partners. One respondent specified, "The idea is fine, but instructions are sometimes obscure and my partners don't stick to the deadline." Another participant highlighted the complexity of project organization (especially time constraints) as one of its main challenges: "I found the project really useful. However, it is difficult to find time for TAPP cooperation and answer emails from my partner regularly because of numerous duties."

When asked specifically about the strong and weak points of their telecollaboration experience, the respondents highlighted a number of points (see Table

12.2). Most of the strong points are similar to those previously mentioned, such as interaction and collaboration with foreign partners who were native speakers of English and the creativity of assignments and organization skills, while the weak points include time zone differences and different levels of commitment on the part of partners.

These perspectives across different respondents sometimes appeared to be slightly contradictory. For example, some students mentioned challenges related to the demands of collaboration (organization, commitment, and time management) and to the short duration and low intensity of the exchange (in collaborations designed to be manageable for students with busy academic lives). One respondent said, "The strong point is the easy communication that we have nowadays, the weak point is the short time project that does not help to obtain a relationship with the TAPP partner." Such contradictory expectations are difficult to meet unless collaborations are designed to be so flexible that they allow for different levels of engagement.

Table 12.2. Strong and weak points of telecollaboration according to students

Strong Points	Weak Points
International, intercultural contact	Different levels of commitment in partners
Improve language skills	Different time zones
Creative, non-routine tasks	Short duration/little intensity

The main points in Table 12.2 are captured in the following response by one of the Polish students who said,

> Strong: getting to know new people, learning about other cultures, improving organization skills. Weak: different time zones can be a problem; The success of cooperation depends on the attitude of people who are taking part in it, whether they are willing to cooperate or not.

The strong points outweigh the weak points, as students show positive perceptions of their telecollaborative projects, seen as a way of developing their competencies and enriching their learning experience, a finding aligned with previous research on telecollaboration (Ferreira et al., 2018; Kohn & Hoffstaedter, 2017) where students show high levels of engagement, building rapport at a distance (Vinagre & Esteban, 2018). Other responses referred to the specifics of the different projects rather than telecollaboration in general. Some engineering students discussed their experience in a collaboration project—a video presenting a technical project—in terms of the challenges that it poses for students who are language learners with different levels of proficiency, or the fact that a communication project is "not real" if compared to the projects developed in other engineering courses which involve the creation and manipulation of tangible objects:

What I didn't like was the short time and that the project just consisted of an idea (nothing real). It would be better to work with other engineering courses to have more technical implications and make something real to share with native speakers.

The same student further elaborates on this idea in response to the question about recommendations for future TAPP projects, suggesting that telecollaboration should be incorporated into "more real" engineering courses (i.e., projects involving tangible objects). This student seems to recognize the value of telecollaboration, although he probably does not fully acknowledge that of technical communication courses: "I would recommend that in the future, projects have to be more than a simple idea, and take it more seriously. I suggest working with other courses like 'project management' in order to do a more serious project."

Another question asked students about their perceived development of professional and technical communication skills. Engineering students mainly reported on language and (technical) vocabulary and increased language awareness (". . . awareness of having to adapt my language to get understood. I noticed what mistakes I made . . . [and saw] how native speakers use their language"). These reflections point to language learning gains in English as a lingua franca as well as to heightened awareness (Arnó-Macià et al., 2019; Helm & Acconcia, 2019).

Probably reflecting on their own specific telecollaboration project, focusing on speaking skills, students mentioned professional speaking skills—and to a lesser extent, writing—as well as "more confidence to speak in English in different situations," as the main learning outcome. The nature of the project (writing-translation partnership) is also reflected in the Polish students' answers, who mentioned writing and translation skills, together with paraphrasing and editing ("paraphrasing the meaning in order to be understandable to my TAPP partners," "how to prepare and translate instruction[s] . . . [and give] feedback on somebody's work").

The fact that some of the engineering students surveyed had participated in several previous TAPP partnerships allowed them to reflect on their overall experience with telecollaboration. For example, one student had participated in two simultaneous TAPP partnerships in the previous term—one on technical writing and the other on technical speaking—and reflected on his development of specific professional communication skills, namely a written genre (instructions manual) and an oral one (the job interview). Another engineering student valued "shar[ing] [their] project with foreign people, from different disciplines and realities" as a positive experience. The best summary of what telecollaboration means for the international employability of graduates is captured in the following answer, pointing out the similarities between the TAPP and the workplace: "You don't get to have a situation where you have to communicate professionally like this one before going into the job market."

These reflections on international employability are made in a context in which most of the students (11 out of 14) consider the possibility of working abroad or for an international company. With this in mind, students were asked for suggestions for future TAPP projects. Apart from the general suggestion of integrating telecollaboration in other engineering courses, specific suggestions were also made regarding the implementation of telecollaboration (see Table 12.3).

Table 12.3. Specific suggestions made by students

(i) clear organization and planning	"To be disciplined, organized, make all the tasks on time, being open-minded for an idea of the other person, More specific information. Also, I suppose that deadlines for performing tasks should be longer."
(ii) communication and participation (more ICT tools used; promoting sustained contact with partners)	"I think it would be great to try to explore different ways of communication, not only by emails. I recommend to keep in contact with his TAPP partners, you can learn more out of the class. Increasing the intensity of contact through, for example, video calls. To involve only the students that are really interested in cooperation."
(iii) diverse task design	"Perhaps less complex tasks would be a good option. Students would find it easier to cooperate then. They should find a different task."

Overall, students suggested very specific guidelines for implementation, involving detailed and clear instructions and deadlines, the use of a wider range of information and communication technology (ICT) tools, better communication between partners, and more clearly designed tasks.

Instructors

Half of the respondents came from the US (and mostly from North Dakota State University (NDSU), a hub of the TAPP), while the other half came from a variety of European universities, thus reflecting typical bilateral partnerships within the TAPP. Specifically, they came from three translation classes in Europe and from TPC classes—in Europe, often termed English for Specific Purposes—from European and U.S. universities. Instructors reported on a wide range of activities developed through TAPP partnerships: (co-)writing different types of texts, translation, editing, storytelling and user experience, and spoken professional communication. Most of the respondents were experienced TAPP lecturers (ranging from 4 to 17 or 18 years of experience), and they had joined the network on their own initiative, although some explained that they had either been invited to participate or that the TAPP was part of the course procedures at their institution.

Instructors discussed their openness towards telecollaboration arising from "the desire to help . . . students understand the skills necessary to function in a globalized work environment." Thus, telecollaboration is regarded as "a means for cultural exchange and develop[ing] the 'additional' proficiencies the networked collaborative medium affords." One of the instructors specifically stressed the potential of telecollaboration in facilitating students' transition into a professional environment: "The first motivation that comes to mind is 'authentic, grounded projects'; that is, I wanted students to participate in work that reflected the transition of an upper-division writing course (student to professional)."

When asked about the skills students can develop as a result of participating in the TAPP, instructors mentioned cooperation, communication, and language. They also mentioned professional competencies, including a broader professional stance and awareness, highlighted in the statement below as "ownership of expertise": "Attention to cultural differences, editing, and professional correspondence. Ownership of their expertise."

One of the U.S. instructors discussed a variety of interrelated competencies she thinks students can develop through telecollaboration:

> Communicating with students who are several time zones away requires students to be mindful of how those time differences affect the transfer of information, which influences project management decisions. When working with students from different culture backgrounds than their own, my students learned to be mindful of how someone else might interpret their words and actions in a way that they did not expect. Students also learned to be mindful of variations between American English and Global Englishes.

The array of professional competencies promoted through the TAPP, including project management, interpersonal skills, and awareness of different varieties of English, also encompasses broader (and probably less well-defined) professional skills, such as "awareness of professional environment diversity." Respondents pointed out that the TAPP becomes a unique global scenario to put such a broad range of competencies into practice: "By participating in the TAPP, students have the opportunity to develop skills in verbal and written intercultural communication, which they likely would not have otherwise."

The value of telecollaboration in teaching technical and professional communication was rated high by all instructors. For instance, one of the U.S. respondents discussed the notion of authenticity by pointing out that telecollaboration projects reach beyond course assignments, as students engage in sustained interaction and collaboration with remote partners. Thus, as telecollaboration projects mirror the authenticity of real-world professional practices, they provide a valuable learning opportunity even when the outcomes are not as satisfactory as expected, which indicates that their learning value lies in the process.

> The collaboration helps to increase the gravitas of projects without ephemeral partnerships (some service learning projects, some

> problem-based pedagogical approaches). Students need practice interacting beyond the classroom without the artificiality of end-of-project presentations to a panel of local business people.... The projects can be an intense interaction between two (or more) student groups, each from a different program, working to complete deliverables. Even when the projects "fall to pieces," it is an instructable moment because "real world" projects often also "fall to pieces"... so learning how to cope with those situations has tremendous value (possibly more than when everything goes "right").

Furthermore, instructors emphasized the role of telecollaboration in preparing students for the globalized workplace. As one said, "Students gain experience working in collaborative situations that closely follow the kinds of international collaborative projects that happen in real-life with many multinational companies."

In line with these answers, the rating given to telecollaboration as a way to enhance global employability is very high, 4.53/5 (SD = 0.64). When asked to explain their rating, a few instructors reported on anecdotal evidence of graduates coming back after having participated in the TAPP and highlighting how it had helped them towards employability. One of these statements is expressed by a U.S. lecturer (who happened to give the lowest rating on this question) as she herself expressed her reluctance to make claims that connect the TAPP with employability, while she pointed out (at the end of the quote below) that this is an aspect that merits further evaluation:

> I do not have available data. I have one anecdote of a student using a TAPP project as part of an interview, and subsequently earning the position. In terms of a rate, one of dozens is not great. However, I do not know if other students mentioned the projects, or if the projects allowed students to set themselves apart from a "crowded" applicant pool. [My] reluctance to be more assertive about this question . . . reflects that I am not comfortable making claims about employability in absence of effort(s) to trace that information. As previous responses make clear, I see value in TAPP. I wish the value was better evaluated, though.

Another point made by this instructor is that boosting employability will depend on how graduates "market" their international experience during a job interview. She said, "The job market is increasingly about how workers can present/market themselves, so being able to draw on interesting experiences like the TAPP can help."

However, most of the instructors made stronger connections with employability based on the development of such skills as international professional communication and collaboration through these projects. Yet other respondents pointed out that greater employability may be affected by several factors, such as whether "it is clearly aligned with related course objectives that are developed before and after the project," "depend[ing] on the individual student's level of engagement," or depending on how successful the project outcome was. This last answer appears to contradict the previous assertion that even less successful projects lead to interesting learning opportunities. Regarding the quality of project outcomes, yet another instructor pointed to the need to show project deliverables, like a professionally produced technical text, as evidence of having acquired the skills needed in the globalized workplace: "Being able to show a potential employer a professional piece of technical writing produced in collaboration with an international partner can provide evidence of skills that not every university student will have."

Finally, telecollaboration is seen as contributing to internationalizing U.S. students' education, opening up new perspectives on internationalization: "Students in the United States, in particular, might not otherwise have the opportunity to work with people outside of the country. . . . This will make them more competitive in the marketplace."

Overall, the instructors interviewed, with a strong motivation towards the TAPP as they took the initiative to join the network, discussed a number of advantages that telecollaboration brings to their teaching (see Table 12.4).

Table 12.4. Advantages of telecollaboration according to instructors

| Development of a variety of skills for the workplace |
| Realism of projects |
| Boosting employability (a connection that merits further research) |
| Acquisition of professional skills |
| Campus internationalization |

Graduates

We were interested in the opinions of former students who had experienced TAPP projects in order to find out whether, in hindsight, they perceive the TAPP to contribute to potentially increased employability. The goal of gathering graduates' views about employability aligned with the responses by instructors in relation to employability, especially the notion that tracing this experience merits further investigation. After contacting former students through social networks and personal contact, we received 11 responses (six from former engineering students at UPC and five from former M.A. and Ph.D. students of English at NDSU). They all had graduated in the past five years and work as engineers or university instructors of English or are pursuing further studies. The engineers had participated in bilateral projects involving the development and review of oral presentations, while the English graduates from the US had participated in both multilateral and bilateral projects. Overall, they expressed a very high level of satisfaction with their participation in the project, with a rating of 4.55 ($SD = 0.52$).

Accordingly, they expressed positive views on the experience, pointing out the advantages of collaboration through technology, the similarity of the learning scenario with the real workplace, and the combination of skills that are practiced in such projects, including project management, communication, collaboration, and, for U.S. students, internationalization. One of the engineers mentioned technical communication in English, from his dual perspective as a former engineering student and a learner of English as a foreign language: "I improve[d] my technical English skills in the technical sector, I had a multidisciplinary experience with another student of different specialties."

We were specifically interested in whether the students had included their TAPP experience on their résumés. All U.S. respondents did, and two of them reported on the interest the project had aroused among search committees (one of them specifically mentioned technical writing). Neither of the Spanish students had included the TAPP experience on their résumés, although one of them referred to an indirect mention (i.e., his technical communication courses featuring the TAPP), and two of them expressed their intention to do it in further résumé updates. This difference in awareness of the potential of the project can probably be explained by the field of study (English) of U.S. respondents, while engineering students (learning English as a foreign language in an optional course) may not be fully aware of the potential of such projects, as they do not come with any additional recognition. With respect to the role of the TAPP in the development of their technical communication skills for the job market, the rating is also quite high (4.18, $SD = 0.75$).

However, in this case there was a difference between U.S. and Spanish graduates, as the former's ratings ranged between 4–5 and the latter's ranged between 3–4. Although the low number of respondents should be taken with caution, what is revealing are the qualitative comments. The two students who had given ratings of 3 justified them in terms of having a low level of proficiency in English at the time of the collaboration and of the short duration of the experience ("Was just one subject! So it was a short time but useful"). In terms of suggested improvements, responses ranged from no suggestions ("I think the project is well organized and I learned a lot about communications skills, so I can't think about improvements") to giving it more visibility, namely by including explicit reflection on skills as part of instruction ("I think the instructions could help us reflect on the skills we've practiced so that we can include them on resumes/CVs"). Other specific suggestions included more focus on professional projects and more creativity in the types of tasks performed. One of the engineering students even suggested matching European students with U.S. companies: "It could be even more interesting if the exchange and telecollaboration project was more focused on collaborating with a real company in the USA. This way, graduates would feel more prepared and value the real needs and tasks as engineers."

The last questions in the survey inquired into professional communication skills that could increase graduates' international employability. Respondents assigned a very high score to the importance of professional communication in the global labor market (4.73, SD = 0.65) and mentioned specific skills that graduates need, like learning about the organization of engineering companies and communication skills for persuasion and transmitting trust: "I think they have to learn how a company works. In engineering it is important to demonstrate that you are able to manage different kinds of situations and scenarios."

Some of the responses were similar to those of students and instructors, especially those that referred to professional communication skills such as technology-mediated intercultural communication and collaboration as well as concise technical writing. Finally, from the respective viewpoints of English native speakers and Spanish learners of English, respondents focused on the development of language skills and greater awareness, which they practiced in their communication between native and non-native speakers, the latter having to accommodate interlocutors with varied levels of proficiency in English. However, what we found especially valuable in the graduates' responses was their connection with the workplace and the views they could provide from the perspective of former students currently in the professional world. Table 12.5 summarizes graduates' views on the advantages of telecollaboration and suggestions for improvement.

Table 12.5. Advantages of telecollaboration and suggestions for improvement according to graduates

Advantages	Suggestions
Collaboration through technology	Connection with companies
Similarity to workplace scenario	Greater visibility/connection with employability (through résumés)
Development of language and communication skills	Greater variety of projects and tasks

Administrators

One of the focuses of this study was the extent to which telecollaboration is recognized by universities, which is why we gathered the perspectives of university administrators. It should be noted that these exchanges are grassroots partnerships initiated by lecturers from different universities. A small number of administrators were approached from NDSU (the hub university) and UPC. A total of four respondents answered the survey, three from UPC and one from NDSU. One of the UPC respondents held a vice-dean's position. The other two were not related to the TAPP: one held an academic management position in the department where the TAPP is implemented and the other was a specialist in quality assurance. The NDSU administrator held a dean's position. The first question asked about participants' knowledge of the TAPP (a question to which two of the UPC respondents and the NDSU respondent answered positively). When asked to describe the project, one of the descriptions was very specific and included an account of its integration in one of the technical communication courses. The same respondent was also specific about how long the university had been involved in the TAPP. Also, the three respondents considered the initiative to be "highly valuable for students as a learning experience," especially for the U.S. students who may have little international experience: "This project connects them directly with classes at a university abroad and provides a framework for working together in a safe and non-threatening manner."

According to the administrators, telecollaboration can help Spanish students of engineering, who are learners of English and novice technical writers, become aware of the complexities of technical writing (which go beyond learning the grammar of a language). Faced with "English-speaking students' difficulties in writing," students discover that writing "is not an easy task, even for [native speakers of the language]."

Two of the respondents said that the TAPP was given visibility/recognition at their university, and more specifically, all three agreed that it deserved institutional support—and even one of them specified the type of support needed, namely greater dissemination and catering for needs derived from TAPP implementation.

While specific TAPP questions were only addressed to those respondents that knew about the project, more general questions about telecollaboration in higher education were asked to all participants. A high rating was given to the question on the value of telecollaboration for teaching professional and technical communication in higher education (4.5, SD = 0.58), and the following benefits were mentioned: (i) creating a global learning network for a globalized workplace, (ii) a more sustainable alternative to physical mobility, and (iii) raising non-native speakers' awareness of technical writing in other contexts. The following quote summarizes these views: "Online collaboration is the only way to go. Physical mobility is expensive, time-intensive, conflicts with other personal and professional obligations, cannot be sustained over long stretches of time. Access to online communication has become ubiquitous."

In terms of boosting students' international employability, respondents mentioned several affordances of telecollaboration, such as experience in international teams, open-mindedness, and writing in different contexts. However, they pointed to several conditions necessary for such telecollaboration to have an impact on students: "One of them is that the contact was sustained and impactful enough to make a difference; another one is that students are made aware of the significance of this particular learning experience so that they are able to express the value of this learning activity in applications and resumes."

This latter reference to including telecollaboration on résumés echoes instructors' reflections on employability and "marketing" students' experience. Regarding the role of telecollaboration in preparing students for the labor market, administrators agree that it integrates multiple skills sought by employers (international communication, collaboration, personal reflection) in a realistic international context: "An essential role, as telecollaboration facilitates contact with people and groups that would be impossible or costly to achieve face to face."

When asked about professional communication skills that can be developed through the TAPP and more generally through technical communication (TC) courses, all administrators agreed on similar answers (self-management, intercultural awareness, adapting to different audiences and situations, problem-solving), which suggests an alignment between telecollaboration practices and objectives set for curricular professional communication courses. Again, differences appear between Spanish respondents' references to a "foreign language" (English) and U.S. respondents' references to adapting to other interlocutors, thus highlighting the role of English as a lingua franca and the benefits of native speaker/non-native speaker communication as preparation for international professional communication contexts. Overall, administrators consider telecollaboration a valuable experience in their universities for a variety of reasons (see Table 12.6). Some argue that telecollaboration is "the way to go," yet they maintain that certain conditions need to be met for this experience to have a lasting impact on students. The administrators' views of telecollaboration as deserving greater support and visibility also points to the need for balancing the flexible, grassroots nature of such partnerships with institutional policies and practices.

Table 12.6. Advantages of telecollaboration according to academic administrators

Safe environment for collaboration
Greater awareness and authenticity of writing
Sustainable internationalization experience for students
Potential to develop a variety of personal, professional, and communication skills

Discussion and Application

In response to the first research question, different stakeholders discussed telecollaboration as a highly valuable tool that helps students strengthen their written and oral communication skills. For students who are non-native speakers of English, it is a way of improving English language proficiency, in addition to practicing intercultural communication, and developing creativity and project organization skills.

Instructors emphasized the potential of telecollaboration in helping students to develop a range of professional competencies. They viewed telecollaboration as a unique opportunity to put students in realistic professional scenarios and to practice a variety of skills related to project management, multidisciplinary work, and communication. Thus, beyond the competencies related to international communication, like interculturality, politeness, and global Englishes, instructors highlighted professional attitudes—less tangible than competencies—such as "awareness of professional environment diversity" and "ownership of expertise." "Authenticity" was also a key word in instructors' responses related to the projects and documents that constitute the output of such collaborations. Authenticity, linked to a learner's agency in the sense of participation and initiative, has also been pointed to as one of the affordances of telecollaboration (Kohn & Hoffstaedter, 2017).

Just like the instructors, the graduate respondents also value the potential of telecollaboration in replicating realistic professional scenarios and in offering useful experiences in project management, collaboration, internationalization, and technical communication in English as a lingua franca. Telecollaboration is equally valued by administrators, who highlight the opportunities it offers students to work in international teams and to write in different contexts. However, they argue that certain conditions should be met for such opportunities to have an impact on students, including a steady contact among the project partners and also the need to raise students' awareness about the value of this experience.

The idea of "marketing" students' international telecollaborative experience as an asset for employability is mentioned by different respondents, which contrasts with the fact that the graduates we surveyed had not included this experience in their job applications. It is assumed that telecollaboration can help bridge the gap between higher education and professional environments, along the lines of the recommendations made by Elspeth Jones (2016). However, to help students make the most of their telecollaboration experience, it is necessary to include an explicit reflection on project participation (O'Dowd, 2015a), which in the TAPP is usually performed through pre- and post-learning reports, to encourage students' reflection on both expectations and learning outcomes (Mousten et al., 2012). The results of this study are in line with previous TAPP studies (e.g., Arnó-Macià et al., 2019) in that as students interact across languages and take on multiple roles in realistic work projects, they show a greater appreciation of language(s), especially English as a lingua franca. Regarding the design of activities, some valuable suggestions were made by students and graduates, mostly (i) simplifying tasks and increasing exposure to communication with foreign partners, (ii) connecting technical communication with other courses, and (iii) connecting with real companies. In terms of planning and implementation, care has to be taken to integrate online and classroom work in a meaningful way (Ferreira et al., 2018).

The second research question focused on a general reflection on the skills needed for international professional communication—as opposed to concrete experience—and was elicited in graduates' and instructors' surveys. Graduates place great value on professional communication in the global labor market, mentioning specific skills that are needed, like learning about the organization of companies and adapting to different situations, as well as specific communication skills related to persuasion and transmitting trust.

Connections can be made between the communication skills that, according to graduates, are needed in the workplace, and students' and instructors' responses about the skills developed through the TAPP. These include intercultural communication and collaboration and technical writing. As English is the language of international professional communication, different perspectives are expressed by native and non-native speakers, with the former learning to accommodate non-native audiences and the latter practicing and improving foreign language skills.

When asked about professional communication skills that can be developed through the TAPP and, more generally, through university courses, administrators give similar answers (self-management, intercultural awareness, adapting to different audiences and situations, and problem-solving). These responses suggest an alignment between telecollaboration practices, and the objectives set for curricular professional communication courses. Like graduates, administrators point to the role of English as a lingua franca in international professional communication contexts from a native vs. non-native speaker perspective (i.e., adapting to non-native audiences vs. improving proficiency in English as a foreign language).

On the last research question, participants' evaluations of their telecollaboration experience were extremely positive, with high ratings by students, instructors, and graduates. their institutional perspective, administrators referred to the potential of the TAPP/telecollaboration initiatives for curriculum (and institutional) internationalization in a sustainable way (Verzella, 2018). The affordances of telecollaboration mentioned by participants in this study, namely interacting and collaborating with foreign partners, developing creative assignments, and developing management and organization skills, were aligned with proposals made in the literature about teaching technical communication, in that it must cover a broader range of skills that are demanded in the workplace (e.g., Brumberger & Lauer, 2015). Thus, telecollaboration projects in TC allow the integration of such broad skills with more specific writing sub-skills, such as paraphrasing or editing, as well as the ability to work with specific genres (usually procedural texts in TAPP technical writing assignments). Such projects bring to the TC classroom the complexity of real workplaces (Maylath et al., 2013b), so that students are faced with the challenges—as confirmed by participants in this study—of managing time zone differences and dealing with different partners that show varied perspectives, skills, and levels of commitment to the project, as well as coping with the demands of multi-tasking and tight deadlines, let alone the linguistic challenges faced by the students with lower proficiency in English.

Overall, the TAPP appears to contribute to students' preparation for international professional communication, although students are cautious in making such a direct link with international employability, as indicated by the rating given on that question (not as high as that for their evaluation of the experience).

This may be due to the fact that either students find the activity enjoyable and different from their everyday academic activity or that the project has a limited impact, probably due to the short-term nature of the collaborations and the lack of engagement of students once they have completed the required work. Despite these limitations, students mention a number of ways in which the TAPP has helped them develop international professional communication skills: negotiation, intercultural communication, use of technology, and especially the opportunity to experience a scenario that replicates the demands of a workplace environment while they are still at university.

Instructors appear as highly motivated and dedicated (not surprisingly, considering the grassroots nature of the project), a result that aligns with previous literature that identifies a broad range of competences and attitudes that characterize instructors engaged in telecollaboration (O'Dowd, 2015b). The motivation among instructors in this study arises from the desire to bring a globalized learning environment—"a networked component"—into their courses, giving students the opportunity to engage in cultural exchange and develop skills related to the complexities of project management in international professional communication (Maylath et al., 2013b). In this regard, the term "authenticity" appears as a key word in participants' responses (authenticity of situations, projects, and especially with purposeful assignments that exceed mere course requirements).

Considering that their opinions combine both the student perspective and that of a post-graduate employee, the responses of graduates were particularly important. In their evaluation of the TAPP experience, they reflect on similar positive points as do students and instructors, namely technology-mediated collaboration, realistic projects, similarity to workplace environments, internationalization, and foreign language learning. However, these highly positive characteristics of telecollaboration are not usually included in their résumés, which indicates that graduates may not always be aware of the value this experience can have for potential employers or do not know how to incorporate this experience into their job application materials.

Furthermore, graduates consider that the TAPP helped them improve their professional communication skills for the job market, with the only hindrances being a low level of proficiency in English and the short duration of the overall project experience. Similarly, administrators who participated in this study expressed a highly positive view of telecollaboration, acknowledging that it needs more visibility and support by universities. It is not surprising that such initiatives are not usually catered to by institutional policies (Helm, 2018), as many TAPP collaborations develop through the arrangements by individual instructors. Apart from praising telecollaboration as a teaching tool, the administrators discuss its advantages as linked to the idea of internationalization at home (Crowther et al., 2000). Such initiatives are a more sustainable alternative to physical mobility, and they contribute to creating international learning networks for globalized work environments.

Respondents' recommendations can help improve future TAPP collaborations focused on TPC. Students' recommendations range from project organization to the types of tasks assigned to extending telecollaboration to subject-specific courses. Instructors' link between the TAPP and increased employability needs further research but encouraging graduates to better "market" their international experience constitutes one step forward. Certain conditions should be met for telecollaboration to fulfill its potential: alignment with broader course objectives and methodology as well as specifying professional outcomes that students can display.

In sum, the telecollaboration initiative analyzed in this chapter is based on students taking on roles that replicate workplace experiences, and it offers enormous potential for increasing the relevance of TC teaching to authentic professional needs. In spite of the limitations of this study in terms of scope and number of respondents, certain implications can be derived for maximizing the effectiveness of telecollaboration: (i) making students aware of the activity so that they can take advantage of its full potential, (ii) organizing the tasks so that they reflect potential professional settings and at the same time are presented to students in a clear and manageable way, and (iii) strengthening the links between telecollaboration and the overall TPC curriculum. Through the practice of intercultural TPC in an experiential way, telecollaboration provides an opportunity for students to go beyond strengthening writing and communication skills by helping them develop broader skill sets to become effective professional communicators for diverse audiences and contexts against the background of an increasingly globalized workplace.

References

Arnó-Macià, E., Isohella, S., Maylath, B., Schell, T., Verzella, M., Minacori, P., Mousten, B., Musacchio, M. T., Palumbo, G. & Vandepitte, S. (2014). Enhancing students' skills in technical writing and LSP translation through tele-collaboration projects: Teaching students in seven nations to manage complexity in multilateral international collaboration. In G. Budin & V. Lušicky (Eds.), *Languages for special purposes in a multilingual, transcultural world. Proceedings of the 19th European Symposium on Languages for Special Purposes, 8–10 July 2013, Vienna, Austria* (pp. 249–259). University of Vienna. http://lsp2013.univie.ac.at/proceedings.

Arnó-Macià, E., Vandepitte, S., Minacori, P., Musacchio, M. T., Hanson, J. & Maylath, B. (2019). A multilingual background for telecollaboration: Practices and policies in European higher education. *European Journal of Language Policy, 11*(2), 235–255. https://doi.org/10.3828/ejlp.2019.14.

Brewer, P. E. & St.Amant, K. (2015). Education and training for globally-distributed virtual teams [Editorial]. *Connexions: International Professional Communication Journal, 3*(1), 3–7. https://connexionsjournal.org/wp-content/uploads/2019/12/brewer-st-amant.pdf.

Brown J. D. (2009). Open-response items in questionnaires. In J. Heigham & R. A. Croker (Eds.), *Qualitative research in applied linguistics: A practical introduction* (pp. 220–219). Palgrave Macmillan.

Brumberger, E. & Lauer, C. (2015). The evolution of technical communication: An analysis of industry job postings. *Technical Communication*, *62*(4), 224–243.

Canagarajah, A. S. (2009). Multilingual strategies of negotiating English: From conversation to writing. *JAC*, *29*(1/2), 17–48. https://www.jstor.org/stable/20866885.

Canagarajah, A. S. (Ed.). (2013a). *Literacy as translingual practice: Between communities and classrooms*. Routledge.

Canagarajah, A. S. (2013b). Negotiating translingual literacy: An enactment. *Research in the Teaching of English*, *48*(1), 40–67. https://www.jstor.org/stable/24398646.

Croker, R. A. (2009). An introduction to qualitative research. In J. Heigham & R. A. Croker (Eds.), *Qualitative research in applied linguistics: A practical introduction* (pp. 3–24). Palgrave Macmillan.

Crowther, P., Joris, M., Otten, M., Nilsson, B., Teekens, H. & Wächter, B. (2000). *Internationalisation at home. A position paper*. The European Association for International Education.

Denzin, N. K. & Lincoln, Y. S. (Eds.). (1994). *Handbook of qualitative research*. Sage.

Ferreira-Lopes, L., Bezanilla, M. J. & Elexpuru, I. (2018). Integrating intercultural competence development into the curriculum through telecollaboration. A task sequence proposal for Higher Education. *Revista de Educación a Distancia (RED)*, *58*. http://www.um.es/ead/red/58/ferreira_et_al.pdf.

Guth, S. & Helm, F. (Eds.). (2010). *Telecollaboration 2.0: Language, literacies and intercultural learning in the 21st century*. Peter Lang.

Hanson, J. R. (2015). Preparing globally distributed virtual team members to bridge boundaries of language difference: A graduate program teaching case. *Connexions: International Professional Communication Journal*, *3*(1), 87–112. https://connexionsjournal.org/wp-content/uploads/2019/12/hanson.pdf.

Helm, F. (2018). The long and winding road… *Journal of Virtual Exchange*, *1*, 41–63. https://doi.org/10.14705/rpnet.2018.jve.3.

Helm, F. & Acconcia, G. (2019). Interculturality and language in Erasmus+ Virtual Exchange. *European Journal of Language Policy*, *11*(2), 211–233. https://doi.org/10.3828/ejlp.2019.13.

Hobbs, V., Matsuo, A. & Payne, M. (2010). Code-switching in Japanese language classrooms: An exploratory investigation of native vs. non-native speaker teacher practice. *Linguistics and Education*, *21*(1), 44–59. https://doi.org/10.1016/j.linged.2009.12.004.

Horner, B., Lu, M.-Z., Royster, J. & Trimbur, J. (2011). Language difference in writing: Towards a translingual approach. *College English*, *73*(3), 303–321. https://www.jstor.org/stable/25790477.

Humbley, J., Maylath, B., Mousten, B., Vandepitte, S. & Veisblat, L. (2005). Learning localization through trans-Atlantic collaboration. In *IPCC 2005. Proceedings. International Professional Communication Conference* (pp. 578–595). IEEE. https://doi.org/10.1109/IPCC.2005.1494227.

Jones, E. (2016). Mobility, graduate employability and local internationalization. In E. Jones, R. Coelen, J. Beelen & H. de Wit (Eds.), *Global and local internationalization* (pp. 107–116). Sense Publishers. https://doi.org/10.1007/978-94-6300-301-8_15.

Kohn, K. & Hoffstaedter, P. (2017). Learner agency and non-native speaker identity in pedagogical lingua franca conversations: Insights from intercultural telecollaboration in foreign language education. *Computer Assisted Language Learning*, *30*(5), 351–367. https://doi.org/10.1080/09588221.2017.1304966.

Lauer, C. & Brumberger, E. (2019). Redefining writing for the responsive workplace. *College Composition and Communication, 70*(4), 634–663.

Maylath, B., King, T. & Arnó-Macià, E. (2013a). Linking engineering students in Spain and technical writing students in the US as coauthors: The challenges and outcomes of subject-matter experts and language specialists collaborating internationally. *Connexions: International Professional Communication Journal, 1*(2), 159–185. https://connexionsjournal.org/wp-content/uploads/2019/12/maylath-king-arno-macia.pdf.

Maylath, B., Vandepitte, S., Minacori, P., Isohella, S., Mousten, B. & Humbley, J. (2013b). Managing complexity: A technical communication translation case study in multilateral international collaboration. *Technical Communication Quarterly, 22*(1), 67–84. https://doi.org/10.1080/10572252.2013.730967.

Maylath, B., Vandepitte, S. & Mousten, B. (2008). Growing grassroots partnerships: Trans-Atlantic collaboration between American instructors and students of technical writing and European instructors and students of translation. In D. Starke-Meyerring & M. Wilson (Eds.), *Designing globally networked learning environments: Visionary partnerships, policies, and pedagogies* (pp. 52–66). Sense Publishers.

Mousten, B., Humbley, J., Maylath, B. & Vandepitte, S. (2012). Communicating pragmatics about content and culture in virtually mediated educational environments. In K. St.Amant & S. Kelsey (Eds.), *Computer-mediated communication across cultures: International interactions in online environments* (pp. 312–327). IGI Global. https://doi.org/10.4018/978-1-60960-833-0.ch020.

Mousten, B., Vandepitte, S., Arnó-Macià, E. & Maylath, B. (2018). Preface. In B. Mousten, S. Vandepitte, E. Arnó-Macià & B. Maylath (Eds.), *Multilingual writing and pedagogical cooperation in virtual learning environments* (pp. xv–xxi). IGI Global.

O'Dowd, R. (2015a). Supporting in-service language educators in learning to telecollaborate. *Language Learning & Technology, 19*(1), 63–82. http://dx.doi.org/10125/44402.

O'Dowd, R. (2015b). The competences of the telecollaborative teacher. *The Language Learning Journal, 43*(2), 194–207. https://doi.org/10.1080/09571736.2013.853374.

O'Dowd, R. (2018). From telecollaboration to virtual exchange: State-of-the-art and the role of UNICollaboration in moving forward. *Journal of Virtual Exchange, 1*, 1–23. https://doi.org/10.14705/rpnet.2018.jve.1.

O'Dowd, R. & Lewis, T. (Eds.). (2016). *Online intercultural exchange: Policy, pedagogy, practice.* Routledge.

Sorensen, K. S., Hammer, S. & Maylath, B. (2015). Synchronous and asynchronous online international collaboration: The Trans-Atlantic & Pacific Project. *Connexions: International Professional Communication Journal, 3*(1), 153–177. https://connexionsjournal.org/wp-content/uploads/2019/12/sorensen-hammer-maylath.pdf.

Tzoannopoulou, M. & Maylath, B. (2018). Virtual networks in English-for-Specific-Purposes education: A translation-reviewing/editing model. In B. Mousten, S. Vandepitte, E. Arnó-Macià & B. Maylath (Eds.), *Multilingual writing and pedagogical cooperation in virtual learning environments* (pp. 318–343). IGI Global. https://doi.org/10.4018/978-1-5225-4154-7.ch013.

Verzella, M. (2018). Virtual exchange between cross-cultural teams: A sustainable path to the internationalization of college courses. *Transformative Dialogues: Teaching and Learning Journal, 11*(3), 1–13. https://kpu.ca/sites/default/files/Transformative%20Dialogues/TD.11.3_Verzella_Virtual_Exchange_between_Cross-Cultural_Teams.pdf.

Vinagre, M. & Esteban, A. C. (2018). Evaluative language for rapport building in virtual collaboration: An analysis of appraisal in computer-mediated interaction. *Language and Intercultural Communication, 18*(3), 335–350. https://doi.org/10.1080/14708477.2017.1378227.

Part Four: Engaging Communities

13. Visual Communication in Community Contexts

Elise Verzosa Hurley
ILLINOIS STATE UNIVERSITY

Abstract: This chapter explores the ways in which critical approaches to visual communication can be fostered through community-based projects in technical and professional communication curricula by discussing a client project between an introductory technical and professional writing class and a local juvenile justice system. I offer a pedagogical approach and course design that integrates visual communication throughout the duration of a technical and professional writing course, arguing for the ways in which visual communication practices are significant not only to technical and professional communication documents, but also within the larger social and cultural contexts in which communication documents are a part.

Keywords: visual communication, client projects, service learning, course design

Key Takeaways:

- Community-based projects in introductory technical and professional writing course can foster an understanding of critical approaches to visual communication.
- Community-based projects offer an understanding of the ways in which visual communication practices are significant not only to technical and professional communication documents, but also within the larger social and cultural contexts in which communication documents are a part.
- This chapter provides a sample course design that integrates visual communication alongside a partnership with a local juvenile justice system.

Over two decades ago, Teresa M. Harrison and Susan M. Katz (1998) called students and teachers of technical communication to "take organizations seriously" by emphasizing the social structures within which organizational cultures "create a world characterized by idiosyncratic knowledge and patterns of symbolic expression" (p. 18). Harrison and Katz suggested that students can learn about organizations both through organizational socialization processes as well as in the classroom and, since then, pedagogical practices such as community-based learning and professional internships that allow students to engage in situated literacy activities specific to a particular profession or organization have emerged (Hayhoe, 1998; Henson & Sutliff, 1998; Huckin, 1997; Sapp & Crabtree, 2002; Savage, 1997).

Today, community-based learning in the technical and professional communication classroom, often in the form of client projects or service learning, is a well-documented pedagogical method of engaging students in real-world situations and rhetorical contexts. While often complex, challenging, and messy—both in its planning and facilitation—recent scholarship provides compelling arguments concerning its benefits, including increased student motivation (Pope-Ruark et al., 2014), exposure to new and unfamiliar communication genres (Willerton, 2012), and authentic opportunities to develop communication skills and negotiate client relationships (Melton & Hicks, 2011), in addition to its potentials in encouraging students to be more engaged civic participants (Dubinsky, 2002; Eble & Gaillet, 2004). Despite the prevalence of community-based learning in technical and professional communication curricula, however, the literature about such projects often focuses primarily on writing projects, traditionally defined, even as visual communication and design in professional, academic, civic, and otherwise public contexts are increasing in prominence.

In the same collection as Harrison and Katz, Kenneth T. Rainey (1998) called for integrating visual communication pedagogies into the technical communication curriculum due to developments in technology that have led to the collapse of discrete communication skills "so that, in many cases, communicator, editor, designer, and producer are the same individual" (p. 231). Fifteen years later, Eva R. Brumberger (2013) noted that teaching visual communication in technical and professional communication courses alongside community-based projects provides students with another layer of "analyzing audience, understanding document conventions, mastering technological tools, and recognizing ethical conflicts" (p. 100). In other words, teaching visual literacies can further enhance community-based projects as well as the broader technical and professional communication curriculum.

This chapter explores the ways in which critical approaches to visual communication can be fostered through community-based projects in technical and professional communication curricula by discussing a client project between an introductory technical and professional writing class taught at the University of Arizona and a local juvenile justice system, the Pima County Juvenile Court Center (PCJCC). I argue that in addition to helping students develop as communicators and thinkers, client projects with a strong visual communication component can allow students to better understand the role of professional documents in organizations, communities, and broader publics, and can encourage students to think of themselves as "citizen designers [who] have the ability to analyze, to respond critically, and to produce visuals in a variety of genres" (Hilligoss & Williams, 2007, p. 230). This chapter proceeds, first, with a review of the literature concerning client projects and service learning in technical and professional communication, followed by a description of the institutional context and background information on the partnership with PCJCC. Then, I describe the course design, focusing on the ways in which visual communication can be integrated throughout the duration of a technical and professional writing course. Finally, I discuss the results of the project and offer insights regarding the added value of teaching visual communication in community-based projects and in the broader technical and professional communication curriculum.[1]

Client Projects and Service Learning in Technical and Professional Communication

Rather than rely solely on case studies, simulations, or textbook assignments, technical and professional communication teacher-scholars have emphasized the benefits of experiential learning because it affords students with opportunities to blend theory and practice by applying their newly learned professional communication skills to real world situations (Blakeslee, 2001; Henson & Sutliff, 1998; Huckin, 1997; Matthews & Zimmerman, 1999; Spears, 1996). Two common experiential learning approaches advocated by teacher-scholars in the field are client projects and service learning. Both approaches ask students to situate their learning beyond the immediate context of the classroom by working with a client (often industry professionals) or community partner (often non-profit organizations), thus providing students with opportunities to communicate within real rhetorical situations, negotiate and problem-solve existing issues, and produce professional documents for an audience beyond the course instructor.

While client projects and service learning have much in common, scholars have also taken care to note how the approaches might differ. Gregory A. Wickliff (1997) argued that client projects are "the most valuable compromise between traditional classroom teaching from cases and the more involved task of designing individualized internships or cooperative educational experiences," and that central to client project pedagogies is that "students do not assume hypothetical roles as members of real or hypothetical organizations" (p. 172). Rather, client projects emphasize authenticity and professionalization by allowing students to experience real organizational cultures and workplace contexts (Kreth, 2005), often by requiring students to develop consultant stances as they produce professional communication deliverables for their client partner.

Although service learning resists a singular definition, scholars have argued that it "involves having students perform a service for a nonprofit organization" (Tucker et al., 1998, p. 89), and that it is "a pedagogical theory and method of experiential education in which students apply their academic skills in ways that both enhance the curriculum and foster a sense of civic responsibility" (Sapp & Crabtree, 2002, p. 411). Service learning, James M. Dubinsky (2002) argues, not only prepares students to "learn the skills they need in the workplace" but also provides a "path toward virtue and can create ideal orators and citizens who put their knowledge and skills

---

1. This study was considered exempt from review by the Institutional Review Board of the University of Arizona. Student and client materials are reproduced by permission.

to work for the common good" (p. 62). Crucial to many service learning approaches, thus, is attunement not only to reciprocity wherein both "community service and classroom learning" are improved (Matthews & Zimmerman, 1999, p. 383) but also to encouraging "students to develop a civic mindset" (Eble & Gaillet, 2004, p. 351). While client projects and service learning may initially seem to have divergent philosophical goals—professionalization versus citizenship—these two types of experiential learning approaches may certainly overlap, depending on the course instructor, the partnering client or community organization, and how they conceive of the partnership in relation to course goals.

Regardless of approach, much has been written about client projects and service learning in technical and professional communication curricula, though less explored are the ways in which visual communication practices can contribute to these two experiential learning pedagogies. Following Michelle F. Eble and Lynée L. Gaillet's (2004) reconfiguring of the term "community intellectual" and pedagogies that not only prepare students for "their chosen professions but also to send them to community organizations and businesses equipped to question community constructions and engage in rhetorical practices" (p. 353), I similarly advocate for a pedagogical view that positions students as citizen designers who not only have the know-how to employ visual communication practices but can do so through a critical lens that takes into account the broader cultural contexts in which visual artifacts circulate.

■ Institutional Context and Background

English 313: Technical and Professional Writing is an upper-division general education course at the University of Arizona that serves a variety of undergraduate majors. Many students who enroll in the course do so in order to fulfill a requirement in their major, and the course is often the only upper-division writing course they will take beyond first-year composition. Moreover, many of the students who enroll in the course have little to no prior experience with technical and professional communication, much less visual communication. A key component of the course is a client project during which students, in teams, collaborate with a local community or campus organization to produce various communication deliverables that fulfill a need of the partnering organization while providing professional development experience for the students.

In this particular case, I actively sought a partnership with the Pima County Juvenile Court Center in the months prior to the beginning of the course, and they were involved throughout the entire length of the project. Together with the presiding judge, the court administrator, the deputy court administrator, the juvenile justice coordinator, and the university's writing program administrator, we determined PCJCC's documentation needs and discussed the necessary course materials to support the project. PCJCC's primary need involved redesigning and creating new documents for each of the court's divisions—family and child services, probation, and detention—in order to better reflect the court's mission of serving the community through outreach and prevention. Like many organizations, PCJCC had limited resources and was unable to update or revise its existing documents, much less create new ones.

PCJCC administrators visited campus several times in order to meet with the students enrolled in the course. These visits included an informative presentation in the first few weeks of the semester to orient the students to the court, its mission, and the populations it serves, as well as later visits to provide students with feedback as they crafted their deliverables. The class also visited and toured the facilities of the Pima County Juvenile Court Center to gain a better understanding of each of the court's divisions and their functions. As the course instructor, I remained involved as the project unfolded, though it is important to note that the nature of the collaboration shifted as the project progressed, with students taking a more active role in scheduling, communicating, and negotiating their proposed projects with PCJCC administrators. At the project's conclusion, the students provided PCJCC with editable digital versions of their project deliverables as well as printed hardcopies for the court to use and revise as necessary. At the end of the semester, I also scheduled a follow-up meeting with PCJCC administrators and solicited their responses and reflections about the collaboration. The success of the collaboration, I believe, can be attributed in part to the involvement and commitment of PCJCC administrators throughout the length of the semester rather than simply during the client project.

▋ Needs Assessment

In many of our planning meetings, PCJCC administrators expressed concern about how their public image was perceived by community members, an image that was constructed, in part, by their existing technical and professional communication documents. As an exemplar of how they did *not* want to be perceived, court administrators shared their limited existing documents, pointing to the ways in which visual representations of the court did not align with their mission, goals, and values. The existing brochure they used, for example, depicted images of youth from minority populations behind bars and in handcuffs, even though handcuffing youth and placing them in barred cells are not practices employed by PCJCC. In his work about organizational narratives, Brenton D. Faber (2002) writes that an organization's image is "constructed and held by audiences of its communication and derives from more sources than just the organization's own communication" (p. 35). Cognizant of the ways in which visual representations affect public perception, PCJCC administrators acknowledged that their existing documents reinforced dangerous societal stereotypes about criminalization and minority youth. The internal organizational narrative of PCJCC as a community institution that prioritizes outreach, prevention, and protective services thus conflicted with its external image as a seemingly punitive institution whose primary purpose is to incarcerate juvenile delinquents. In other words, PCJCC administrators were concerned with what Faber calls *image-power*, or the ability of an organization to "reproduce, alter, create, or otherwise influence the way other people perceive images" (p. 123). In our joint discussions about pedagogical and curricular support for the project, PCJCC administrators emphasized that it was important for students to understand how the court's primary audience—youth and their families—tend to be constructed as either victims, or more commonly, as delinquents, a binary often reinforced by other cultural representations such as those in films, news media, and popular music. PCJCC administrators, thus, also expressed the desire to destabilize dominant assumptions regarding victimization and delinquency in their technical and professional communication documents.

In addition to issues of visual representation, PCJCC administrators also voiced concerns about the functionality of their existing documents, noting that inattention to visible document features often hindered readability and usability. For example, they pointed to an existing handbook for detention and probation that was designed with large blocks of text and little white space and lacked any semblance of information architecture. Further, the informative content in many of their other documents was written for adult audiences presumed to be familiar with legal vocabulary, rather than for youth with little knowledge and understanding of legalese. In sum, the client project with PCJCC required a pedagogical framework that included explicit instruction about the rhetorical consequences of visual representation as well as the visible features of document design.

■ Course Design, Goals, and Classroom Activities

Given the numerous learning objectives that comprise most introductory technical and professional writing courses, it can be challenging to find time to devote instruction solely to visual communication and design. Although the client project with PCJCC accounted for much of the sixteen-week semester, the curriculum also included three other major projects. In her survey of the teaching of visual rhetoric in the professional communication curriculum, Brumberger (2005) noted that visual communication tends to be treated as a "unit," "discrete entity," or "add-on" to the broader technical and professional communication curriculum rather than as essential or foundational to it (p. 324). In order to support the needs of PCJCC and the complexity of the client project, however, I integrated visual communication components throughout the semester in order to lay the groundwork for the visual literacies that would be required of the students to complete the client project.

■ Visual Representation

In preparation for Unit I, where students are asked to assess and analyze their skills and experiences for their future professions by completing a professional inventory sheet and researching potential professional positions for which they might apply upon graduation (Hea, 2005), I also asked students to find a visual image that best represented their target professions. Students had the option of finding an image that, in their opinion, reflected their perceptions of what their chosen professions might be like, or an image that communicated how the profession is characterized. Students located a variety of images, reflecting the range of the students' majors and professional interests. During the first two weeks of the semester—the duration of the job analysis unit—we spent the first few minutes of each class session discussing the images the students brought to class. Although this was an informal activity, our classroom conversations helped students to begin talking and thinking about visual texts critically by connecting them to prevailing cultural assumptions and perceptions about their chosen profession. For example, a female non-traditional student enrolled in the university's race track industry (RTI) program shared a photograph of jockeys on horses and commented on her difficulties in finding images of the industry that included women, thus prompting a brief discussion about gender representation in RTI. Brief informal discussions such as this allowed students to explore how images and visual representations function, prompting them to consider the relationships among dominant images about various professions, how those images affect public perception, and how such perceptions and assumptions are related to broader social and cultural concerns.

Document Design

During Unit II, where students are asked to produce and design job materials such as a résumé and cover letter for an actual professional position to which they might apply upon graduating, several class sessions were spent exclusively on the visible features of document design, covering topics ranging from Gestalt design principles, color theory, typography, and information architecture (Baker, 2006; Campbell, 2006). By the time we were ready to begin the client project, students already had some understanding of visual rhetoric, visual representation, and visual communication, all of which would inform their production of various deliverables for PCJCC.

Visual Culture

After PCJCC's initial visit to our classroom, we determined that the development of informative packets consisting of three to four separate deliverables addressing information needs specific to the three divisions of the court would be the most viable in meeting the court's document needs. In self-selected teams of three, students proposed to create the deliverables outlined in Table 13.1.

Prior to drafting their proposals outlining the specific deliverables each team would create, students researched PCJCC's local context by conducting primary research in the form of interviews with court staff and site visits, in addition to researching the ways in which other juvenile courts used various technical and professional communication documents. As part of their investigation, students also conducted image searches on the web to better understand the ways in which juvenile court systems use visual communication principles. Unsurprisingly, search results on different search engines turned up numerous images of youth dressed in prison overalls, often in handcuffs or behind bars. These image searches also yielded several flow charts detailing the intake process, as well as various kinds of graphs displaying statistics about youth and crime. Students quickly pointed out that the underlying mission of PCJCC and those of other juvenile courts, with their focus on prevention and outreach, contradict how youth are typically represented in the documents provided by PCJCC and in those they found through web research. Thus, in the first few weeks of the semester and in the initial research stages of the client project, students were immersed in the local context of their community partner while also becoming more familiar with visual communication practices. This immersion process, I believe, was a necessary step for the students before they could even contemplate their proposed project deliverables since many parts of it—including both the written informative content as well as the visual design—were dependent on their understanding of the visual culture of juvenile justice systems.

Table 13.1. Project deliverables

Family and Child Services	Probation	Detention
<ul><li>Community resource directory</li><li>Family and child welfare brochure</li></ul>	<ul><li>Probation brochure</li><li>FAQ Sheet</li></ul>	<ul><li>Handbook for youth</li><li>Handbook for parents</li></ul>

Bringing It All Together: Multiple Interpretative Visual Frameworks

Because it was necessary for the students not only to learn how to analyze visuals but also to begin thinking about how their analyses can inform their own production practices, I employed what Candice Welhausen (2009) calls *visual topoi*—or visual commonplaces—which I tailored to the specific context of the client project as both an analytic and a heuristic for generating design ideas (see Appendix A). According to Welhausen, using visual commonplaces can enhance visual communication pedagogies because they "link [visual] analysis and production, link visual invention to classical theory," and provide students with a means of "drawing from a common body of cultural knowledge that allows us to construct visual knowledge in particular ways" (pp. 182–183). Beginning with visual analysis, Welhausen suggests that adapting multiple interpretive visual frameworks can allow students to tease out the ways in which visual texts are constructed both materially (form, mode of delivery, usability, etc.) and culturally (content, ideology, rhetorical purpose, etc.). Moreover, using multiple interpretive visual frameworks builds from Rainey's (1998) assertion that instructors should teach the following principles of visual communication: selection, design, position, production, and cost. Using the heuristic, I prompted students to analyze PCJCC's primary brochure, which depicted images of forlorn youth behind bars, similar to the images they found during their initial Google search.

By engaging with the interpretive visual frameworks outlined in the heuristic, students were able to gain a more nuanced understanding of the ways in which the brochure may be read and interpreted as a primarily visual document. Moreover, this approach allowed them to gain a richer understanding of visual design as it informed the specific local context of PCJCC. In relation to graphic design theories and formal elements of design, for example, students noted that PCJCC's brochure was a relatively well-designed document because it made good use of page layout, had balanced and aligned design elements, and used typography and spacing in a manner that created a clear organizational hierarchy. As students moved through the heuristic questions, however, they also noted that the photographs depicted on the brochure reinforced a division between those depicted as having power (authority figures such as law enforcement and attorneys) and those depicted without it (children and youth). Further, students were quick to point out that the children and teenagers included in the brochure were nearly all people of color, while the authority figures were not, thus reinforcing dangerous stereotypes regarding criminality and minority youth.

In relation to the heuristic questions pertaining to rhetorical context, students mentioned that while the PCJCC brochure states that it is for both parents and youth, parents were not represented in the brochure at all; further, the language employed in the brochure only addresses youth, although students noticed that the tone may be inappropriate, noting that the use of bolded typeface and exclamation points on the back of the brochure could be interpreted as condescending. Guided by the interpretive visual frameworks listed on the heuristic, students concluded that PCJCC's brochure did not communicate a philosophy of outreach and prevention but projected instead a punitive image—one that was in stark contrast with PCJCC's mission and goals.

Integrating critical approaches to visual communication and design within the context of PCJCC's existing communication documents not only helped students gain a more nuanced understanding of the court's goals in relation to visual representation, but it also helped student teams to articulate and frame specific goals they wanted to achieve during both the written proposal and production stages of the project. Doing this exercise prior to the written proposal allowed students to connect their analyses of the existing documents to PCJCC's goals and then convey their rationales for their proposed deliverables in ways that explained and justified the reasoning behind their visual design decisions.

Stakeholder Theory and Reflection

As the literature states, community-based pedagogies should extend beyond simply producing deliverables to meet a community partner's needs and should also prompt students to think critically about the social problems that give rise to the exigencies surrounding various community organizations (Huckin, 1997; Sapp & Crabtree, 2002; Scott, 2006). One strategy instructors have used to prompt such awareness includes teaching students to identify the stakeholders invested in community-based projects. Stakeholder theory not only helps students negotiate various stakeholder relationships (Hea, 2005), but also has the potential to help students make conscious decisions about the deliverables they are to design when integrated alongside a visual communication project. During the research and proposal stages of the project, students were asked to complete a stakeholder chart (see Appendix B). They identified a range of stakeholders, including youth and families in the local community, PCJCC administrators and staff, social workers, law enforcement, attorneys, other community organizations, themselves, and me, among others. Once students identified stakeholders, ascertaining and problematizing the exigencies and social issues that contribute to the need for juvenile justice systems became much clearer.

For example, one student team tasked to produce deliverables for the family and child services division noted that one reason why youth may enter the juvenile justice system might, in part, be due to the lack of available support systems in the home and in the community. Reflecting on the stakeholders they identified led them to consider how external social factors, such as the working conditions of parents and guardians, access to information about after-school and outreach programs, and funding limitations to support such programs, might all contribute to issues affecting the stakeholders of the family and court services division.

As one of their deliverables, the student team then proposed to create a community resource directory that could be distributed locally in order to raise awareness about various organizations and outreach programs available to the community. In addition to providing important information to youth and their families, the student team also considered the possibility that the design and distribution of the directory might increase the visibility of the community programs and organizations included in it, which could then potentially lead to additional funding and/or volunteer resources for the organizations. Mapping the stakes of stakeholders invested in the client project, thus, helped the student team to identify and articulate social issues that affect the stakeholders they identified which, in turn, guided their design decisions.

The stakeholder charts became useful, again, after the client project was completed and the class moved on to the final reflective report. As many scholars have confirmed, reflection is an important part of experiential learning and, when built into a course design, can allow students the opportunity for metacognitive awareness about course learning concepts and applications (Mahin & Kruggel, 2006; O'Toole, 2007). Students returned to their stakeholder charts as they considered how their finished deliverables might impact the stakeholders they initially identified. For example, one student whose team produced deliverables for the probation division used this final assignment as an opportunity to critique his team's design, noting that the needs of some stakeholders were featured more prominently than the needs of others. The student's reflective report acknowledged the ways in which his team's deliverables fulfilled the wishes of their community partner, while also allowing him to consider the implications of his team's design on other stakeholders. As Harrison and Katz argued, "students must also understand how their actions contribute to the construction, maintenance, and potential transformation of the organization and its culture" (p. 28), and returning to the stakeholder charts in the reflection process allowed students to complicate their ideas, assumptions, and experiences about community partnerships, professional relationships, and visual representation, highlighting the ways in which professional practices must always be refigured and reconsidered.

■ Outcomes

Many of the students wrote about their positive experiences during the client project, noting especially the ways in which the partnership allowed them to gain actual experience that would be beneficial to their future professional identities and beyond, while also allowing them to understand the significance of visual communication as a situated practice. Moreover, PCJCC administrators similarly noted increased awareness to reflect on and consider the role of their communication documents in the community. In the feedback form completed by PCJCC administrators at the end of the semester, they noted that "the students gave us a lot to think about in terms of the current documents we have available for the local community. The court has felt listened to as evidenced by the materials presented to us. Students understood our goal of being more engaging, professional, compassionate, and sensitive in our materials." Additionally, PCJCC administrators shared their monthly communication bulletin featuring the students' deliverables with our class. For the students, PCJCC's public recognition of their work allowed them to witness the ways in which their deliverables were being implemented to effect change in their own local community, further emphasizing the impact of technical and professional communication documents as visual texts beyond the classroom context.

While the community partnership was ultimately successful in meeting our joint goals of designing usable deliverables for PCJCC, the project was certainly not without challenges. Instructors who work with community partners must take care to familiarize students with communication practices that are conducive to the partnering organization's schedule. For example, as students began to brainstorm ideas after our initial meeting with PCJCC administrators, the student teams were eager to get started, and individual students from the same team emailed our PCJCC contact multiple questions, several times a day, resulting in a deluge of emails for PCJCC's administrative assistant. Because I was copied on the emails, I was quickly able to address the issue of respecting our community partner contact's time and labor by suggesting that each team brainstorm a list of questions and then send a single email to PCJCC's administrative assistant. As Amy C. Kimme Hea and Rachael Wendler Shah (2016) note, "introducing students to ever-complex interpretations of partner contexts, rhetorical situations, and civic responsibility" (p. 64) requires that instructors work carefully to understand the stakes of community partners in tandem with the community partner's organizational social context. Thus, as Lee-Ann M. Kastman Breuch (2001) argued, instructors must also anticipate that working with clients or community partners also requires explicitly teaching students more intangible skills concerning communication, interaction, and engagement.

Conclusion

The course design and activities discussed in this chapter offer just a few ways of attuning instructors to the added benefits of teaching visual communication in community-based projects, further opening the possibilities for integrating visual learning in the larger technical and professional communication curriculum. In so doing, students enrolled in our courses can better understand not only the significance of visual communication in technical and professional communication documents, but also the ways in which visual practices participate within larger social and cultural contexts, including the ways in which such practices are often mirrored in organizations. This understanding helps students develop abilities to potentially transform organizational discourses by being attentive to the social and cultural implications of the documents they produce currently, as students, and in the future, as communication professionals.

References

Baker, W. H. (2006). Visual communication: Integrating visual instruction into business communication courses. *Business Communication Quarterly*, 69, 403–407.

Blakeslee, A. M. (2001). Bridging the workplace and the academy: Teaching professional genres through classroom-workplace collaborations. *Technical Communication Quarterly*, 10, 169–192.

Breuch, L-A. M. K. (2001). The overruled dust mite: Preparing technical communication students to interact with clients. *Technical Communication Quarterly*, 10, 193–210.

Brumberger, E. R. (2005). Visual rhetoric in the curriculum: Pedagogy for a multimodal workplace. *Business Communication Quarterly*, 68, 318–333.

Brumberger, E. R. (2013). Teaching visual communication through community-based projects. In E. R. Brumberger & K. M. Northcut (Eds.), *Designing texts: Teaching visual communication* (pp. 99–115). Baywood.

Campbell, N. (2006). Communicating visually: Incorporating document design in writing tasks. *Business Communication Quarterly, 69*, 399–403.

Dubinsky, J. M. (2002). Service-learning as a path to virtue: The ideal orator in professional communication. *Michigan Journal of Community Service Learning, 8*, 61–74.

Eble, M. F. & Gaillet, L. L. (2004). Educating 'community intellectuals': Rhetoric, moral philosophy, and civic engagement. *Technical Communication Quarterly, 13*, 341–354.

Faber, B. D. (2002). *Community action and organizational change: Image, narrative, identity*. Southern Illinois University Press.

Harrison, T. M. & Katz, S. M. (1998). On taking organizations seriously: Organization as social contexts for technical communication. In K. Staples & C. Ornatowski (Eds.), *Foundations for teaching technical communication: Theory, practice, and program design* (pp. 17–29). Ablex.

Hayhoe, G. F. (1998). The academe-industry partnership: What's in it for all of us? *Technical Communication, 45*(1), 19–20.

Hea, A. C. K. (2005). What's at stake? Developing stakeholder relationships. *Reflections: Writing, Service-Learning, and Community Literacy, 3*, 54–76.

Hea, A. C. K. & Shah, R. W. (2016). Silent partners: Developing a critical understanding of community partners in technical communication service-learning pedagogies. *Technical Communication Quarterly, 25*, 48–66.

Henson, L. & Sutliff, K. (1998). A service learning approach to business and technical writing instruction. *Journal of Technical Writing and Communication, 28*, 189–205.

Hilligoss, S. & Williams, S. (2007). Composition meets visual communication. In H. A. McKee & D. N. DeVoss (Eds.), *Digital writing research: Technologies, methodologies, and ethical issues* (pp. 229–248). Hampton.

Huckin, T. N. (1997). Technical writing and community service. *Journal of Business and Technical Communication, 11*, 49–59.

Kreth, M. L. (2005). A small-scale client project for business writing students: Developing a guide for first-time home buyers. *Business Communication Quarterly, 68*, 53–59.

Mahin, L. & Kruggel, T. G. (2006). Facilitation and assessment of student learning in business communication. *Business Communication Quarterly, 69*, 323–337.

Matthews, C. & Zimmerman, B. B. (1999). Integrating service learning and technical communication: Benefits and challenges. *Technical Communication Quarterly, 8*, 383–404.

Melton, J. & Hicks, N. (2011). Integrating social and traditional media in the client project. *Business and Professional Communication Quarterly, 74*, 494–504.

O'Toole, K. (2007). Assessment in experiential learning. *Education Research & Perspectives, 34*(2), 51–62.

Pope-Ruark, R., Ransbury, P., Brady, M. & Fishman, R. (2014). Student and faculty perspectives on motivation to collaborate in a service-learning course. *Business and Professional Communication Quarterly, 77*, 129–149.

Rainey, K. T. (1998). Visual communication: The expanding role of technical communicators. In K. Staples & C. Ornatowski (Eds.), *Foundations for teaching technical communication: Theory, practice, and program design* (pp. 231–242). Ablex.

Sapp, D. A. & Crabtree, R. D. (2002). A laboratory in citizenship: Service learning in the technical communication classroom. *Technical Communication Quarterly, 11*, 411–432.

Savage, G. J. (1997). Doing unto others through technical communication internship programs. *Journal of Technical Writing & Communication, 27*(4), 401–415.

Scott, J. B. (2006). Extending service-learning's critical reflection and action: Contributions of cultural studies. In J. B. Scott, B. Longo & K. V. Wills (Eds.), *Critical power tools: Technical communication and cultural studies* (pp. 241–258). State University of New York Press.

Spears, L. (1996). Adopt-a-nonprofit: A project in persuasion and collaboration. *Business Communication Quarterly, 59*, 21–28.

Tucker, M. L., McCarthy, A. M., Hoxmeier, J. A. & Lenk, M. M. (1998). Community service learning increases communication skills across the curriculum. *Business Communication Quarterly, 61*, 88–99.

Welhausen, C. A. (2009). *Toward a visual paideia: Visual rhetoric in undergraduate writing programs* [Doctoral dissertation, University of New Mexico]. University of New Mexico Digital Repository. https://digitalrepository.unm.edu/engl_etds/6/.

Wickliff, G. A. (1997). Assessing the value of client-based group projects in an introductory technical communication course. *Journal of Business and Technical Communication, 11*, 170–191.

Willerton, R. (2012). Teaching white papers through client projects. *Business and Professional Communication Quarterly, 76*, 105–113.

Appendix A: Sample Heuristic for Integrating Multiple Visual Frameworks

These heuristic questions help students analyze the visible features of a document from a plurality of visual theories relevant to the parameters of the client project and serve as a starting point for generating design ideas. Instructors can choose to revise the visual frameworks to support the specific needs of their partnering organizations in relation to the specific context of a design project.

Graphic Design

- What elements are emphasized in the existing client brochure? How does the alignment and layout contribute to the brochure's purpose and meaning? How does repetition, proximity, and contrast enhance/detract from the brochure's purpose, meaning, and likelihood for usability?
- How is line, shape, texture, typeface, color, and space being used? What do these elements suggest and how do they contribute to the purpose of the document?

Social Semiotics

- What sign systems are represented in the existing brochure (icons, indexes, symbols)? How realistic/abstract are they? What meanings are associated with them? Why?

Visual Culture

- How are audiences positioned to view this brochure? What viewpoint is privileged? How? How does this contribute to the overall purpose and meaning of the brochure?

- Focus on the photographs in the existing brochure. Who is represented/not represented? What angles/point of views are depicted? What meanings are associated with them? Why?

Rhetorical Context
- Why might the client need to produce documents such as this? What are the potentials and constraints of the brochure genre?
- For whom is this brochure created? When and where might they encounter this brochure? What might they need to know from a brochure such as this?

Drawing on your answers to the above questions, what conclusions might you draw about how the client is represented? Keeping in mind the various interpretative visual frameworks listed above, what are some other alternatives for representing the client, their mission, and the populations they serve through a visual document like a brochure? *Adapted from Welhausen (2009).

■ Appendix B: Sample Stakeholder Chart

A stakeholder chart allows students to (1) identify multiple stakeholders involved in a client project, (2) identify each of the stakeholder's goals/investments in relationship to the project, (3) identify each of the stakeholder's needs in relationship to their goals/investments, and (4) identify external factors that can impact each stakeholder's goals, investments, and needs.

Stakeholders	Goals/Investments in the Project	Needs in Relation to the Project	External Factors to Consider

*Adapted from Hea (2005).

14. Competing Mentalities: Situating Scientific Content Literacy Within Technical Communication Pedagogy

Lisa DeTora
HOFSTRA UNIVERSITY

Abstract: Technical communication students may need to learn how to manage scientific information for a general audience because documents written by and for theoretical and applied scientists, as well as data outputs, provide essential source material for many technical communicators. Hence, technical communication instructors and students might benefit from understanding the production and pedagogy of such scientific textual materials. A key disjunction was identified in the late twentieth century by Carolyn R. Miller and Charles Bazerman (among others): scientific writing does not look, or operate, like humanistic writing, which opens the possibility for criticism and critique rather than understanding and respect. By recognizing that reading practice in the sciences consists largely of scanning for key information and assembling useful datasets, the logic of scientific writing pedagogies becomes more apparent. Technical communication pedagogies can leverage existing advice in the sciences to help students gain fluency in reading texts that are intentionally constructed using difficult jargon in order to maintain the integrity of scientific information.

Keywords: scientific writing, technical communication pedagogy, layered literacies

Key Takeaways:

- Scientific literacy relies on content mastery and statistical reasoning rather than more writerly concerns, which represent competing mentalities.
- Technical communicators can enrich their pedagogical practices by fostering a deeper understanding of scientific materials and their modes of production, producing more effectively layered literacies.
- Students may benefit from understanding the subtleties of scientific writing within expert discourses.

Competing Mentalities

Tom Johnson's (2019b) keynote address at the Symposium on Communicating Complex Information suggested that certain types of technical writing jobs are declining relative to the growth of the software industry (a field informed by theoretical and applied sciences), largely as a result of rapidly increasing technical specialization. In other words, technical communication programs may not be keeping up with increasingly sophisticated technology, which may indicate that their pedagogies could better address the complexity of scientific and technical information.

Johnson's observations highlight a site of intellectual tension often embodied in a dispute between Carolyn R. Miller's (1979) case for technical communication as a humanistic discipline and work by those who see technical writing as an expression of proficiency and scientific literacy. Miller (1979) calls for readers to understand both humanities and science as motivated by rational practices. Current readers will, no doubt, be aware that Miller's comments enjoy huge traction in technical communication pedagogy. Yet few current scholars refer to Miller's cogent remarks on the "communal rationality" (p. 617) of the science that underpins technical communication. In fact, many current scholars behave as though Miller's remarks begin and end in considering technical communication to be humanistic. Thus, pragmatically-grounded technical or scientific praxis is now often considered a type of "contextless logic" (p. 617), even though Miller clearly debunks the idea that science operates in this intellectually limited way. What is lost in this maneuver is a connection to the materially grounded practices of workplaces—laboratory, office, pilot plant, hospital, research center—or what Paul R. Meyer and Stephen A. Bernhardt (1997) refer to as "workplace literacy" (p. 86). I suggest that the type of scientific knowledge that is recognizable to scientists is an essential underpinning of successful technical communication that should be better incorporated into its pedagogies.

This chapter presents some advice for technical communication pedagogies to encourage a better understanding of scientific content and writing practices on their own terms, that is, considering the ways that people who participate in various scientific discourse communities constitute what they consider to be good communication. Current technical communication pedagogies might benefit from considering scientific inquiry as invention, following Miller, and also how writing pedagogy by scientists tends to approach specific, practical problems of composition and critique. Thus, the argument below is organized in response to a specific question: Given that scientific texts often provide an essential body of knowledge for technical communication professionals, how can understanding pedagogies and textual production in scientific disciplines strengthen technical communication pedagogy? One answer was suggested by Kevin Garrison (2014) in "The Scientist, Philosopher, and Rhetorician: The Three Dimensions of Technical Communication and Technology": to balance the "competing mentalities" (p. 359) he cites in the title of his paper in order to form a better foundation for technical communication pedagogies. Garrison cautioned readers that his framework required ongoing examination and revision, and this paper is one such follow-on. In the subsequent pages, I will explain why I think that increased attention to scientific material is warranted, justify my central question, and describe some underexamined scientific materials that could be of use for technical communicators, all to inform a heuristic for technical communication pedagogies that accounts for both humanistic and scientific understanding.

The Need for Scientific Literacy in Technical Communication

An opening question might be what scientific information has to do with technical communication pedagogies. One reason is that Miller's observation that scientific information might be subject to intervention and analysis via technical communication scholarship continues to hold true. Organizations such as the Special Interest Group on Design of Communication (SIGDOC) of the Association for Computing Machinery (sigdoc.acm.org), the Society for Technical Communication (stc.org), or tekom (technical-communication.org) define technical communication as an attempt to convey useful information, either technical or instructional, in traditional textual or electronic formats. In other words, technical communication *translates* complex scientific or technical information for people who might need to use it. Thus, technical communication differs from specialist scientific and highly complex technical writing, such as regulatory documentation (Benau, 2020) in its purposes and audiences. Technical communication scholars may not know how scientists, who often see their inquiry as necessarily interdisciplinary, adapt their work for varied audiences and genres. This circumstance presents challenges for those who need to present complex information to the general public by troubling the ability to clearly define a boundary between scientific fields or even between those fields and applied practice.

The primary ethos of much technical communication centers on usability and user experiences (UX), which evolved from earlier practices such as alpha and beta testing and user-friendliness in computing design (Seffah & Metzker, 2004). For example, computer specialists Sari Kujala and colleagues' (2011) scientific discussion of UX as centering on adoption rather than use, notes a lack of useful definitions, but merely cautions the reader not to conflate different usages. In contrast, technical communication scholars like Lisa Melonçon (2017) and Kirk St.Amant (2017) suggest both practical solutions and useful common vocabulary when they promote patient-based UX design (PXD) and intercultural PXD (I-PXD) in healthcare contexts. When Kujala and colleagues identify deficits in UX terminology without correcting them, this signals a tolerance for discursive problems, while St.Amant and Melonçon display attention to language rather than a tacit acceptance of an unclear general literature. Technical communication, thus, attempts to displace responsibility for comprehension from the reader to the writer, a pattern of behavior consistent with the humanistic values Miller describes.

Of course, what may be lost in a focus on humanistic values is the practical and scientific content of technical writing. And this point is well taken, given the proliferation of scientific information and technologies since the late twentieth century. Richard Van Noorden (2014a) observed that the global scientific publication output doubles every nine years, and this information has increased in complexity as well as volume. As Johnson (2019b) comments, subject-matter expertise, much more than writing and thinking, is highly valued in many technical writing settings. Technical communication pedagogies, thus, should help students manage both the increasing complexity of scientific and technical information as well as follow through on an enhanced attention to user experiences.

■ The Need for Layered Literacies

A key hurdle in translating scientific and technical information for users is understanding this information in the first place, or what might be considered a type of content literacy that augments general and workplace literacies. For example, Johnson (2019a) advises technical training to offset the tendency for technical communication degree programs to "drift" toward the humanities. Such drift may occur even in pedagogy that seeks to account for multiple literacies. For example, Kelli Cargile Cook's (2002) model of "layered literacies" (p. 5) for technical writers calls on instructors to impart models of understanding rhetorical, technological, ethical, and critical content and approaches. Cargile Cook's model omits specific attention to scientific literacy, situating technological literacy in social, rather than pragmatic, terms. Her reference list reflects a strong trend toward literary and social theory, supporting its participation in the production of new humanistic knowledge. Similarly, J. Harrison Carpenter's (2011) update of layered literacies for scientific writing concentrates not on the acquisition of scientific knowledge but on graphical, technological, sociocultural, and communicative values in science. Hence, layered literacy approaches ultimately position themselves as liberal arts , remaining within only one of the types of competing mentalities Garrison describes. Neither Carpenter nor Cargile Cook consider how to layer scientific literacy as understood by scientists, or even other workplace literacies as described by Meyer and Bernhardt, into technical communication pedagogies. I believe that encouraging scientific and workplace literacies is essential for many sites of technical communication pedagogy and should be an added layer in this milieu of competing mentalities.

Layered literacies should require an articulation of specific disciplinary knowledge that highlights the function of such contents within specific rhetorical activities. In "Articulation: A Working Paper on Rhetoric and *Taxis*,"

Nathan Stormer (2004) explains the historically constituted and performative nature of rhetorical constructions. For Stormer, articulation is a means of understanding the material practices of rhetoric as arising from "shared acts" (p. 257) as well as a means of "bringing together the material world, language, and spatial arrangement in one act" (p. 263). Attention to the specific means of building and ordering information may provide a framework in which layered understandings might operate more effectively. Two features of Stormer's argument are germane here: first, that articulation is historically situated and second, that language scholars should attend to the arrangement and ordering of elements within rhetorical activities. I suggest that an inadvertent omission of scientific literacies in technical communication pedagogies could contribute not only to the "drift" Johnson (2019a) describes but also to an apparent loss of job share in the technical sector (Johnson, 2019b). One way of remedying this problem is by enhancing our understanding of how scientific literacy plays out in scientific writing advice and teaching by and for scientists. Below, I present a heuristic for technical communication pedagogies before discussing what scientists—as opposed to humanists or even social scientists—might mean by scientific literacy and how those differing understandings may be used to enrich technical communication pedagogies.

In Table 14.1, I offer some practical suggestions for operationalizing science as an element of layered literacy and a competing mentality in the technical communication classroom. This model is based in part on existing work, such as Melody A. Bowdon and J. Blake Scott's (2003) volume on service learning in technical communication, which already advises technical communicators to consider the various positions and needs of users, readers, and writers. Layered literacies and a recognition of competing mentalities are excellent models for technical communication pedagogies, as long as teachers and students understand various modes of ordering textual and conceptual elements as a type of performance, as Stormer indicates. The heuristic below is compatible with the advice of the technical communication scholars quoted above as well as information shared within scientific education communities, creating a site for effecting layered literacies that better account for scientific knowledge. Each of these activities may be analyzed as a type of rhetorical performance.

Technical communicators can use scientific information and scientific writing pedagogies to improve teaching and practice in order to better inform and develop activities like those in Table 14.1. It is important to note that the activities in Table 14.1 are not intended to replace the work already being done in the field. In other words, these exercises are intended to enhance and develop scientific literacy and to enable teachers and students to articulate scientific literacy more effectively into existing technical communication pedagogies. The following sections identify obstacles to scientific literacy that can impede the work of technical communicators and offer information to help situate humanistic and scientific approaches to writing study.

Table 14.1. A heuristic for technical communication pedagogies using scientific information

Step to Take	Reason for Taking the Step	Example Activities
Follow scientific advice for reading scientific materials (Pain, 2016): scan for main points, take notes, review tables and figures carefully.	Scientists construct documents to be managed in specific ways; practicing these skills enhances knowledge.	Build annotated bibliographies of scientific works, including key tables or figures, before reflecting on their contents. Read scientific papers from generalist and specialist journals to identify differences in placement of key information.
Closely examine table and figure legends and footnotes. Consider how the placement of titles legends, captions, and footnotes functions in different discursive situations (manuscripts, slides, posters, for example).	Scientists rely heavily on these types of text to understand data.	Use figure legends from scientific papers to identify elements of study design and key results, including statistical analysis. Review guidelines for figure legend, caption, and table heading composition from various journals or sources. Compare placement of titles, captions, legends, and footnotes in different journals.
Examine the role of mathematical and quantitative literacies in technical communication.	Scientists distinguish between mathematical and quantitative reasoning; the former is a strong predictor for success in science.	Ask students to define mathematical and quantitative competencies using specific examples from the scientific literature. Build a heuristic of mathematical versus quantitative literacy in a specific field, discipline, or technical setting.
Review uses of jargon/ pull new copies of papers often.	Scientific work undergoes constant revision; terminology may drift over time.	Identify different uses of the same term by authors over time. Identify different uses of the same term in different fields.
Understand scientific context by examining citation practices.	Scientific conversations occur over multiple papers.	Identify scientists in a field, their affiliations and citation habits. Develop a "citation map" of thinkers who cite one another's work.

Step to Take	Reason for Taking the Step	Example Activities
Think about plain language.	Scientific writing is not intended for mainstream audiences and needs translation to be useful.	Translate key elements of a paper—like figure and table legends—into plain language versions. Analyze what might be lost in translating technical or scientific content into plain language.
Recognize the wisdom in Miller's 1979 paper (and other key works in technical communication).	Miller finds ways to value both humanistic and scientific modes of thinking.	Identify evidence of communal rationality in a group of scientific papers.
Identify the components needed to support an argument across multiple scientific papers.	Humanistic values emphasize argumentation; identifying the needed components to make a humanistic-type argument can help students understand scientific writing genres and their role in communication.	Have students find a literature review and then read several of the cited papers to identify how they were adapted for the purposes of review.
Identify historical/chronological relationships between texts and ideas.	Scientific communication and humanistic studies of scientific discourses are ongoing conversations; students will benefit from understanding how ideas build on one another.	Build a timeline of key works about a scientific topic, then build a parallel timeline of work in technical communication over the same time period.

■ Scientific Literacy in Scientific Terms

A major obstacle to developing scientific literacy is a primary disjunction between scientific and humanistic habits of mind—competing mentalities that might impede the project of developing layered literacies. These patterns of thought inform the accepted standards for logic and convincing evidence. For example, mathematical aptitude and training predict success in science majors, even in disciplines like biology, that require relatively little mathematical training (Shapka et al., 2006). Kyla Flanagan and Jillian Einarson (2017) identified mathematical confidence as a more critical factor for success in college biology than "grit" (p. 1) or tenacity: as students gained mathematical confidence, their overall performance increased regardless of stick-to-it-iveness. Importantly, confidence was strongly associated with actual mathematical skills, which meant that students had low confidence *because* they lacked certain skills, as reflected in exam results (Flanagan & Einarson, 2017). Thus, mathematical skills might be a valuable literacy to layer into technical communication pedagogies that could be incorporated into the activities in Table 14.1.

Yet, according to the American Association for the Advancement of Science (AAAS; 1990), mathematical skills are only one important habit of mind that characterizes scientific literacy. Computation and estimation skills must be augmented by curiosity, openness to new ideas, informed skepticism, and material practices like manipulation and observation, as well as effective communication (AAAS, 1990). Since AAAS initiatives are aimed at primary and secondary schoolchildren, the AAAS values might be compatible with the aims of technical communication by enhancing the ability of all Americans to understand scientific information. Unfortunately, another stumbling block emerges here. While it may seem to humanists that the habits of mind described by the AAAS already characterize their own engagement with technical or scientific materials, scientific discourses reveal fundamental differences.

Such differences might derive from what scientific writing expert Scott L. Montgomery (2017) identifies as a contrast between writing training in the humanities and in the sciences in *The Chicago Guide to Communicating Science*. Montgomery, clearly addressing what he views as an audience of fellow scientists, notes that "a major difference between the humanities and sciences is that composing, critiquing and revising papers forms a central part of learning in the former, while in science it does not" (p. 5). Montgomery explains that scientists are "supposed to pick up" good habits of writing "either from a course or two in technical writing while at school, or through osmosis after entering the caffeine-riddled world of professional research" (p. 5). Montgomery reiterates a common truth for the culture of science, which Miller (1979) comments on as well, that writing is often understood as an obstacle to true science, an "opponent" (p. 6) that competes with content knowledge.

Montgomery (2017) also contrasts patterns of reading in humanities and the sciences, noting that attention to historical texts is a hallmark of the humanities but not the sciences. Of course, scientific findings often have a short self-life, being displaced quickly in the light of new discoveries, and even current reading is very demanding. As Allen H. Renear and Carole L. Palmer (2009) observe, scientific reading has long been tactically complex, requiring "strategically working with many articles simultaneously to search, filter, scan, link, annotate, and analyze fragments of content" (p. 828). And because scientists glean "fragments of content" (p. 828) for varied purposes, the writer cannot presume to dictate to the reader how or when to make use of the information provided. In effect, the process of reading is constructed as an act of scientific discovery, which might help explain Kujala and colleagues' acceptance of unclear terminology—they assume the caution will be enough because of the way they view reading.

Van Noorden (2014b) also notes that scientific work is continually subject to revision, even once published, creating a burden for readers to go back and double-check specifics. Further, given that training in scientific reading reinforces the skills Renear and Palmer cite, it is not only possible, but likely, that humanistic reading expectations are not a strong driver for scientific writing. To recur to Stormer's model of articulation and taxis, then, the ordering of items in scientific texts might be seen to function not as the formation of a specific argument so much as to allow other scientists to glean useful fragments for their own research. Scientific writing pedagogies like Bruce Schulte's (2003) "Parallel Hourglass Structure in Form and Content," hence, emphasize students' ability to place information where other researchers can expect to find it. Technical communication pedagogies should, at the least, acknowledge this reality.

■ Humanistic Critique and Scientific Literacy

A challenge to understanding scientific habits of communication and an obstacle to completing some of the activities in Table 14.1 may arise from humanistic reader expectations. Humanistic studies of scientific writing do not see the fragmentation and continuous revision of scientific materials that Montgomery, Renear, Palmer, or Van Noorden describe as value-neutral. This is a significant site of competing mentalities that has significant implications for developing scientific literacy. For example, in *Shaping Written Knowledge: The Genre and Activity of the Experimental Article in Science*, a critical text in writing studies and rhetoric of science, Charles Bazerman (1988) comments on the shift from argumentative to structured papers by the American Psychological Association as cause for complaint. For Bazerman, the fragmentation of argument not only across sections of a paper, but across multiple papers, increases reading burdens that should be undertaken by the author. And Bazerman further questions authors' knowledge in discontinuous narratives that present an introduction, methods, results, and discussion because "the author escapes the need for transitions to demonstrate the coherence of the enterprise" (p. 260). For Bazerman, certain rhetorical formulae are necessary to prove coherence and soundness of thinking, suggesting a fundamental disparity between his position and that of scientists well-acculturated to the reading practices Renear and Palmer or Montgomery describe. As Schulte (2003) explains, a certain logic informs the presentation of an introductory rationale for a study, its methods and results, followed by a discussion that highlights successes and failures and suggests next steps. So, Bazerman calls for specific modes of rhetorical articulation that provide a "complete" argument, which highlights the expectations that inform his humanistic mentality in such reading. Fostering this sort of expectation would limit the ability of students to develop scientific reading literacy on its own terms. And while Bazerman's book appeared several decades ago, it remains a foundational text in humanistic studies of scientific information, continuing to influence new generations of thinkers.

Bazerman (1988) also takes an approach Montgomery (2017) describes as characterizing humanities approaches to writing, as previously observed in "Owning Our Limits: Composition and the Discourse of Science"(DeTora, 2012). Beginning with the 1665 *Philosophical Transactions of the Royal Society*, Bazerman traces the development of the structured scientific format. More recent texts by rhetoric and writing studies experts like Alan Gross (2006), Jeanne Fahnestock (2005), or Michael Zerbe (2007) also present a history of scientific writing through the works of famous scientists like René Descartes, Sir Isaac Newton, and Crick and Watson, historical works that also feature in seminal linguistic studies by M. A. K. Halliday and J. R. Martin (1993). These scholars articulate a coherent, linear history that scientists might call into question. Such humanistic studies of scientific writing might also, like Zerbe's and Bazerman's, go on to criticize the rhetorical shortcomings of structured formats.[1] These studies drift from Miller's (1979) earlier remarks, casting scientific writing as positivist and instrumental rather than as an independent intellectual endeavor that intentionally articulates its writing practices in certain ways. Fahnestock characterized such moves as a "desire to dethrone science" (p. 272), calling for greater understanding to enrich rhetorical studies of science. Students in technical communication programs might benefit from reading this work in the context of the work of scientists as distinct modes of articulation and rhetorical performance rather than as a corrective.

Another generative pedagogical approach for technical communicators might be to explain and examine both the utility and the limitations of the works briefly reviewed here. For example, the retrospective historical progression of scientific writing manufactures an independent historical discourse of science that does not fully account for intellectual conditions before the disciplines became differentiated in the nineteenth century. As also previously noted (DeTora, 2012), critical discussions about the establishment of scientific education by figures such as Matthew Arnold and T. H. Huxley explicitly addressed the relationship between the sciences and humanities. Another limitation of historical progressions of scientific writing is a tendency to group works intended for popular and scientific audiences. For instance, Halliday and Martin (1993) use *Scientific American* and other popular texts in their linguistic analyses, which limits their applicability for scientists engaged in highly technical discourses of the types Montgomery, Van Noorden, Renear, and Palmer describe. Analyzing this tendency could provide better insights into work respected within technical communication contexts and create a model for understanding scientific discourses on their own terms.

Finally (and perhaps surprisingly for technical communication students), the current structured scientific format, again as Montgomery (2017) indicates,

---

1. Most scholars in writing studies, rhetoric, and technical communication would more strongly differentiate the authors I have grouped here. For example, Zerbe describes how to use rhetorical studies in freshman composition pedagogies, while Gross' and Fahnestock's works are more commonly read by scholars and students of rhetoric. Yet each of these authors, with the possible exception of Zerbe, can be seen to have influenced early discussions in technical communication, especially insofar as studies of the scientific format are concerned.

is only one dominant and longstanding communication model for scientists. Letters, brief communications, white papers, editorials, perspectives, and reviews each have an important place in the overall milieu of scientific writing, and all antedate the structured scientific format as currently published. While Bazerman (1988) describes differences across these genres as ongoing "innovations" (p. 319), Montgomery sees these same works as adhering to specific conventions and expectations that are grounded in a longer history. For example, *Science* specifies various article formats:

- Peer-reviewed research articles, reports, or reviews
- Commentaries
- Perspectives
- Book and media reviews
- Policy forums
- Letters
- eLetters
- Technical comments

Each of these formats follows specific aims, scope, and word counts, as well as the maximum number of tables, figures, and references. Many other journals share these formats and expectations. This circumstance suggests that what Bazerman (1988) views as invention in scientific writing formats actually follows fairly proscriptive rules. In fact, the structured research paper is intended as an aid to allow targeted reading by always presenting the same type of information in the same place. Editorials and reviews and perspectives, which gather information broadly, are vehicles for more complete arguments of the type Bazerman values. It could also be that similar opportunities for invention exist within structured formats but are more difficult to perceive for those less fluent in such communications. Thus, scientific writing literacy might be articulated not merely through humanistic understanding but also through the material and textual practices described by the AAAS, Montgomery, Schulte, Renear, and Palmer. These practices could be understood as one of what Garrison (2014) might call "competing mentalities."

How Scientists Construct Literacies

As Montgomery, Renear, Palmer, and others have noted, textual expectations among scientific audiences rely on certain habits of mind, which foster particular reading practices. Advice for students learning to read scientific literature often provides a heuristic for gleaning needed information with minimal expenditure of time and effort (Pain, 2016). Such heuristics often advise reading the abstract in a database to decide whether to review the full paper. And when reading a paper, tables and figures are often most worthy of initial review, making captions and legends crucially important. Discussions, results, methods, and introductions are less critical unless a reader is trying to replicate an experiment or use the data in some other way. These reading habits might seem alien to those used to reading linear narrative arguments. Of course, even the iterative and recursive modes of reading suggested in these forms are only stepping stones to fluent scientific readership: true expert readers can manage dozens of publications at once, as Renear and Palmer (2009) observe. A further challenge for developing fluency in scientific reading is managing vocabularies that vary from paper to paper. An effect of being left to pick up good practice, as Montgomery (2017) indicates, is a proliferation of vocabulary.

Scientists also develop novel vocabularies for writing about writing. For example, George D. Gopen and Judith Swan (1990)—whose "The Science of Scientific Writing" is widely used as a teaching tool in humanities-based science writing classes—introduce a vocabulary, borrowed from linguistics, for describing the functions of "units" of scientific discourse and the concept of "stress positions." They comment that even grammatically correct sentences can resist reading if they contain too much information or place details in a counterintuitive order. Ultimately, Gopen and Swan propose three essential "rhetorical principles": "grammatical subjects should be followed as soon as possible by their verbs," "every unit of discourse . . . should serve a single function or make a single point," and "information intended to be emphasized should appear at points of syntactic closure." They also discuss reader expectations: presenting what is known before what has been discovered, for example. Yet, Gopen and Swan also aim to retain jargon and complexity. Indeed, Gopen and Swan see "plain English" for "the general public" as a means of diluting science, an idea that runs counter to prevailing humanistic notions in technical communication.

A recent example of writing pedagogy by scientists is Tracy Ruscetti, Katherine Krueger, and Christelle Sabatier's (2018) "Improving Quantitative Writing One Sentence at a Time." This work exemplifies a trend in evidence-based scientific writing instruction that links writing success to specific content measures and/or test scores (see, for example, Morgan et al., 2011). The authors, teaching biologists, quantify the quality of quantitative statements in student laboratory reports, then use calculations to identify specific shortfalls for each student. Ruscetti et al. concluded that targeted feedback improved writing quality, that student writing quality decreased as content complexity increased, and that science teachers must adjust writing instruction for more complex conceptual tasks.

Of note and in contrast to Gopen and Swan (1990), Ruscetti et al. (2018) used writing studies texts as a means of contextualizing their findings. This indicates that scientists may seek to triangulate their findings by as many means as are available to them, which offers a vantage point for technical communication interventions. Morgan and colleagues' (2011) scientist/writing studies collaboration found that greater content comprehension translated into better student writing, indicating that the anecdotal scientific perspective that good writing stems from strong science mastery is not incorrect. What remains is to offer some specific means of translating this wisdom into pedagogical practice in technical communication.

While Table 14.1 has some suggestions, these are only a starting point. The most effective pedagogies might be those where students are offered models like those in Table 14.1 and asked to develop their own ideas as to how they might best leverage scientific knowledge in technical communication projects. In other words, another option for using Table 14.1 is as pre-work or preparation for specific projects.

Conclusion

Humanistic and scientific interpretations of the same genre conventions differ profoundly, which is a symptom of what Garrison (2014) might have termed competing mentalities. Thus, significant work is required to create a layered literacy model that includes scientific content literacy. Significant evidence supports the idea that humanists want to understand scientific textual practice in the same terms as they understand belletristic or critical texts, while scientists understand the same materials quite differently. This creates a fundamental disjunction that speaks to Meyer and Bernhardt's (1997) ideal of workplace literacy. Since technical communication often aims to translate complex information, like scientific data, into suitable forms to meet user needs, recognizing the disjunction between scientific textual expectations and humanistic ones is an important first step in meeting practical and pedagogical user needs. In other words, the scientific literacy that should be layered into these activities requires a recognition of the basic modes of scientific expression. The goal of technical communication to reconstruct desired humanistic formats from scientific ones can only be furthered by understanding source texts. Since calls for better attention to workplace realities are not possible by current technical communication pedagogies without recourse to the humanistic values that now inform the field, reconciling these competing mentalities would be an important first step.

Technical communication constituted as a humanistic major must provide pedagogical solutions for students to manage scientific information for a general audience. Documents written by and for theoretical and applied scientists, as well as data outputs, provide essential source material for many technical communicators. Hence, technical communication pedagogies might benefit from understanding the production and pedagogy of such scientific textual materials. A key disjunction in this process was identified in the late twentieth century by Miller (1979) and Bazerman (1988): scientific writing does not look, or operate, like humanistic writing, which opens the possibility for criticism and critique rather than understanding and respect. By recognizing that reading practice in the sciences, as Renear and Palmer (2009) indicate, consists largely of scanning for key information and assembling useful datasets, the logic of scientific writing pedagogies becomes more apparent. Technical communication pedagogies can leverage existing advice in the sciences, like Pain's (2016) model for scientific reading, to help students gain fluency in reading texts that, as indicated by Gopen and Swan (1990), are intentionally constructed using difficult jargon in order to maintain the integrity of scientific information.

■ References

American Association for the Advancement of Science. (1990). *Science for all Americans online.* http://www.project2061.org/publications/sfaa/online/sfaatoc.htm.

Bazerman, C. (1988). *Shaping written knowledge: The genre and activity of the experimental article in science.* Wisconsin University Press.

Benau, D. (2020). An overview of medical and regulatory writing. In L. DeTora (Ed.), *Regulatory writing: An overview* (2nd ed., pp. 1–14). Regulatory Affairs Professionals Society.

Bowdon, M. A. & Scott, J. B. (2003). *Service-learning in technical and professional communication.* Longman.

Cargile Cook, K. (2002). Layered literacies: A theoretical frame for technical communication pedagogy. *Technical Communication Quarterly, 11*(1), 5–29.

Carpenter, J. H. (2011). A "layered literacies" framework for scientific writing pedagogy. *Currents in Teaching and Learning, 4*(1), 17–33.

DeTora, L. (2012). Owning our limits: Composition and the discourse of science. In J. Rich & E. D. Lay (Eds.), *Who speaks for writing: Stewardship in writing studies in the 21st century* (pp. 49–60). Peter Lang.

Fahnestock, J. (2005). Rhetoric of science: Enriching the discipline. *Technical Communication Quarterly, 14*(1), 279–286.

Flanagan, K. M. & Einarson, J. (2017). Gender, math confidence, and grit: Relationships with quantitative skills and performance in an undergraduate biology course. *CBE life sciences education, 16*(3), ar47. https://doi.org/10.1187/cbe.16-08-0253.

Garrison, K. (2014). The scientist, philosopher, and rhetorician: The three dimensions of technical communication and technology. *Journal of Technical Writing and Communication, 44*(4), 359–380.

Gopen, G. D. & Swan, J. (1990). The science of scientific writing. *American Scientist.* https://cseweb.ucsd.edu/~swanson/papers/science-of-writing.pdf.

Gross, A. G. (2006). *Starring the text: The place of rhetoric in science studies.* Southern Illinois University Press.

Halliday, M. A. K. & Martin, J. R. (1993). *Writing science: Literacy and discursive power.* University of Pittsburgh Press.

Johnson, T. (2019a). *Should you get a degree in a tech comm program? Two considerations to keep in mind.* I'd Rather Be Writing. https://idratherbewriting.com/2019/03/13/should-you-enroll-in-tech-comm-program/#tech-comm-programs-are-a-popular-topic.

Johnson, T. (2019b). *Tech comm trends: Providing value as a generalist in a sea of specialists* [Keynote address]. Symposium on Communicating Complex Information, Shreveport, LA, United States. https://idratherbewriting.com/2019/02/24/slides-for-trends-preso-symposium-for-communicating-complex-info/#my-presentation-on-trends.

Kujala, S., Roto, V., Väänänen-Vainio-Mattila, K., Karapanos, E. & Sinneläa, A. (2011). UX curve: A method for evaluating long-term user experience. *Interacting With Computers, 23*(5), 473–483.

Melançon, L. (2017). Patient experience design: Expanding usability methodologies for healthcare. *Communication Design Quarterly*, *5*(2), 19–28.

Meyer, P. R. & Bernhardt, S. A. (1997). Workplace realities and the technical communication curriculum: A call for change. In C. Staples & C. Ornatowski (Eds.), *Foundations for teaching technical communication: Theory, practice, and program design* (pp. 85–98). Ablex.

Miller, C. R. (1979). A humanistic rationale for technical writing. *College English*, *40*(6), 610–17. https://www.jstor.org/stable/375964.

Montgomery, S. L. (2017). *The Chicago guide to communicating science* (2nd ed.). Chicago University Press.

Morgan, W., Fraga, D. & Macauley, W. (2011). An integrated approach to improve the scientific writing of introductory biology students. *The American Biology Teacher*, *73*(3), 149–153.

Pain, E. (2016, March 21). How to (seriously) read a scientific paper. *Science*. https://doi.org/10.1126/science.caredit.a1600047.

Renear, A. H. & Palmer, C. A. (2009). Strategic reading, ontologies, and the future of scientific publishing. Science, 325(5942), 828–832. https://doi.org/10.1126/science.1157784.

Ruscetti, T., Krueger, K. & Sabatier, C. (2018). Improving quantitative writing one sentence at a time. *PLOS ONE*. https://doi.org/10.1371/journal.pone.0203109.

Schulte, B. A. (2003). Parallel hourglass structure in form and content. *The American Biology Teacher*, *65*(8), 591–594.

Seffah, A. & Metzker, E. (2004). The obstacles and myths of usability and software engineering. *Communications of the ACM*, *47*(12), 71–76.

Shapka, J. D., Domene, J. F. & Keating, D. P. (2006). Trajectories of career aspirations through adolescence and young adulthood: Early math achievement as a critical filter. *Educational Research and Evaluation*, *12*(4), 347–358.

St.Amant, K. (2017). The cultural context of care in international communication design: A heuristic for addressing usability in international health and medical communication. *Communication Design Quarterly Review*, *5*(2), 62–70. https://doi.org/10.1145/3131201.3131207.

Stormer, N. (2004). Articulation: A working paper on rhetoric and *taxis*. *Quarterly Journal of Speech*, *90*(3), 257–284.

Van Noorden, R. (2014a, May 4). *Global scientific output doubles every nine years*. Nature newsblog. http://blogs.nature.com/news/2014/05/global-scientific-output-doubles-every-nine-years.html.

Van Noorden, R. (2014b, February 3). *Scientists may be reaching a peak in reading habits*. Nature newsblog. https://www.nature.com/news/scientists-may-be-reaching-a-peak-in-reading-habits-1.14658.

Zerbe, M. (2007). *Composition and the rhetoric of science: Engaging the dominant discourse*. Southern Illinois University Press.

15. Technical Communication Pedagogy and Layered Literacies in Workplace Training Courses

Elizabeth L. Angeli
MARQUETTE UNIVERSITY

Abstract: Technical communication scholarship explores how students who are enrolled in university courses acquire and transfer rhetorical skills and literacies into the workplace (Beaufort, 2008; Brent, 2011; Cargile Cook, 2002; Haas, 1996; Russell, 2007; Winsor, 1996). Absent from these discussions are workplace writers who may not attend college and instead complete non-academic training courses. Often these writers are expected to complete technical documents without receiving formal writing instruction (Amidon, 2014; Angeli, 2015, 2019). The widely used layered literacy framework (Cargile Cook, 2002) offers a way in which to understand how writers learn multiple literacies in the hybrid workplace-classroom context of workplace training programs. As such, this chapter uses the layered literacy framework to better understand how workplace communicators learn multiple literacies outside of the technical communication classroom and inside workplace training. In turn, this chapter shows how the layered literacies framework informs other actions related to writing, including decision making and synthesizing data, and how the framework is strained when applied to workplace training contexts.

Keywords: workplace education, adult learners, EMS report writing, EMS education, fire department training

Key Takeaways:

- The concept of layered literacies informs other actions related to writing, especially decision making and synthesizing data, which students can develop in the technical communication (TC) classroom.
- Workplace communicators learn multiple literacies outside of the TC classroom and inside workplace training.
- The workplace training classroom suggests that literacies, especially embodied and multisensory, are more than layered; they are symbiotic and in tension.

Technical communication (TC) scholarship has explored how students enrolled in university courses acquire rhetorical skills and literacies they learn in academic TC courses and transfer them into the workplace (Bay, 2006; Beaufort, 2008; Brent, 2011; Cargile Cook, 2002; Haas, 1996; Munger, 2006; Russell, 2007;

Winsor, 1996). As a part of this discussion, scholarship focuses on how we as TC teachers can bridge the gap between academia and the workplace, pointing to internships, service learning, co-ops, and client-based projects as ways to prepare students for the classroom-to-workplace transition.

Absent from these discussions, though, are courses and training programs developed by and offered through workplaces. Research on workplace training education related to writing or literacy has been conducted outside of the United States (Matthews, 1999; Taylor, 2000), and research conducted in the United States is at least ten years old (Bogert, 1989; Hollenbeck, 1993; Karlson, 1991; Thrush & Hooper, 2006). Despite this gap in scholarship, the workplace training context promises to teach TC teachers and scholars much about how we might prepare our students for workplace writing.

In the United States, workplace training programs are often designed by employers and immerse their employees in workplace-specific skills and the workplace environment. As such, workplace training programs are a hybrid space because they are both workplaces and classrooms. The programs are not typically offered at colleges, and they are not just workplace activities—they are on-the-job training, developed by workplace instructors who may or may not have pedagogical training. These training programs, then, can provide insight into how workplace instructors prepare their employees for their local workplace. In turn, university TC teachers might better understand how to prepare college students for the workplace.

The profile of participants in these workplace training programs varies in a few ways. Age range varies, from participants who are directly out of high school to participants who have decades of workplace experience. They might be current employees who are in the program in order to complete continuing education, or they might be new to the workplace with limited workplace experience. In turn, education levels vary, ranging from a high school diploma to a few college credits to bachelor's degrees and completed certifications. Programs vary in length too; continuing education programs can last a few hours, while workplace training programs that are designed for new employees can span a few days, weeks, months, or years.

The training program that I'm currently studying is a three-year firefighter and paramedic training program that enrolls recent high school graduates. In this way, these students, called "cadets," are similar to students who are enrolled in our university TC courses with one exception: the cadets are expected to complete high-stakes workplace documentation every day on the job with about only two hours of formal writing instruction, which is significantly less time than we spend with our TC students who might not be entering such high-stakes workplaces.

In this chapter, I contribute to workplace training education scholarship, specifically workplace writing, by examining how the TC theory of "layered literacies" manifests in and is strained by workplace training contexts. In doing so, I aim to better understand how students develop the requisite TC skills that have been shown to be fundamental to the workplace. Thus, this chapter answers two questions:

1. How do we account for teaching TC in different contexts with current theoretical and/or pedagogical knowledge, specifically layered literacies?
2. How do students in workplace training programs prepare for employment opportunities that include managing, reporting, and sharing diverse types of data to multiple audiences in various discourse communities?

To answer these questions, I draw on the first two years of a six-year longitudinal research project with my research site, an urban fire department's three-year training course, known as the Training Academy. The Academy trains recent high school graduates to become firefighters and paramedics through rigorous physical training, classroom-based teaching and testing, and field training. Although the cadets may not see themselves as technical communicators, TC is fundamental to their workplace (Angeli, 2019). It is the lifeblood that allows them to achieve workplace goals, and as part of this work, the cadets are expected to develop layered literacies to care for patients and document medical decisions.

▪ Layered Literacies and TC Pedagogy

Scholars have argued that technical communicators' work is often invisible in the workplace (Brady & Schreiber, 2013), and part of this work involves "layered literacies." As developed by Kelli Cargile Cook (2002), layered literacies refer to a framework that structures the six interrelated "key literacies" of TC pedagogy: "basic, rhetorical, social, technological, ethical, and critical" (p. 7). This framework has been applied to a number of studies that range from research on the design of informed consent forms (Wright, 2012) and TC certificate programs (Turner & Rainey, 2004) to assessment (Brinkman & van der Geest, 2003; Thomas & McShane, 2007), service-learning and civic engagement (Dush, 2014; Eble & Gaillet, 2004; Turnley, 2007), and civic and ethical literacies (Batova, 2013; Hannah, 2010; Kienzler & David, 2003).

Recently, scholars have added two literacies to this framework: embodied and multisensory. Embodied literacy is a way that students develop a "critical awareness of one's own embodied positions, as the technical communicator, in relation to users and technical documents" (Swacha, 2018, p. 264). Citing Jay Timothy Dolmage, Kathryn Y. Swacha notes that the body is rhetorical and informs the technical communicator's writing processes and practices. Likewise, in healthcare contexts, such as prehospital emergency medicine, the patient's body is rhetorical; it communicates information through various pathways, including vital signs when hooked up to a telemetry machine, changes in skin color and temperature, and pupil dilation (Angeli, 2019; Fountain, 2014; Melonçon, 2017).

Focusing further on healthcare contexts, "multisensory literacies" are a way to understand healthcare's embodied, mediated experience and how expert and non-expert users engage their senses in medical settings (Bivens et al., 2018). For paramedics, the patient's body and the patient's environment cue them into treatment decisions. For example, if a patient has an altered mental status due to low blood sugar, paramedics will follow treatment protocols to increase the patient's blood sugar. Then, to ensure the patient's blood sugar remains at a healthy level, paramedics might look to see if the patient has food in the house. If the patient does not, the paramedic might suggest to the patient that they transport the patient to the hospital to ensure the patient's blood sugar doesn't drop again. To do so, paramedics will draw on their rhetorical persuasive skills to convince the patient to go to the hospital.

Despite the wide range of studies that apply the layered literacies framework, less explored is how this framework translates into training courses outside of the university. In these contexts, students are expected to have knowledge of multiple literacies: basic, technological, rhetorical, ethical, critical, and social. My research site, the Academy, offers an opportunity to understand how layered literacies manifest in non-academic training courses. In turn, the Academy teaches us about two newer, interrelated literacies, embodied and multisensory, which are fundamental to the healthcare workplace. Additionally, this site shows how the workplace-classroom hybrid space strains parts of the layered literacy framework, thus illustrating how it risks being unstainable (Lawrence & Hutter, this collection).

■ The Research Site

This chapter is drawn from a six-year, multi-cohort longitudinal research project with the Milwaukee Fire Department's (MFD) Cadet Training Academy (IRB Protocol #HR-3332, approved 5/19/2017). In the Academy, cadets earn four licenses, each demonstrating a higher level of firefighting skill and medical care: EMT (Emergency Medical Technician), Fire 1, Fire 2, and Paramedic. Cadets also take national exams to earn their National Registry EMT and Paramedic certifications. To apply their skills in the field, cadets complete three rounds of ride-alongs: 4-hour EMT shifts, 8-hour paramedic shifts, and 12-hour paramedic shifts. During these shifts, they must care for a certain number of patients and complete a certain number of hours, and as they move through their shifts from EMT to paramedic, their medical and writing responsibilities change and increase in complexity. For example, during EMT-BASIC, cadets handwrite patient care narratives on paper and include basic information, such as patient vitals and transport decisions. Then, as cadets start their paramedic training, their medical responsibilities and decisions become more complex, and they document all decisions and interventions electronically, requiring them to remember more information and synthesize it into a patient care narrative.

My larger research project answers two questions:

1. If workplace communicators do not attend college, how do they develop effective technical workplace communication skills?
2. In contexts in which college-aged students are immersed in the discipline (e.g., fire science) instead of taking university courses about the discipline (e.g., "Writing about Fire Science"), what curricula can promote technical writing skills required to succeed in the workplace?

The first three years of this longitudinal study were dedicated to a pilot study. In the pilot, I conducted classroom observations, collected survey responses from 18 cadets, and worked with six focal participants. Focal participants completed one 40-minute one-on-one interview with me and two 30-minute focus groups. Additionally, I collected all of the writing these participants completed, which includes their entrance exams, class notes, clinical notes, and practice patient care reports (PCRs), which documents the decisions made and treatments provided during a 911 response. Additionally, I observed each of them for eight hours during their paramedic field training where they work under MFD providers (who are paramedics). I rode along with the cadets and their supervising providers as they responded to 911 calls. This involved a total of 48 hours of field observations.

As such, collected data includes completed surveys, field notes from observations, audio recording and transcripts from interviews, and student writing (class notes, completed writing assignments, and practice PCRs). Collected data also includes completed PCRs that current MFD providers submit, which provide a point of comparison for writing the cadets will be expected to complete once they graduate from the Academy. This chapter is informed specifically by my work with six focal participants whom I have been working with for two years at the time of writing.

▌ EMS Providers are Technical Communicators

First responders, who include Emergency Medical Service (EMS) providers, are rhetoricians and technical communicators in many respects. In their workplace, written documentation skills and visual literacy play a significant legal and medical role (Amidon, 2014; Angeli, 2015, 2019; Helferich, 2016; Seawright, 2017). As technical communicators, EMS providers translate medical language to lay people throughout a response, most importantly to the patient. Their writing responsibilities include completing a PCR for every 911 response in which they integrate their observations, actions, and memory from the 911 response into a narrative-style summary. Ultimately, the PCR persuades various audiences that the actions EMS took were appropriate and effective, and it allows those audiences to continue patient care. Specifically, the PCR audiences include

- quality assurance professionals who review quality of medical care provided in the field;

- medical directors and medical examiners who also review quality of medical care and investigate high-profile responses, like a shooting or multi-vehicle car accident;
- physicians and nurses who continue patient care in the hospital;
- insurance and billing companies who determine medical coverage; and
- lawyers who litigate law suits.

The consequences of ineffective, inaccurate, unpersuasive PCRs can be dire. Patients may be left with large medical bills, EMS agencies may not be reimbursed for supplies and expenses, and if called to testify, an EMS provider may lose her license if the PCR does not persuade a lawyer, judge, or jury that effective care was provided to a patient. As a persuasive document, the PCR narrative is the culmination of a complex writing process that, I argue, draws on layered literacies.

EMS Writing and Pedagogy

Despite its importance and complexity, writing in the EMS workplace is given little attention compared to the other critical skills EMS providers learn. For example, EMS providers learn clinical skills, like intubation, and they practice these skills many times to ensure their skills result in effective patient care. However, they might not use this skill every day on the job; they may use it once a week, once a month, or once every three months depending on their community's medical needs. The PCR, on the other hand, is completed at the end of every 911 response. As such, if an EMS provider cares for seven patients on her shift, she will write seven reports that day.

Turning to my research site, MFD's primary role in Milwaukee is to provide the community with emergency medical care. In 2017, 81 percent of all MFD fire and EMS responses were EMS responses (Milwaukee Fire Department, 2017), and at the end of each response, MFD employees are obligated to write a PCR. Despite this prominence of report writing in their workplace, MFD's Training Academy, like most EMS training programs in the United States, lacks a formal mechanism to teach PCR writing. Based on the first 24 months of my pilot study and my near ten years as an EMS writing researcher, I have learned that writing is taught as a product instead of the complex process we in TC understand writing involves. These workplace training programs tend to use what writing scholars would recognize as "the inoculation model" of writing instruction: students are introduced to the basics of documentation in a few hours and then practice it during their field training, and if students are given feedback, feedback strategies may not follow best practices. EMS textbooks also reinforce this model; a typical textbook is around 1,500 pages, and about five pages (~0.003%) are spent on documentation.

The inoculation model does not reflect the writing and literacy practices of EMS providers. Instead, it shortchanges these literacies, all of which influence their ability to practice prehospital medicine and document patient care. The layered literacies framework offers one way we can teach students how to gather, manage, report, and distribute data to a variety of stakeholders and audiences.

I should note that the Academy does not use language from the layered literacy framework. Rather, I am making the implicit work that cadets engage in explicit. In doing so, we might learn how to tend to students' layered literacies, including embodied and multisensory literacies, inside and outside the TC classroom.

▪ Coding Scheme

For this chapter, I draw on my corpus of data collected from working with six focal participants who were Academy cadets. This corpus includes their completed surveys, field notes from my classroom and field observations, audio recording and transcripts from interviews and focus groups, and cadets' writing (class notes, completed writing assignments, and practice PCRs). Using layered literacies as codes, I identified areas in my corpus where students applied or enacted these literacies (see Table 15.1).

Table 15.1. Coding scheme of how layered literacies manifest in the study's corpus, including focal participants' writing, my field notes, and interview and focus group transcripts

Literacy	Definition	Example
Basic	"The ability to read and write" (Cargile Cook, 2009, p. 8)	Cadets read and write during class to take notes and during ride-alongs when they re-read protocols and documents.
Technological	Learn and use technology; understand how technology facilitates social interaction and action.	Cadets interact with medical technology during ride-alongs and refer to information gathered through technology, such as cardiac rhythms, when writing up reports.
Rhetorical	"[A]nalyze, evaluate, and employ various invention and writing strategies based upon [students'] knowledge of audience, purpose, writing situation, research methods, genre, style, and delivery techniques and media" (Cargile Cook, 2009, p. 10).	Cadets discuss the audiences and purposes for writing during ride-alongs and address them in their practice PCRs.

Literacy	Definition	Example
Social	"Collaborate and work well with others" (Cargile Cook, 2009, p. 11).	Cadets collaborate with one another to practice skills and to write the practice narrative reports. They asked providers for help, too. During field training, they eat meals and care for the fire stations with MFD providers.
Ethical	"Knowledge of professional ethical standards ... [and the ability] to consider all stakeholders involved in a writing situation" (Cargile Cook, 2009, p. 15)	Cadets focus on providing effective, appropriate, professional patient care and communication. In their PCRs, cadets learn how to document objective information ("The patient struggled to walk to the stretcher, falling twice.") instead of subjective information ("The patient was obviously drunk.").
Critical	"Ability to recognize and consider ideological stances and power structures and the willingness to take action and assist those in need" (Cargile Cook, 2009, p. 16)	Cadets understand where they fall in the hierarchies of patient-healthcare provider and between cadet-MFD provider. They knelt next to patients when providing treatment and let patients know what they were doing, verbalizing all moments of care: "Sir, I'm going to listen to your heart now." They are also aware of the necessary hierarchy in the firehouse during ride-alongs, for example, referring to providers as "sir" and "ma'am" and standing at attention when MFD providers were talking.
Embodied and multisensory	"[T]he ability to understand how bodies and embodied experiences affect and are affected by how users interact with technologies and texts in varied physical, material ways" (Swacha, 2018, p. 261) "[A]ural, tactile, and visual experiences" that inform interaction of users and health (Bivens et al., 2018)	Cadets learn how to gather information from the environment and the patient's body, and how their environment impacts their ability to provide treatment and, thus, document. One cadet was frustrated by not being able to start an IV because the ride in the ambulance was bumpy due to pot holes in the road. The provider offered feedback, sharing that the cadet needs to trust his skills despite a bumpy ride. The providers assured the cadet that, with time, he'll learn which roads are smoother and conducive to starting an IV; even though providers can't tell which roads they're on in the back of an ambulance, they learn where they are by how the ride feels and by the number of turns the ambulance takes.

Analysis of pilot study data suggests that cadets are expected, although not explicitly, to develop layered literacies (see Table 15.1). For example, they engage their embodied and multisensory literacies by completing high intensity workouts to prepare their bodies for and to mimic the intensely physical scene of a fire rescue. In doing so, cadets better understand how to perform in high-stress, potentially dangerous situations. Of most importance to TC, though, is that cadets integrate all of these literacies to write a PCR that justifies their decisions and actions during a 911 response. To complete this technical document, cadets draw on multiple literacies to integrate their observations, actions, and memory from the 911 response. Overall, layered literacies manifested in their learning process, specifically during their field training and in their written products, especially their practice PCR narratives.

■ Literacies Are Interrelated and in Tension

Although all the literacies informed cadets' learning and medical practice, the literacies were interrelated and, perhaps most notably, in tension. This tension is part of the interrelated relationship of layered literacies (Bay & Blackmon, 2016), and Cargile Cook (2002) notes that this tension is at the core of how layered literacies work in TC.

Focal participants navigated this tension as they developed their literacies, which was most noticeable during focus groups when they shared their experiences with documenting in the field. During paramedic training, cadets completed their practice PCRs electronically, and to facilitate their writing and to help them remember information, participants reported that they took notes by hand during a response, and their notes were chronological, following the order of steps and actions taken during a call, from start to finish. When they transferred these notes to the computer, participants noted they felt frustrated because the organization of the computer report did not follow their note-taking methods in the field. This genre shift "interrupted" their writing process, impacting their ability to recall details and present a cohesive, synthesized narrative. Participants shared they preferred to write by hand because the technology was limiting and did not reflect how they organized their ideas—the technology, in other words, did not mirror their writing process. Participants were concerned about this tension because they knew they would be required to write PCRs electronically once they became MFD providers.

In this way, literacies are in tension. Learning technological literacy bumps up against their basic, multisensory, and embodied literacies in that technology does not facilitate the information gathered through and created by these literacies. When participants wrote in the field, they shared that they walk through the response in their head to recall the care they provided (embodied, multisensory) and the information gathered through medical technology (technological). When they wrote with technology, they reported that their writing ability was negatively impacted.

In sharing this experience, though, participants demonstrated their growing awareness of their literacies, specifically their embodied and rhetorical literacies. Their responses indicate they are aware that their physical interaction with technology is in relationship with their writing and documentation abilities. Likewise, by sharing that their writing-by-hand process allows them to follow the chronological order of the call and better recall details, they demonstrate their rhetorical literacy: they know that their documentation needs to include details, because one of the main purposes of their documentation is to re-create the 911 response in writing. In achieving this purpose, they allow other stakeholders to act accordingly and appropriately, which is a key element of social and ethical literacies.

■ Literacies Manifest in the PCR Narrative

Although cadets did not use the term "layered literacies," they noted and demonstrated that they were developing each literacy by gathering information for and by writing their practice PCRs. I observed these developments during ride-alongs, in interviews, and in focus groups, and these literacies also manifested in their practice PCRs. In doing so, cadets demonstrated layered literacies as they integrated actions, observations, and memories into their writing.

The following two PCR narratives were written by two participants, Gary and Sophie, who were completing ride-alongs together during their EMT training. In this response, they were called to the scene of a motor vehicle accident with two cars and one patient, and following their training requirements, they hand-wrote their practice PCR narratives after the response. I have included in brackets where the literacies are present in their narratives.

> Gary's practice PCR narrative:
>
> Dispatched to special case. We arrived on scene to find a 4 door sedan that had been T-boned [multisensory]. Intrusion on the vehicle was about 20 inches [multisensory]. Patient was loaded into MED and 2 IVs were placed [technological, embodied, ethical]. Patient complained of pelvic pain but pelvis was stable on palpation [embodied, rhetorical]. A c-collar was applied and phentynole was administered IV [technological, embodied, ethical]. Patient was initially hypertensive. After placing patient in Trendelenburg position, BP and vitals stabilized [embodied, multisensory]. Tenderness was found in the RLQ on palpation [embodied].
>
> Sophie's practice PCR narrative:
>
> Med X was dispatched to the scene of a car accident with 1 patient. Upon arrival, the patient was standing next to the car and reported that she got out through the passenger side after her side, the driver side, was impacted [multisensory, rhetorical]. Since there was an intrusion of about 20 inches, a major injury is possible [multisensory, rhetorical, critical, ethical, embodied]. The patient reports

pelvic and back pain. After lying on the cot, she was moved into the med unit for further evaluation. A cervical collar was applied and warm IV fluids were administered via 18 gauge [technological]. With a pain level of 8, pain meds were administered which improved the pain to a level 2 [technological, rhetorical, ethical]. No DCAP-BTLS was found except tenderness in the abdomen and pain in the pelvis and back [embodied]. Rapid transport was given to HOSPITAL and care was transferred to the ER staff [rhetorical, social, ethical].

These practice narratives suggest that participants integrate the information gathered through literacies into their writing, thus demonstrating their skills and achieving the rhetorical purpose of the narratives. The detailing of appropriate medical treatment decisions indicates participants' ethical literacies, as they are tending to their patient's needs in effective ways. They understand how technology, like IVs and c-collars, contributes to patient care, and they include how their senses and the patient's body informs treatment decisions and medical care, highlighting their embodied and multisensory literacies.

Additionally, participants' rhetorical literacies are visible, most notably in the second cadet's narrative. In this narrative, Sophie highlights her understanding that a legal audience might read her document: "Upon arrival, the patient was standing next to the car and reported that she got out through the passenger side after her side, the driver side, was impacted." Sophie relies on her observation skills to detail this part of the response, and she also demonstrates her rhetorical skills, especially her audience awareness. In this excerpt, Sophie details that the patient was standing next to the car upon arrival, and in this enthymemic statement, the missing detail is that the responding EMS crew did not move her—Sophie reported that the patient moved herself by exiting from the passenger side. This move may have required the patient to potentially twist and turn, which could implicate underlying injuries. If, after the response, the patient experienced medical problems that were potentially related to this car accident, like paralysis, and if the patient sued the EMS agency, a lawyer would read this narrative to learn more about the scene. The lawyer might consider questions like "Did the EMS crew move the patient? Were those movements appropriate? Might they have caused the patient to become paralyzed?" By noting that the patient was "standing next to the car upon arrival," Sophie speaks to a potential legal audience and answers these questions, with the subtext, "Our crew did not move her; she moved herself. We took care to treat her pain and ensure she did not have imminent life threats. We secured her neck to prevent any spinal cord damage. If she becomes paralyzed or has related issues in the future, it was not due to our care." To develop this subtext, Sophie drew on her rhetorical literacy, and to convince her potential legal audience, she integrated details and evidence drawn from her embodied and multisensory literacies.

Embodied and Multisensory Literacies Are Interdependent

Findings suggest that multisensory and embodied literacies are interdependent. During coding, many of the references that I coded "multisensory" also applied to "embodied." This dual-coding suggests that these literacies have a symbiotic relationship in the healthcare workplace training environment, and they are so intertwined that they might actually be one literacy instead of two in this setting. Likewise, this finding also suggests that the linear nature of layered literacies might not be sustainable (Lawrence & Hutter, this collection).

I hesitate to combine these two literacies or choose between them for two reasons. First, I do not want to contribute to the layered literacy framework's expansiveness that Lawrence and Hutter discuss in this collection. Second, reducing these two literacies to one risks making it fit the framework's linear structure and removes any nuanced exploration of the literacies' symbiotic relationship. The following paragraphs are my attempt at this exploration.

To understand how their bodies and patient's bodies influence the writing, data-gathering, and medical care process (embodied literacies), participants gathered information from the environment through their own individual senses (multisensory literacy). They needed to focus on where they were in the response—in relation to the patient or other pertinent objects, like a car in a car accident—to gather data and respond accordingly. In doing so, participants developed their situational and sensory awareness (Angeli, 2019), which informs embodied and multisensory literacies.

For example, during classroom observations, cadets learned about toxicology and how to treat and transport patients who were intoxicated or overdosed. The instructor emphasized that, especially when responding to these situations, cadets need to "know your surroundings" because "you pretty much know if they're [the patient] poisoned, intoxicated, or overdosed" upon visual assessment: "look at the bottle, the date prescribed, today's date, the number of pills left, and the dosage amount; put your police hat on." The instructor verbally described what a scene could look like, noting that cadets need to pay careful attention to a patient's environment because a patient might report that she has not been drinking, but "six empty beer cans" may be on the front porch, indicating that *someone* has been drinking. At that point, the responding EMS provider must use his senses to quickly determine if the patient is lying or if someone else drank the beer, because that information will inform an EMS provider's treatment decisions.

During this lecture, cadets were expected to develop their content expertise in treating an intoxicated or overdosed patient. In doing so, cadets were encouraged to develop their sensory and environmental awareness, fine tuning both their multisensory and embodied literacies. Without the information they gathered from embodied and multisensory literacies, they would not be able to enact their basic and rhetorical literacies to write an effective PCR.

In a way, embodied and multisensory literacies serve as data collecting literacies in this training course—they are methods by which participants gathered information so that they could write, communicate, and use technology to provide patient care. As such, one part of embodied literacy that the EMS workplace helps us better understand is the "interaction between technical documents and bodily, material experience" (Swacha, 2018, p. 263). The technical document—the PCR—reflects, accounts for, and captures the bodily, material experience of a 911 response. It is an artifact that captures a moment in time that helps other people act and continue patient care. Without a well-written, detailed PCR, other stakeholders of a 911 response, like lawyers or other healthcare providers, cannot act.

■ Takeaways and Future Directions

By examining the layered literacies framework in workplace training courses, we can see how this framework applies outside of the TC classroom and to the workplace writing process and product. In turn, we learn that the concept of layered literacies informs other actions related to writing, including decision making and synthesizing data. For example, layered literacies remind us that workplace writing is an ethical activity, and along with it, then, decision making and synthesizing data must be seen as ethical activities that require careful reasoning. In other words, when gathering data through the senses, as Academy cadets do, this data must be gathered ethically and appropriately in order for the corresponding written product—the PCR—to be ethical.

Likewise, literacies are visible in more spaces than a written product and must be tended to throughout the learning process. This process extends beyond one course, and the Academy demonstrates this learning trajectory. Cadets develop their literacies over three years and in different contexts, which include the classroom and various field contexts on ambulances, in simulation training, and in fire stations. Looking at the TC classroom, layered literacies might be scaffolded throughout TC courses and curricula and tended to in a variety of spaces, including in the classroom, service-learning, co-ops, and internships.

Additionally, this research raises important questions for TC instructors to consider, especially related to accessibility and layered literacies. The Academy and the EMS and fire service workplace require cadets and providers to be physically fit, to be capable of completing demanding work, and to be able to hear, see, and speak so they can complete required tasks, like lifting and carrying patients to safety. This physical work is important to their embodied and multisensory literacies, as outlined above, and it requires cadets to be physically able to complete this work. How might we engage students' physically in the classroom while also tending to accessibility? Might embodied and multisensory literacies be developed in ways that do not privilege all senses and physical abilities? How might we integrate sensory and situational awareness into the design and writing process while also tending to accessibility? What might a multisensory, embodied, and accessible TC classroom look like?

Ultimately, this piece highlights the complex literacy work required in a first responder training course, and it presents a few ways that training courses facilitate layered literacies development. In doing so, it also demonstrates how two newer literacies, embodied and multisensory, inform TC workplace practice.

■ References

Amidon, T. R. (2014). *Firefighters' multimodal literacy practices* [Doctoral dissertation, The University of Rhode Island]. DigitalCommons@URI. https://digitalcommons.uri.edu/dissertations/AAI3619422/.

Angeli, E. L. (2015). Three types of memory in emergency medical services communication. *Written Communication, 32*(1), 3–38.

Angeli, E. L. (2019). *Rhetorical work in emergency medical services: Communicating in the unpredictable workplace.* Routledge.

Batova, T. (2013). Legal literacy for multilingual technical communication projects. In K. St.Amant & M. Rife (Eds.), *Legal issues in global contexts* (pp. 83–101). Routledge.

Bay, J. (2006). Preparing undergraduates for careers: An argument for the internship Practicum. *College English, 69*(2), 134–141.

Bay, J. L. & Blackmon, S. (2016). Inhabiting professional writing: Exploring rhetoric, play, and community: Second Life. In J. DeWinter & R. M. Moeller (Eds.), *Computer games and technical communication: Critical methods and applications at the intersection* (pp. 211–232). Routledge.

Beaufort, A. (2008). Writing in the professions. In C. Bazerman (Ed.), *Handbook of research on writing: History, society, school, individual, text* (pp. 269–289). Routledge.

Bivens, K. M., Arduser, L., Welhausen, C. A. & Faris, M. J. (2018). A multisensory literacy approach to biomedical healthcare technologies: Aural, tactile, and visual layered health literacies. *Kairos, (22)*2. http://kairos.technorhetoric.net/22.2/topoi/bivens-et-al/index.html#1.

Bogert, J. (1989). Improving the quality of writing in the workplace: A case study. *Management Communication Quarterly, 2*(3), 328–356.

Brady, M. A. & Schreiber, J. (2013). Static to dynamic: Professional identity as inventory, invention, and performance in classrooms and workplaces. *Technical Communication Quarterly, 22*(4), 343–362.

Brent, D. (2011). Transfer, transformation, and rhetorical knowledge: Insights from transfer theory. *Journal of Business and Technical Communication, 25*(4), 396–420.

Brinkman, G. W. & van der Geest, T. M. (2003). Assessment of communication competencies in engineering design projects. *Technical Communication Quarterly, 12*(1), 67–81.

Cargile Cook, K. (2002). Layered literacies: A theoretical frame for technical communication pedagogy. *Technical Communication Quarterly, 11*(1), 5–29.

Corbin, J. & Strauss, A. (2008). *Basics of qualitative research: Techniques and procedures for developing grounded theory* (3rd ed.). SAGE.

Dush, L. (2014). Building the capacity of organizations for rhetorical action with new media: An approach to service learning. *Computers and Composition, 34*, 11–22.

Eble, M. F. & Gaillet, L. L. (2004). Educating "community intellectuals": Rhetoric, moral philosophy, and civic engagement. *Technical Communication Quarterly, 13*(3), 341–354.

Farkas, K. R. H. & Hass, C. (2012). A grounded theory approach for studying writing and literacy. In K. M. Powell & P. Takayoshi (Eds.), *Practicing research in writing studies: Reflexive and ethically responsible research* (pp. 81–95). Hampton Press.

Fountain, T. K. (2014). *Rhetoric in the flesh: Trained vision, technical expertise, and the gross anatomy lab*. Routledge.

Haas, C. (1996). *Writing technology: Studies in the materiality of writing*. Lawrence Erlbaum.

Hannah, M. A. (2010). Legal literacy: Coproducing the law in technical communication. *Technical Communication Quarterly, 20*(1), 5–24.

Helferich, G. (2016, November 15). How to write good medical reports, part 3: Justify medical interventions. *Journal of Emergency Medical Services*. https://www.jems.com/ems-insider/how-to-write-good-patient-care-reports-part-3-justify-medical-interventions/.

Henschel, S. & Melonçon, L. (2014). Of horsemen and layered literacies: Assessment instructions for aligning technical and professional communication undergraduate curricula with professional expectations. *Programmatic Perspectives, 6*(1), 3–26.

Hollenbeck, K. (1993). *Classrooms in the workplace: Workplace literacy programs in small- and medium-sized firms*. E. Upjohn Institute for Employment Research.

Karlson, K. J. (1991). Writing training: Collaboration between academy and government agency. *Technical Communication, 38*(4), 493–497. https://www.jstor.org/stable/43095824.

Kienzler, D. & David, C. (2003). After Enron: Integrating ethics into the professional communication curriculum. *Journal of Business and Technical Communication, 17*(4), 474–489.

Lawrence, H. & Hutter, L. (2021). Confronting methodological stasis: Re-examining approaches to technical communication pedagogical frameworks. In M. J. Klein (Ed.), *Effective teaching of technical communication: Theory, practice, and application* (pp. 91–110). The WAC Clearinghouse; University Press of Colorado.

Lindlof, T. R. & Taylor, B. C. (2002). *Qualitative communication research methods* (2nd ed.). SAGE.

Matthews, P. (1999). Workplace learning: Developing an holistic model. *The Learning Organization, 6*(1), 18–29.

Melonçon, L. (2017). Bringing the body back through performative phenomenology. In L. Melonçon & J. B. Scott (Eds.), *Methodologies for the rhetoric of health & medicine* (pp. 96–114). Routledge.

Milwaukee Fire Department. (2017). *Milwaukee Fire Department annual report '17*. City of Milwaukee. https://city.milwaukee.gov/ImageLibrary/Groups/mfdAuthors/media/2017AnnualReport.pdf.

Munger, R. (2006). Participating in a technical communication internship. *Technical Communication, 53*(3), 326–338.

Russell, D. R. (2007). Rethinking the articulation between business and technical communication and writing in the disciplines: Useful avenues for teaching and research. *Journal of Business and Technical Communication, 21*(3), 248–277.

Seawright, L. (2017). *Genre of power: Police report writers and readers in the justice system*. National Council of Teachers of English.

Swacha, K. Y. (2018). "Bridging the gap between food pantries and the kitchen table": Teaching embodied literacy in the technical communication classroom. *Technical Communication Quarterly, 27*(3), 261–282.

Taylor, M. C. (2000). Transfer of learning in workplace literacy programs. *Adult Basic Education, 10*(1), 3.

Thomas, S. & McShane, B. J. (2007). Skills and literacies for the 21st century: Assessing an undergraduate professional and technical writing program. *Technical Communication, 54*(4), 412–423.

Thrush, E. A. & Hooper, L. (2006). Industry and the academy: How team-teaching brings two worlds together. *Technical Communication, 53*(3), 308–316.

Turner, R. K. & Rainey, K. T. (2004). Certification in technical communication. *Technical Communication Quarterly, 13*(2), 211–234.

Turnley, M. (2007). Integrating critical approaches to technology and service-learning projects. *Technical Communication Quarterly, 16*(1), 103–123.

Winsor, D. (1996). Writing well as a form of social knowledge. In A. H. Duin & C. J. Hansen (Eds.), *Nonacademic writing: Social theory and technology* (pp. 157–172). Routledge.

Winsor, D. (2006). Using writing to structure agency: An examination of engineers' practice. *Technical Communication Quarterly, 15*(4), 411–430.

Wright, D. (2012). Redesigning informed consent tools for specific research. *Technical Communication Quarterly, 21*(2), 145–167.

16. Hidden Arguments: Rhetoric and Persuasion in Diverse Forms of Technical Communication

Jessica McCaughey
GEORGE WASHINGTON UNIVERSITY

Brian Fitzpatrick
GEORGE MASON UNIVERSITY

Abstract: This chapter explores how persuasion and rhetoric appear within technical writing forms in nuanced and unexpected ways—and in ways that new graduates are not prepared to successfully navigate. Even though the field agrees that technical writing is rhetorical, this isn't always the case in discipline-specific courses that ask students to perform technical writing forms. This qualitative, IRB-approved investigation explores interviews from three professionals who perform technical writing daily—a physician assistant, a CPA, and a labor and delivery nurse. We find that even within practical forms of writing, such as medical records and accounting documents, hidden arguments exist, and that these professionals recount that learning to write persuasively in these forms was a complex, disorienting process that took place entirely on the job. Drawing from these results, this chapter argues that by producing writing prompts and instruction centered on detailed, realistic case study situations and problem-solving, instructors can diminish the disparity between abstract classroom audiences and stakes and concrete workplace audiences and stakes, better preparing our students for real-world writing contexts.

Keywords: workplace writing, writing in the professions, writing transfer, persuasion, technical forms.

Key Takeaways:

- Persuasive and rhetorical writing are embedded within positions traditionally perceived as objective in their communications.
- Producing writing prompts and instruction centered on detailed, realistic case study situations and problem-solving can diminish the disparity between abstract classroom audiences and concrete workplace audiences.
- Courses and instructors should extend rhetorical thinking and teaching alongside the genres and modes of writing.

Cezar M. Ornatowski's (1997) chapter "Technical Communication and Rhetoric" in *Foundations for Teaching Technical Communication*, the text to which this collection is a response, outlines how technical communication is rhetorical and considers the ways in which we might view these rhetorical moves in a classical framework. This chapter reflects and builds upon Ornatowski's exploration, focusing on how persuasion and rhetoric appear within technical communication in nuanced and unexpected ways in the workplace—ways in which new graduates are not prepared to successfully navigate.

Inauthentic classroom experience ill equips newly graduated workplace writers with a generic sense of audience and purpose. The initial writing role of a physician assistant (PA), for instance, might look, even to themselves and their colleagues, like straightforward, templated writing. However, within this writing mode is a hidden requirement for the PA to be persuasive and convince doctors to take a case. Through failed communications and continuous interactions in their daily tasks, new workplace writers eventually develop rhetorical competencies and learn to cater their writing to their hidden roles.

This chapter offers insights about persuasion in unexpected and often subtle forms, representing a selection of data from a larger study. This qualitative, IRB-approved investigation examines the inherent reconceptualization efforts, particularly in the context of industry-specific technical writing that is typically viewed as informational and straightforward. The approach we've taken is rooted in a grounded theory methodology, which emphasizes intensive interviewing with an increased flexibility that allows for follow-up questions and, perhaps more practically, memoing, a "constant-comparison" method, and more than one round of coding (initial and focused; Charmaz, 2014). Using this methodology, we ask the following questions: In what unexpected/unexplored ways does persuasive writing happen in technical fields, and how do communicators learn to make these arguments? Further, how can writing instructors better prepare students for these particular writing situations?

It's important to note that this study is about workplace writers who do *not* identify as technical or professional writers. They see their primary job functions as something other than writing, but they perform, like almost all professionals, a significant amount of writing for their work. The study contains, at this time, 48 interviews with working professionals and examines the university-to-workplace transition. Their job titles range from lawyer to foreign affairs officer to lab manager, as a small sampling, and their time on the job ranges from one year to thirty years. Participants were asked a set of 18 questions, with occasional follow-up questions. These questions centered on the writing they perform on the job and their perceptions about how they developed the skills necessary to succeed in this writing. For instance, we ask, "Could you walk us through the process for one specific recent project or type of project?" and "Can you describe a time in your career that you felt unprepared as a writer at work?"

In this study, we see narratives of students transitioning into professionals, struggling to acquire the skills necessary to thrive as they encounter hidden or unexpected persuasive opportunities. This chapter examines interviews with three such professionals (a certified public accountant [CPA], a labor and delivery nurse, and a neurosurgery physician assistant [PA]), each of whom perform technical writing daily and whose experiences stand out as examples of the ways in which persuasive and rhetorical writing are embedded within positions traditionally perceived as objective in their communications.

These concrete examples add to the body of research on transfer, or the ways individuals gain one kind of knowledge—in this case, technical writing skills—in a particular context and then apply the related skills in a different context. We explore not only the nuanced persuasive writing happening in these technical documents but also how professionals perceive this skill acquisition. The professionals reflect on the roles of university learning and on-the-job experiences in their growth as communicators as they transitioned into the workplace, considering the ways in which they were unprepared and expressing what they wish they'd learned in their writing classes.

Changing Views of Technical Communication

Carolyn R. Miller's seminal 1979 article, "A Humanistic Rationale for Technical Writing," addressed the problematic and then-common view that "science and rhetoric are mutually exclusive" (p. 611). She knew, and many technical writing teachers and scholars since then have agreed, that technical writing isn't—or shouldn't be—simply documenting or instructing. She acknowledged, however, the extreme challenge of teaching scientific or technical writing as more complicated and rhetorical. Miller (1979) claims that good technical writing should be "a persuasive version of experience" (p. 616). She asks, "Why has it been so difficult in a technical writing class to talk about the relationship between writer and readers and the reasons for saying anything about a subject in the first place?" (p. 615). This question can extend to include any workplace writers performing "technical communication." Whether in the classroom or on the job, why has it been so difficult to talk with nuance about the complicated persuasive work that we perform in our documents?

For many years, technical communication was primarily thought of and taught as expository and informative, rules- and conventions-based. Early technical writing textbooks were decidedly anti-rhetorical thinking. Typical excerpts from the 1970s include "[t]echnical writing is expected to be objective, scientifically impartial, utterly clear, and unemotional . . . concerned with facts" and "[t]echnical writing has one certain, clear purpose: to convey information and ideas accurately and efficiently" (Miller, 1979, p. 611). Not until 1993 did Jennifer D. Slack and colleagues describe the transmission, translation, and articulation views of technical communicators. Even in the late 1990s, nuanced persuasive skills were not generally thought of as skills crucial to technical communication (Grice, 1997). More recently, though, several scholars—most notably Erin A. Frost and Michelle F. Eble (2015)—have argued that technical communication is "implicitly persuasive" and encouraged technical communicators to be more open, particularly to public audiences, about "the persuasive nature of technical communication" (p. 1). In some technical writing classrooms, students are learning about promoting social justice through technical communication and about equality or accessibility as a responsibility of the technical communicator. Those pursuing technical writing as a profession are now receiving instruction about the rhetorical nature of what might have been called "neutral" or "objective" 20 or 30 years ago.

We know that a lot of teachers are successfully taking this scholarly view and putting it into practice in their classrooms. For instance, many look towards concepts such as rhetorical problem solving as a way of grounding technical communication pedagogy (Karatsolis et al., 2016). Still, when instructors frame technical communication as rhetorical and persuasive, many do so only around explicitly persuasive forms, such as proposals and other "public" technical texts (Frost & Eble, 2015)—implying that traditional technical writing (which can, of course, mean many things) is not argumentative. In many textbooks we still see what might only be called a "nod" to rhetorical awareness. Even many recent popular technical, professional, and workplace writing textbooks speak only broadly to "considering the needs of your audience" (Alred et al., 2015, p. 3) and ask writers to "try to see from another person's point of view, beyond your own personal concerns" (Marsen, 2013, p. 203). Others encourage writers to work towards "understanding" or "considering" purpose, that a document "must be tailored to its intended audience; otherwise, it probably won't achieve the desired results" (Searles, 2013, p. 3). Those texts that explicitly have sections like "Principles of Persuasion" do not go beyond extremely simple concepts, such as "[d]on't get bogged down in unnecessary details or arguments" (Blake & Bly, 2000, p. 86) and "[c]onsider Whether Your Views Will Make Problems for Readers" (Ewing, 2010, p. 231). Examinations of situations that call for "persuasive" writing always use examples that are explicitly so (Lannon & Gurak, 2013, p. 13).

We see even less of this rhetorical thinking about texts at play in discipline-specific writing textbooks. A text on writing for accountants emphasizes, alongside standard organization and conciseness, a generic nod towards "communication skills" and "appropriate style" (May, 2014). Likewise, many medical texts discuss audience only in terms of tone and expected length and structure of each document type. When these texts do engage in more explicit rhetorical discussions ("Who is going to read this? What do I know about my audience? How do I make decisions about language?"), it is generally either in the abstract (unapplied to any particular technical form) or applied to modes of writing not applicable to the technical discipline itself (advertising/marketing, poetry, press releases). Few, if any, texts apply the persuasive/rhetorical lens to forms seen as explicitly practical and expository or that genuinely offer scenarios or strategies related to the specific profession for complex audience analysis. These textbooks for professionals performing technical writing across industries and roles brush up against rhetorical thinking but rarely go beyond a superficial treatment. In discipline-specific writing situations particularly, we do our students a disservice by not acknowledging and teaching towards the persuasive and rhetorical writing they can expect to encounter within technical forms. Specifically, in fields not focused on writing but in which professionals must write technical documents consistently, graduates do not seem to be examining the rhetorical purpose of their field's writing. While many of our students will not be technical writers formally, they will be asked—sometimes daily or even hourly, depending on their profession—to perform technical writing. It's important, then, to make a key distinction between teaching technical communication and teaching the technical writing that happens in "the professions."

▮ Forms, Documentation, and Persuasion

Professional writing adheres to certain conventions created by the organization or industry (Winsor, 1989), and at times, persuasive writing might be a part of those conventions. Robert I. Williams (1983) argues that technical writing forms, in and of themselves, are persuasive, and that "the very conventionality of format works in the writer's favor" (p. 11) when technical communicators are working in traditional forms. The "message of the standard format is that this is a sound document" (Williams, 1983, p. 12). Moving beyond the form itself reveals even more hidden moments of argument and persuasion in the practical execution of these documents.

We can see such moments of hidden persuasive communication clearly in the interviews of our three professionals: the CPA, the neurology PA, and the labor and delivery nurse. All describe finding themselves, immediately upon entering the workforce, encountering forms of writing thought of as (and sometimes explicitly stated to be) straightforward and objective, as many organizations still adhere to the "transmission" view, or the idea that technical communicators are simply "the neutral vehicle" conveying meaning from one place to another (Slack et al., 1993, p. 14). The CPA, for instance, writes a letter to the IRS that one might assume is essentially a form letter; the two medical professionals work within templates constrained by the medical records system. They are asked to "fill in the blanks" in some sense, to write down what they and others around them refer to as strictly factual documentation. They are taking notes about things that must be "recorded" and "documented," as though their understanding of everything that occurs in their work is a fact, with no grey area. These forms appear, at first, to leave little room for individual perception, opinion, or motive.

One clear example of this can be seen in our interview with a neurosurgery PA who works in an emergency room. She describes her writing in this way:

> Documentation is pretty important in medicine. [These documents] are electronic, typed consultation notes or history and

> physicals. Also, daily progress notes . . . and physical exams. And for OR procedures, a brief summary of the procedure itself. . . . Most of our writing is actually in template form. So it doesn't really take too much time, and most documentation will include a summary of the patient themselves and their background, specifically their past medical history and things that are pertinent to their hospital stay. . . . They will also include a physical exam, so my exam of the patient and a plan. So plans for all of the diagnoses that the patient has, and documentation that my attendings and surgeons have agreed to plans that I'm making.

Likewise, when discussing the kinds of writing performed on the job, the labor and delivery nurse repeatedly references "documentation" and the constrained and objective nature of these forms, even when they are narrative or descriptive:

> So most of the writing that takes place as a staff nurse is on an electronic medical record, where we joke that it's an elaborate billing system, because it is, but they try to make it as easy for the billers to use as possible, and as easy for you to not get yourself in trouble as possible. So they do a lot of like, selecting options for charting, so it's like a column where you select options, you can type in things like, you know, blood pressures, or temperatures, and then you can select options for pain levels, or assessment findings, like color of the skin, they'll give you options like, "appropriate for ethnicity, warm, dry, clammy, red, hot, weeping," like tons of different options. . . . there's also notes you write that are more narrative. . . . And it's really easy when you're using click boxes to fill in your answers to, if you're not being careful, just fill in like your normal answers, like the standards, and then if you write something different in a note, and it contradicts what you already charted, it makes it look like you're not competent. So you're trying to make sure that you're being consistent with what you're writing unless it's actually discussing a change.

A standard form of communication for nearly all medical professionals, electronic medical records are, again, generally thought of as strictly objective documentation. In the classroom, too, such communication is framed consistently by instructors and by textbooks as "neutral," "template-based," or strictly informational or factual. The idea of simply "recording information" is so ingrained that the interviewees still explain their work in this way despite offering examples of particularly complex, persuasive compositions. The CPA, for example, talks about "being able to keep it neutral" in much of her writing. The PA consistently describes her writing as simply "documentation." The labor and delivery nurse describes the electronic medical record, jokingly, as "an elaborate billing system" and a tool that makes it as easy as possible for you to "not get yourself in trouble." She acknowledges the possibility of a situation in which a failure to document something she did could have an impact on care, but generally trivializes her charting, saying it's "mostly going to be proof that I've followed up on things, and acknowledged things, and noticed changes." She defines success as a writer in her position as having "more to do with how quickly and efficiently can I say the bare minimum to show that I did my job."

To these writers, even complex rhetorical moves seem to feel mechanical, supporting the assertion that "[w]orkplace writers are far more skilled and accomplished than they themselves or their managers acknowledge" (Dias et al., 1999, p. 233). For example, when asked to describe a specific writing experience, it becomes clear that at least some of the writing the PA performs requires rhetorical flexibility and persuasion:

> I think the way that I approach it is, how do I shape his story into something that's going to catch someone's attention? So most of us in medicine, like if I get a call of a consult to say, "Hey this patient has some kind of an issue and it looks like they have a fracture in a bone near the ear," I'm immediately checked out thinking, "Why are you calling a neurosurgeon for this? We don't take care of this. I'm not interested." So same thing if I'm trying to talk to a [medical] doctor. I'm trying to frame my note that would be appealing to them to say, "Hey this is exactly why we need you, and this is why we hope that you're going to accept our patient," because there is still, you know, some procedure in the hospital involved once a surgical problem is managed and taken care of, you want to transfer your patient to a doctor that can better take care of their medical needs, things that I don't really manage myself. So you want to try to kind of frame the patient as, "Oh this is a really interesting medical patient now that we're done with the surgical part of things." So having to write something in a way that's going to make it relevant to other people and catch their attention is a big challenge in writing and I think it's a challenge that's kind of fun to try to do.

The PA states that she wasn't taught to write compelling narratives about her patients to convince doctors to take them on; this work is implicit and unspoken. Both she and the labor and delivery nurse describe their respective typical writing forms as objective record-keeping and documentation of fact. Interestingly though, both describe situations in which the authentic professional and medical circumstances require this writing to serve other purposes. It's important to note that such forms are not limited to accounting and medical professions. Content management (CM) systems, for instance, mimic similar constraints; their form presents a "simple" way to create and store information or text. Many argue that for the sake of "ease," such systems strip rhetorical agency from communicators, who are then "relegated to working within the confines of . . . systems that, in most cases, others have designed. These writers are not tasked with making situated rhetorical decisions" (Andersen, 2014, p. 121). But these writers do make rhetorical decisions, even within the confines of limited text boxes. These forms, while generally perceived as templated or mechanical, do in fact require more complex awareness and writing skill from professionals who understand well their audiences, purposes, and larger organizational contexts. Persuasive writing happens, then, at both the textual and the metatextual levels.

The genre of the technical writing form itself, according to Williams (1983), is persuasion of authority; working successfully within the constraints and expectations of a form argues to the audience that this "is a sound document." The nurse who successfully records the state of her patient shows her understanding of not only the software but also of her role, particularly in the context of staying "in her own lane," as we'll see below. Through the structural and formal constraints of its genre, the form actually argues against its own persuasion. Consider again, the patient chart: its checkmarks and small text boxes demand shorthand expository writing, using objectively clinical and fragmented language. The stated goal for these documents is to serve as records and for the writer to get to the point efficiently and accurately. However, if the workplace writer explicitly follows the constraints of pure objectivity and simple documentation that the template itself requests, the writer ultimately cannot succeed in the real world. For example, the labor and delivery nurse recounts navigating these hidden moments of persuasion:

> I remember having to sit there and write notes with people, and you would always seek out like someone you felt comfortable with and saying, "Can you help me write this note? This difficult thing happened." Like generally then, it had to do with pain management, and you couldn't get anesthesia to get there on time, or something like that, right? Patient's in pain, you're out of pain medicine, anesthesia isn't coming, it took an hour, your patient hates you now, you know, something like that [laughter], and you have to be careful not to write, "I called anesthesia a hundred million times and they didn't want to come, because they didn't like the page," like, you can't write that, right? So, it's like going back in time and someone you know, teaching you how to write, okay, write a note for the first time that you notified anesthesia. And then write another note that says, "notified anesthesia." Write another note that says, "notified anesthesia, anesthesia now in rounds," you know, and you write it that way. Like these little one-line notes that say, "Hey, I did it. Hey, I did it. Hey, I did it." And as someone showing you, instead of writing one long note, it shows this persistence, for example. . . .

> [I]f I was concerned about a patient, let's say she had chest pain after delivery, and I was concerned, and I took some vital signs and everything was normal, and her bleeding was all normal, and everything was great. But I'm still going to . . . let the physician know, "Hey, she's having chest pain. This is her blood pressure, this is her heart rate, this is her temp., this is what her bleeding is like." And if they say like, "I'm not worried about it." And then I'm like, "Well, don't you want an EKG?" If the provider's like, "No, I don't." Okay, so I don't want to write a note that says that exactly, because it makes them look like they're not doing their job, even if I feel that way. So, I have to write, for example, . . . "Patient complained of chest pain." I might like list the vital signs, [and] "Provider notified, no new orders."

In both cases, adhering strictly to the perceived objectivity and documentation of the form would undercut a colleague's authority or result in reduced quality of patient care. The new workplace writer mistakes the form and its constraints as signals that the required writing must be not just objective but also explicitly not persuasive. The form, by way of its clinical template, actively hides its opportunities for persuasion; only through understanding and resolving metatextual issues—in this case, collegial relationships—can a writer recognize the hidden demands for persuasion and resolve such discrepancies to succeed in their writing.

We see this play out differently for the CPA in her own understanding of the need for persuasion within what is generally referred to as a "form letter" that she prepares and sends to the IRS on behalf of a client. She initially describes the letter as usually "kind of standard," in that she records and reports to the government required details and information on a client's finances. However, she acknowledges that her purpose for writing is to make a case for forgiveness for her client:

> I'd [ask the client] what are the circumstances, and then you write the saddest story that the truth will allow. . . . If you're requesting like an abatement of a penalty or something, . . . it's "my dog died, my wife got sick, my car broke down, and I ran out of" . . . you know, it's just whatever the circumstances are, you write it in a way that's like, "it was so sad and it was so awful and they couldn't possibly have filed that day. But look! They did it two days later and it will never happen again and they have reached out to a professional to ensure [that] and we're on top of it." And that's sort of how you write these things.

She is, of course, not simply writing about the numbers in this form letter, as we might expect. The phrase "the saddest story that the truth will allow" inherently describes the contradiction between the objective form she sees herself writing in and the authentic goals of her correspondence. It is the truth, framed purposefully and rhetorically as advocacy on behalf of her client. The CPA, the PA, and the labor and delivery nurse all write as "advocates" on behalf of their clients or patients; rarely, though, do they frame their technical documentation as championing.

To that end, for all three of these workplace writers, efficacy is determined by two factors: successful advocacy on behalf of their patients/clients and learning the rules wherein the former becomes possible. Each of these writers progressed from failed writing attempts to successful ones via either their own or witnessed failed attempts. They each can point to these past failures in very tangible, practical ways. As the CPA states, she learns based on what works: "I guess if it's like an appeal . . . whether it's successful or not." Similarly, the PA learned through experience that patients not presented as "interesting" enough for the neurologist are not chosen to be seen by the specialist. Meanwhile, the labor and delivery nurse cautions against explicitly indicting a colleague in a document of record. The stakes and possible repercussions are not limited to just the workplace writer. Even more compellingly, perhaps, they extend to the writers' clients and patients, for whom the stakes—financial or medical—are even higher.

Each workplace writer defines success as their writing contributing to the desired outcome for the people for whom they are advocating. The gravity of their charges' situations clearly motivates the writers, although they may feel or express this only in the abstract. And yet concretely, it seems as though a fully developed awareness of this context informs the mechanics of how they approach their texts.

While we as instructors cannot authentically replicate the empathy and desire to write well on behalf of a medical patient, we can at least model for our students the mechanism for this kind of successful writing—a strong foundation in rhetorical awareness and decision making. We can also show them very clear, real-world examples that model the authentic situation and what it might entail. We can't, metaphorically speaking, take them to Mount Everest to prepare them to climb it. However, we can expose them to the real experiences of those who have climbed it and the lessons they've learned. We can offer them practice in the specific conditions they can expect not just on any mountain, but on the one they will face. Preparing students for writing targeted to their specific goal workplaces is a clear improvement from generically preparing students for writing in the workplace.

■ Application

The tensions Katherine Staples and Cezar M. Ornatowski (1997) articulate between the university and the workplace still exist. Though our study catalogs the experiences of workplace writers after they have left the classroom, the interviews we've highlighted from it clearly have pedagogical implications for the classroom. This final section will examine how instructors rely (often necessarily) on generic and inauthentic forms (Anson & Forsberg, 1990) and how we must change these patterns in order to reinforce to students the opportunities—and needs, in some cases—for persuasive, rhetorically complex workplace writing.

The limited nature of the classroom itself challenges the application of what we understand about classroom teaching and transfer of writing skills into real-world technical writing spaces. Chris M. Anson and L. Lee Forsberg (1990) assert that students struggle in their transition because of classroom writing's rhetorical limitations and its tendency to focus on "generic skills," rather than on "developing strategies for social and intellectual adaptation," applicable across both academic and professional writing contexts (p. 201). They highlight the gap between the phases of *Expectation* and *Disorientation*, wherein a writer with a series of past successful assignments, writing to hypothetical, generic audiences with abstract stakes, is jarred when their writing expectations become incongruous with the new and newly elevated stakes of the authentic workplace. The *Disorientation* phase (Anson & Forsberg, 1990) makes adapting to specific workplace contexts difficult for students, contributing to what Rebecca S. Nowacek (2011) categorizes as challenging "reconceptualization" efforts once graduates begin writing in the workplace.

The broad view of technical writing as a homogenous field emphasizes rote recall of forms and conventions over what we argue are more valuable and versatile skills. Based on the findings above, we argue that students must be immersed in genuinely authentic forms and examine and practice the nuanced persuasive writing types we see in the three example writers' experiences. If we better understand not only the work these communicators perform, but also how they perceive their acquisition of their skills, then we can build into our courses transfer-based activities and ways of thinking (Elon, 2015) so that students can take what they've learned in highly specific texts and apply them in very different, although equally nuanced, persuasive writing situations.

All too often, unless students take explicit technical writing courses, the explicit teaching of rhetoric and persuasion falls to the wayside once students have passed beyond first-year composition (FYC). In the types of major classes professionals like those in this study take, they may see acknowledgments of generic or implied rhetoric (formal vs. informal tone, active vs. passive voice), but a lack of access to authentic forms and the limitations of the classroom make it a struggle to teach concrete and authentic strategies of persuasion. Imagine, for example, a pre-med/nursing student being told that "your audience is a doctor." To a student without the proper context (and industry experience) of the authentic situation, this might mean no more than "your audience is formal, so don't use contractions," and perhaps "use technical jargon." But this audience is, of course, more rhetorically complex than this quick assessment suggests. Consider the following from an interview with the PA:

> [The doctor] want[s] something interesting . . . [They are] also protective of their workload because, you know, I'm maybe seeing 30 patients on my service and then I'm saying, "Oh my god," I'm having five other people try to give me other patients so I'm thinking, "Okay, do I really want this patient that's not . . . really relevant to me? Or do I want the ones that are specifically neurosurgery? Yes, I can do something to help you," that kind of thing.

The difference between the generic and abstract idea of "doctor" as audience versus a concrete touchstone is significant. While the FYC student may have a generic sense of audience and in turn a generic rhetorical response to that prompt, our workplace writer, a PA with the authentic experience and workplace context, understands her audience with additional, crucial nuance. Her audience insights—doctors are overworked, "protective of their workload" and reputation, and also want to serve patients with cases related to their research interests—lead to a set of focused and tangible rhetorical decisions that can be made and seen in her writing.

The first chapter of Michael A. Arntfield and James W. Johnston's 2016 textbook *Healthcare Writing: A Practical Guide to Professional Success* covers "in-patient writing," including charting, reporting, and writing notes in what they designate as "expository prose." This text does a better job than most acknowledging how complicated the field's various audiences are and the rhetorical nature of medical documentation. Sub-sections like "Rhetoric and Rhetorical Modes" acknowledge audience, context, and purpose as key factors in small-space medical charts. The text offers the following examples, in a chapter regarding clear and concise technical writing of patient notes, charts, and reports:

> A 19-year-old female patient came into the emergency room with a friend who drove her, and complained in detail that she had been feeling very sick to her stomach for several days, having begun vomiting earlier in the afternoon. The patient appeared pale and disheveled and explained that this has never happened to her before, and that she hadn't been able to hold down as much as a glass of water since yesterday morning. (Arntfield & Johnston, 2016, p. 20)

An improved excerpt they offer reads:

> 19-year old female patient attended the emergency room complaining and presenting with symptoms of severe nausea with vomiting, and with no known cause. No previous history disclosed and patient unable to ingest food or water in over 24 hours. (Arntfield & Johnston, 2016, p. 20)

A textbook example like this, advocating for fewer words and direct, catalogued language, seems like common-sense best practice. Without the understanding that these technical professionals are, in fact, writing nuanced rhetorically complex persuasive documents, and without access to authentic and detailed real-world experiences, this example—focusing primarily on economy of language and direct prose—represents successful technical documentation. In a hospital setting, there is genuine value in concision and efficient use of already constrained time and writing space. As authentic as this example may appear on the surface, the reality is that textbooks operate under the same constraints that classrooms do, including the constraint of inauthentic situation. The example implies that the critical skill of this writing style is capturing the message in as few words as possible, but it doesn't acknowledge, as the PA and labor and delivery nurse do, the varied contextual factors at play in determining successful writing in these situations. So, while this text serves as a model for employing concision in technical writing when physical writing space is constrained, like many other textbooks, it operates under one of two assumptions: every situation these workplace writers will encounter is similarly straightforward and will require only "transmissive" documentation (Slack et al., 1993, p. 14), or the student-reader will know which information is important (without genuine experience in the profession or direct access to professionals modeling the complexities of these detailed situations, they cannot). The interviews with healthcare professionals we present here both emphasize the importance of thorough documentation; the question for this field, then, becomes "How does a new workplace writer in this situation learn to make the distinction between comprehensive documentation and relevant documentation?"

Similarly, as teachers, we are forced to teach a typical approach to writing forms—understood best approaches or the "middle-of the road." Sometimes even strong writing assignments still teach either to an abstract understanding of form or to an audience or purpose without authentic stakes (beyond a student's grade). We might teach certain writing decisions as "typical" for proposals, résumés, reports—that students should use a certain kind of language or keep certain kinds of audiences' concerns in mind. Without real stakes or a genuine, knowable audience, though, these ideas do not develop beyond the hypothetical ideal. Textbook examples and classroom assignments are often devoid of the authentic stakes and audiences established in real workplace writing. Further, students taught to take a best-average approach to writing forms will likely struggle, once in the workplace, to diverge from those same abstract norms.

Training a student to continuously navigate to the middle of a figurative road in their writing takes for granted their future flexibility when the road itself becomes fully realized and fraught with obstacles. In successful transfer, a student enters a workplace with the ability to recognize the average of those modal practices and the awareness to shift direction as the authentic audience, genre, and purpose require. When we present technical writing modes only at face value, as documentation, we obscure their complexity and create hidden moments of persuasion that require rhetorical maneuvering to address a complex and knowable audience or make choices about factors such as audience, genre, and purpose. Without acknowledging that these technical writing forms are and will continue to be persuasive when real audiences are introduced, students won't know which choices to make, nor recognize when exactly those opportunities for choices even arise.

How do we fix what sounds like an impossible problem with only imperfect solutions? While there are always limitations to how disciplinary classrooms and their texts can close the gap of authenticity, continuously emphasizing and explicitly teaching rhetoric when teaching technical forms and writing can fortify a student's rhetorical flexibility and capacity to navigate the remaining gaps and complications they'll encounter in workplace writing. Courses and instructors should extend rhetorical thinking and teaching alongside the genres and modes of writing. This exercise becomes more complex and focused as it relates to specific workplace writing expectations and focuses on solving the authentic problems that workplace writers encounter.

Between FYC and upper-level technical writing courses, the focus often shifts from rhetorical awareness and persuasive modes to elements like disciplinary structure, style, and diction. It's arguably more important, though, as students progress deeper into the complex pockets of disciplinary technical writing, that they continue their rhetorical educations. Having an instructor familiar with the technical and formal expectations of the field's forms, audiences, and purposes will also mean having an instructor capable of helping students apply their broad understanding of rhetoric in these more specific and authentic writing spaces.

Producing writing prompts and instruction centered on detailed, realistic case study situations can diminish (but likely not eliminate) the disparity between abstract classroom audiences and stakes and concrete workplace audiences and stakes. Rather than relying on generic business or technical writing practices and modes, or sometimes outdated forms untethered to specific disciplines or industries (memo, business letter, or generic proposal; Dias et al., 1999), we must do our best to tailor writing assignments as closely as possible to the real-world writing contexts in which our students will be asked to perform. For instance, an assignment constructed around the case study of our physician assistant might provide not only information about the purpose of the piece of writing, but also a knowable and researchable audience—doctor, overworked, protective of their caseload, etc. In fact, the writing a student is asked to perform may not even mimic traditional correspondence with the doctor, but rather take the form of an audience assessment that helps them to understand the rhetorical choices necessary in such a correspondence in their future careers. An assignment for accounting students should go beyond what we think of as traditional best practices in the field (accuracy, formality, objectivity) and consider the nuanced persuasive writing that takes place when advocating to the IRS on behalf of a client, particularly when such advocacy might require less expected tactics, such as emotional appeals. In nursing programs, students should consider not only issues of conciseness in writing medical records, but also how potentially complex their narrative and persuasive language choices can be in a medical chart for real patients.

In each of these examples, the focus is ultimately on making these moments of hidden persuasion visible and explicit for the developing writer.

We must find inventive ways to effectively capture complex rhetorical situations and have students interrogate and work through them. This can be achieved through guest speakers, case studies, or other points of access to the truly specific and complex ways in which workplace writers are asked to write. Modeling and/or creating these more authentic situations will enable our students to first observe and imitate examination of and response to these technical forms, and eventually, with more experience, uncover the complexities themselves. In understanding the hidden demands of audience and purpose within these forms, the student-writer grows into the workplace-writer role. Imitation falls away and the challenge of writing within the authentic context resolves.

■ References

Alred, G., Oliu, W. E. & Brusaw, C. T. (2015). *The business writer's handbook* (11th ed.). Bedford; St. Martin's.

Andersen, R. (2014). Rhetorical work in the age of content management: Implications for the field of technical communication. *Journal of Business and Technical Communication, 28*(2), 115–157. https://doi.org/10.1177/1050651913513904.

Anson, C. M. & Forsberg, L. L. (1990). Moving beyond the academic community transitional stages in professional writing. *Written Communication, 7*(2), 200–231.

Arntfield, M. & Johnston, J. (2016). *Healthcare writing: A practical guide to professional success*. Broadview Press.

Blake, G. & Bly, R. W. (2000). *Elements of technical writing: The essential guide to writing clear, concise proposals, reports, manuals, letters, memos, and other documents in every technical field*. Pearson.

Charmaz, K. (2014). *Constructing grounded theory* (2nd ed.). Sage.

Dias, P., Freedman, A., Medway, P. & Pare, A. (1999). *Worlds apart: Acting and writing in academic and workplace contexts*. Lawrence Erlbaum Associates.

Elon Center for Engaged Learning. (2015). *Elon statement on writing transfer*. http://cel.wpengine.com/wp-content/uploads/2014/09/Elon-Statement-Writing-Transfer-2015.pdf.

Ewing, D. (2010). Strategies of persuasion. In K. J. Harty (Ed.), *Strategies for business and technical writing* (7th ed., pp. 231–242). Longman.

Frost, E. A. & Eble, M. F. (2015). Technical rhetorics: Making specialized persuasion apparent to public audiences. *Present Tense, 4*(2). https://www.presenttensejournal.org/volume-4/technical-rhetorics-making-specialized-persuasion-apparent-to-public-audiences/.

Grice, R. A. (1997). Professional roles: Technical writer. In K. Staples & C. Ornatowski (Eds.), *Foundations for teaching technical communication: Theory, practice, and program design* (pp. 209–220). Ablex.

Karatsolis, A., Ishizaki, S., Lovett, M., Rohrbach, S. & Kaufer, M. (2016). Supporting technical professionals' metacognitive development in technical communication through contrasting rhetorical problem solving. *Technical Communication Quarterly, 25*(4), 244–259. https://doi.org/10.1080/10572252.2016.1221141.

Lannon, J. M. & Gurak, L. (2013). *Technical communication* (3rd ed.). Pearson.

Marsen, S. (2013). *Palgrave study skills: Professional writing* (3rd ed.). Red Globe Press.

May, G. S. (2014). *Effective writing—a handbook for accountants* (10th ed.). Pearson Education.

Miller, C. (1979). A humanistic rationale for technical writing. *College English, 40*(6), 610–617. https://doi.org/10.2307/375964.

Nowacek, R. S. (2011). *Agents of integration: Understanding transfer as a rhetorical act.* Southern Illinois University Press.

Ornatowski, C. M. (1997). Technical communication and rhetoric. In K. Staples & C. Ornatowski (Eds.), *Foundations for teaching technical communication: Theory, practice, and program design* (pp. 31–52). Ablex.

Searles, G. J. (2013). *Workplace communication* (6th ed.). Pearson.

Slack, J. D., Miller, D. J. & Doak, J. (1993). The technical communicator as author: Meaning, power, authority. *Journal of Business and Technical Communication, 7*(12), 12–36.

Staples, K. & Ornatowski, C. M. (Eds.). (1997). *Foundations for teaching technical communication: Theory, practice, and program design*. Ablex.

Williams, R. I. (1983). Playing with format, style, and reader assumptions. *Technical Communication, 30*, 11–13.

Winsor, D. A. (1989). An engineer's writing and the corporate construction of knowledge. *Written Communication, 6*(5), 270–285.

Contributors

Elizabeth L. Angeli is Associate Professor of English at Marquette University, where she studies technical communication and writing education in healthcare. Liz is a leading expert in documentation practices and training for first responders, and her award-winning book, *Rhetorical Work in Emergency Medical Services* (Routledge), is the first book-length work to examine how first responders harness rhetoric's power to document patient care. Her work has also been published in academic and practitioner venues, including *Written Communication*, *Communication Design Quarterly*, *Rhetoric of Health and Medicine*, and *EMS1*.

Elisabet Arnó-Macià is Associate Professor of Technical Communication at Universitat Politècnica de Catalunya (BarcelonaTech, Barcelona, Spain). Her research interests include technical communication and English for Specific Purposes, Internationalization in higher education, and the role of technology in language education, especially intercultural virtual exchange. She has co-authored two textbooks on technical and academic English and has co-edited two research volumes on the use of technology in *Languages for Specific Purposes* and *virtual exchange*, respectively. Her work has appeared in journals such as *The Modern Language Journal*, *Journal of English for Academic Purposes*, *English for Specific Purposes*, and *Language Learning Journal*, among others.

Joseph Bartolotta is Assistant Professor in the Department of Writing Studies & Rhetoric at Hofstra University. His work examines the training and application of usability and user experience principles in writing programs and for students in TPC. He further explores the ways schools and industry organizations define best practices, competencies, and ethics in their respective contexts and looks for ways to bring both together for generative discussions.

Jennifer L. Bay is Associate Professor of English and Director of Professional Writing at Purdue University, where she teaches courses in rhetorical theory, professional writing, feminist rhetorics, and community engagement. Her work has appeared in journals such as *Technical Communication Quarterly*, *College English*, and *Community Literacy Journal*, as well as in edited collections.

Chen Chen is Assistant Professor of English at Winthrop University in Rock Hill, South Carolina, where she teaches first-year writing and professional and technical communication courses. She received her Ph.D. in Communication, Rhetoric, and Digital Media from North Carolina State University. She researches technical and professional communication pedagogical practices. She also studies how graduate students professionalize into the field of rhetoric and composition across different disciplinary spaces, and Chinese feminist rhetorics.

Lisa DeTora is Associate Professor and the Director of STEM Writing at Hofstra University. She teaches courses in writing studies, rhetoric, English, chemistry, biology, disability studies, sports science, and medical humanities. Her scholarship and service bridge various intellectual and practical domains, including biomedical publication ethics, regulatory documentation practice, medical humanities, the rhetorics of health and medicine, technical communication, and graphic narrative research. Upcoming publications include work on the representation of gender and science in graphic narratives, considerations of quantum physics as a metaphor for fan experience, and a volume on "graphic embodiment," co-edited with Jodi Cressman (University of Leuven Press).

Kira Dreher is Assistant Teaching Professor at Carnegie Mellon University's campus in Doha, Qatar, where she teaches courses in first-year writing, rhetoric, and technical writing. She received her Ph.D. in Rhetoric and Scientific and Technical Communication from University of Minnesota- Twin Cities. Her research currently focuses on plain language in technical communication and the history of "plainness" in rhetoric, medicine, and other traditions. She has published in journals such as *IEEE Transactions on Professional Communication*, in various edited collections, and as a co-author of *Arab Women in Arab News* (Bloomsbury, 2012).

Ann Hill Duin is Professor of Writing Studies and Graduate-Professional Distinguished Teaching Professor at the University of Minnesota, where her research and teaching focus on collaboration, digital literacy, analytics, and writing futures. She served 15 years in higher education administrative roles, including Vice Provost and Associate Vice President for Information Technology. Her most recent scholarship appears in *Computers and Composition*, *Communication Design Quarterly*, *IEEE Transactions on Professional Communication*, *Technical Communication Quarterly*, and the edited collection *Content Strategy in Technical Communication*. Her international collaboration includes research cluster leadership in the Digital Life Institute at Ontario Tech University.

Brian Fitzpatrick is Assistant Professor at George Mason University, where he teaches composition. His research is primarily focused on professional and workplace writing, as well as online and hybrid pedagogies. He is the co-founder of the Archive of Workplace Writing Experiences and, along with his co-author, was recipient of the Conference on College Composition and Communication's Emergent Researcher Award in 2017.

David M. Grant is Associate Professor of Writing Studies in the Department of Languages & Literatures at the University of Northern Iowa, where he teaches courses in science communication, environmental rhetoric, and indigenous North American rhetorics, and helps build critical literacy awareness at his institution, with his students, and in the community. He researches at the intersection of posthumanisms and decolonization, and his work has appeared in *College Composition and Communication*, *enculturation*, *Rhetoric Review*, *Kairos*, and *Pre/Text*, among other venues.

Elise Verzosa Hurley is Associate Professor of Rhetoric, Composition, and Technical Communication at Illinois State University and editor of *Rhetoric Review*. Her research and teaching interests include visual and spatial rhetorics, professional and technical communication theory and pedagogy, multimodal composition, public rhetorics, feminist rhetorics, and civic and community engagement. Her scholarship has been published in *Technical Communication Quarterly*; *Kairos: A Journal of Rhetoric, Technology, and Pedagogy*; *Rhetoric Review*; *Res Rhetorica*; and various edited collections.

Liz Hutter is Assistant Professor in technical communication in the English department at the University of Dayton. Her teaching and research interests include STEM pedagogy; rhetoric of health, medicine, and science; and medical humanities. Her work has been published in *Journal of Technical Writing and Communication*, *Communication Design Quarterly*, *Computers and Composition*, *World Medical Health Policy*, and *Configurations*.

Michael J. Klein is Associate Professor of Writing, Rhetoric and Technical Communication at James Madison University in Harrisonburg, Virginia. He directs the Cohen Center for the Humanities, a university center focused on humanistic inquiry across disciplines, and is also the founder and coordinator of the interdisciplinary minor in medical humanities, which comprises 12 academic units across three colleges. He teaches courses in technical communication, scientific and medical communication, and writing in the health sciences. His recent scholarship has focused on medical narratives and intercultural communication, and the creation of graphic embodiment memoirs in an interdisciplinary writing course.

Adrienne Lamberti is Associate Professor of Languages and Literatures and Coordinator of the Professional Writing Program at the University of Northern Iowa, where she researches and teaches rhetoric and writing in the disciplines. Her pedagogical work focuses on community engagement and other forms of service learning. Dr. Lamberti's scholarly work focuses on boundary and Othered communications, including discourses on the production side of agriculture, as well as conflict and crisis communications. Her most recent publications include *Cultivating Spheres: Agriculture, Technical Communication, and the Publics* and (with Anne R. Richards) *Communication and Conflict Studies: Disciplinary Connections, Research Directions*.

Liz Lane is Assistant Professor of English in the Writing, Rhetoric, and Technical Communication concentration at the University of Memphis. Her research explores the intersections of activism in digital spaces, feminism, and technical communication. Her work has appeared in *Peitho*; *Computers and Composition*; *Composition Studies*; *Ada: A Journal of Gender, New Media, and Technology*; and various edited collections. She is also the co-managing editor of *Spark: A 4C4Equality Journal*, an open-access, peer-reviewed journal of activist rhetorics in writing studies, which can be found at sparkactivism.com.

Halcyon M. Lawrence is Assistant Professor of Technical Communication and Information Design at Towson University. She has over 20 years of professional experience as a technical trainer, writer, and usability practitioner. Her research focuses on speech intelligibility and the design of speech interactions for voice technologies, particularly for under-represented user populations. She holds a Ph.D. in Technical Communication from the Illinois Institute of Technology.

Jessica McCaughey is Assistant Professor at George Washington (GW) University, where she teaches academic, professional, and technical writing. She transitioned to this role after more than a decade of business writing and editing in corporate and non-profit organizations. In her role at GW, Professor McCaughey directs the Professional Writing Program, which consists of workshops, assessment, and coaching to help organizations improve the quality of their writing. Her research focuses on the transfer of writing skills from the university to the professional realm. In 2017, with her co-author, she won the Conference on College Composition and Communication's Emergent Researcher Award.

Julianne Newmark teaches technical and professional communication (TPC) at the University of New Mexico, where she is Assistant Chair for Core Writing and Coordinator of the TPC program. Her publications have considered the school-to-work transition, multimodal community creation in online classrooms, and usability/UX/UCD. She also teaches, conducts research, and publishes in Indigenous studies, particularly concerning early twentieth-century Native activist writers' rhetorically impactful bureaucratic writing, especially in Bureau of Indian Affairs contexts. She is currently at work on her second book, *Reports of Agency: Retrieving Indigenous Professional Communication in Dawes Era Indian Bureau Documents* and continues to serve as editor-in-chief of *Xchanges*, a writing studies ejournal.

Dr. Isabel Pedersen is Canada Research Chair in Digital Life, Media, and Culture and Professor of Communication Studies at Ontario Tech University. She is Founder and Director of the Digital Life Institute. She studies the rhetorical, ethical, and political challenges posed by technological change on communication and digital literacy practices. She is co-editor of *Embodied Computing: Wearables, Implantables, Embeddables, Ingestibles* (2020, MIT Press). She is published in many academic journals, including the *Journal of Information, Communication and Ethics in Society*; *International Journal of Cultural Studies*; and the *Journal on Computing and Cultural Heritage*.

Derek G. Ross is Professor in the Master of Technical and Professional Communication Program at Auburn University, where he teaches courses in technical communication, document design, environmental rhetoric, and ethics, among others. His research interests include perceptions of environment-related rhetoric, document design, ethics, and audience analysis. His work has appeared in *Technical Communication Quarterly*, *Technical Communication*, *Written Communication*, *IEEE Transactions on Professional Communication*, and *The Journal of Technical Writing and Communication*, among others. He is the editor of *Topic-Driven Environmental Rhetoric*, Editor in Chief of *Communication Design Quarterly*, and Co-director of Auburn's Lab for Usability, Accessibility, Communication, and Interaction (LUCIA).

Tatjana Schell is an independent scholar based in Germany. She holds a Ph.D. in Rhetoric, Writing and Culture from North Dakota State University. Her research focuses on professional and technical communication, college composition—specifically, when it comes to ESL writing pedagogy—archival research practices, and histories of rhetoric. Her work has appeared in *Rethinking Post-Communist Rhetoric: Perspectives on Rhetoric, Writing, and Professional Communication in Post-Soviet Spaces*, edited by Pavel Zemliansky and Kirk St. Amant, which won the 2017 Conference on College Composition and Communication's Best Original Collection of Essays in Technical or Scientific Communication Award.

Jason Tham (Ph.D., University of Minnesota) is Assistant Professor of Technical Communication and Rhetoric at Texas Tech University, where he co-directs the User Experience Research Lab. His scholarship has appeared in *Technical Communication Quarterly, Journal of Business and Technical Communication,* and *IEEE Transactions on Professional Communication.*

Luke Thominet is Assistant Professor in the English Department at Florida International University. His research interests include technical communication, user experience, and design thinking. His recent projects have examined user experience research practices in video game development, the rhetorical construction of academic job market correspondence, the use of design thinking in academic program design, and the deliberative discourse of institutional review boards. His work has appeared in *The Journal of Business and Technical Communication, Communication Design Quarterly*, and *Technical Communication Quarterly*.

Julie Watts is Professor of English with the Department of English and Philosophy at the University of Wisconsin-Stout. She served as founding director of the online M.S. program in Technical and Professional Communication for 12 years. She teaches courses in composition, business communication, document design, and theory and research methods. She was awarded the 2019 Conference on College Composition and Communication's Best Article on Pedagogy or Curriculum in Technical or Scientific Communication. Her research interests focus on program assessment as well as the communicative dynamics and culture of the classroom learning community and what instructors can do to facilitate student learning.

Index

A

academy, 7, 14, 151, 157, 175, 176
accessibility, 3, 67, 106, 123, 207, 299, 306
applied learning, 16, 139
assessment, 93, 108, 117, 133, 147, 152, 157, 158, 160, 289, 298, 308, 313, 316, 322, 323
   ecological, 143
audience analysis, 35, 58, 306, 322

C

case study, 7, 68, 81, 96, 98, 209, 216, 222, 257, 303, 316
case study method, 22
client projects, 258
clients, working with, 6, 126, 134, 139, 181, 266
collaboration, 4, 6, 15, 18, 20, 24, 32, 33, 35, 41, 68, 104, 114, 122, 126, 127, 142, 186, 195, 200, 211, 220, 222, 223, 232, 233, 259, 282, 320
   international, 249
   scientific, 177, 178
   workplace, 175, 177, 181
collaborative writing, 175, 178, 214, 222, 223
community-based learning. *See* service learning
community-based projects, 7, 257, 264, 266
community of inquiry (COI), 6, 202, 207, 208
community of practice, 15
constructivism, 33, 170, 212
contract grading, 117
Cook, Kelli Cargile, 29, 92, 94, 95, 96, 97, 98, 99, 100, 274, 289
COVID-19, 4, 20, 23, 126, 143
critical thinking, 13, 14, 20, 24, 30, 35, 38, 143, 195, 196

curricula. *See* curriculum
curriculum, 4, 5, 23, 91, 98, 113, 116, 120, 127, 142, 169, 170, 198, 199, 201, 202, 247, 249, 258, 260, 266, 323
   development, 127, 148, 164

D

design, 47, 49, 52, 67, 78, 79, 123, 152, 174, 181, 186, 201, 206, 234, 238, 256, 260, 262, 264, 268, 273, 299, 321
   communication, 36, 171
   course, 7, 93, 96, 97, 111, 112, 113, 114, 116, 118, 120, 163, 223, 231
   document, 29, 46, 58, 59, 94, 100, 104, 114, 116, 117, 121, 131, 133, 217, 260, 261, 289, 322, 323
   information, 46, 50, 80, 126, 181
   interstitial, 5, 41
   organizational, 155
   principles, 41, 155, 261
   social, 37
   theory, 5, 29, 37
   user-centered, 29, 114, 163
   visual, 104, 155, 262, 263
discourse communities, 3, 232, 272, 289
distance education. *See* online learning
distributed technologies, 14, 18
diversity, 20, 22, 24, 36, 106, 114, 123, 148, 149, 150, 177, 184, 239, 246

E

ethical models, 5, 67, 70, 72, 74, 78, 80, 82, 83, 84
ethics, 13, 14, 24, 39, 53, 84, 95, 114, 116, 119, 126, 143, 171, 172, 186, 319, 320, 322
   anthropocentric, 69
   Aristotelian, 72, 73
   biocentric, 69

of care, 72, 78, 81
code, 70
communication, 79
decision-making, 5, 84
ecocentric, 69
feminist, 72, 77, 78, 82
Kantian, 72, 74
non-anthropocentric, 69
scholarship, 68
utilitarianism, 72, 75, 81
virtue, 69
workplace, 67
zoocentric, 69
ethos, 68, 171, 172, 220, 273
experiential learning, 13, 15, 16, 257, 258, 264. *See also* service learning

F

feedback, 18, 35, 41, 122, 151, 154, 181, 196, 197, 200, 201, 208, 212, 213, 214, 222, 259, 265, 282, 292, 294
   from students, 20, 38, 122, 127, 157, 221, 235, 237
Foundations for Teaching Technical Communication, 3, 131, 173, 303

G

gender, 22, 72, 74, 77, 78, 80, 123, 150, 172, 173, 218, 261, 320
genre, 5, 19, 29, 31, 32, 38, 41, 93, 105, 106, 126, 136, 148, 155, 163, 214, 217, 232, 237, 247, 256, 269, 273, 281, 283, 293, 295, 310, 315, 316
global communication practices, 18

I

images, 84, 182, 259, 261, 262, 263
inclusivity, 3, 35, 116
independent study, 19, 24
informed consent, 118, 289
intercultural communication, 231, 232, 233, 235, 239, 243, 245, 247, 248, 321
interdisciplinary, 29, 30, 31, 32, 33, 35, 37, 40, 58, 133, 140, 142, 175, 182, 273, 321

internships, 24, 122
invention, 171, 172, 186, 262, 272, 281, 293

L

Lave, Jean, 14, 15, 16, 17, 174
layered literacies, 6, 7, 29, 91, 92, 98, 194, 271, 274, 275, 277, 299
learning community, 193, 195, 323
literacy frameworks, 91, 93, 99, 100, 101, 107

M

Miller, Carolyn R., 56, 271, 272, 305

O

online learning, 6, 19, 202
open-access tools, 6
organizational dynamics, 7

P

persuasion, 7, 55, 79, 155, 243, 247, 303, 304, 307, 309, 310, 311, 313, 315
persuasive. *See* persuasion
plain language, 5, 66, 79, 80, 320
practicum, 13, 14, 15, 17, 19, 24, 157
problem-solving, 6, 30, 34, 95, 111, 117, 126, 245, 247, 303
professionalism, 18, 82, 137, 152

R

race, 22, 54, 72, 150, 261
racial, 14, 18, 20, 22, 37, 62
readability formulas, 46, 49, 54
reflection, 18, 36, 54, 92, 97, 98, 124, 140, 157, 159, 160, 162, 178, 184, 195, 220, 222, 237, 242, 245, 246, 259, 264
   internship experience, 22, 23
rhetorical, 5, 24, 49, 68, 78, 94, 97, 100, 125, 142, 153, 173, 182, 260, 274, 275, 279, 280, 282, 313, 314, 322, 323
   analysis, 114
   aspects of technical communication, 136, 305, 306, 310
   awareness, 306, 312, 316

context, 256, 263
literacies, 290, 299
purpose, 21, 263, 297, 307
situation, 19, 116, 121, 137, 154, 173, 257, 266, 317
skills, 13, 14, 19, 21, 24, 116, 121, 126, 134, 136, 183, 274, 287, 290, 297
strategy, 21, 49, 155, 206, 222, 309, 315
studies, 56, 280
tradition, 49, 55, 143, 164, 176, 185
rubric, 96, 97, 142, 160, 162

S

service courses, 93, 209, 211
service learning, 125, 239, 258, 275, 288, 321
situated learning, 5, 14, 15, 16, 24
social justice, 3, 5, 6, 29, 31, 40, 41, 45, 46, 47, 49, 54, 57, 67, 78, 106, 111, 118, 119, 127, 147, 150, 154, 157, 161, 162, 164, 306
social justice turn, 3, 5, 22, 29, 37, 40, 83, 116, 127, 150
social media, 4, 79, 149, 178, 179, 180, 186, 216
Society for Technical Communication (STC), 70, 81, 82, 147, 148, 150, 153, 174, 273
soft skills, 5, 24
student-learning outcomes (SLOs), 6, 164
study abroad, 18, 23

T

team-based learning (TBL), 6, 171, 223
teams, 38, 132, 136, 150, 154, 156, 160, 261, 263, 266
    student, 140, 141, 173, 174, 185
team science, 171, 177, 186

technological literacy, 95, 97, 98, 99, 274, 295
technology, 171
    and collaboration, 173, 186
telecollaboration, 7, 249
templates, 100, 307
Trans-Atlantic and Pacific Project (TAPP), 6, 184, 185, 249
translation, 177, 184, 231, 233, 237, 238, 305

U

universal design, 32, 36
usability studies, 32, 33, 35
usability testing, 122, 123, 233

V

visual communication, 7, 155, 266
visual literacy, 95, 98, 291
visual rhetoric, 32, 114, 260, 261

W

workplace, 4, 7, 31, 36, 68, 70, 78, 81, 84, 98, 106, 117, 119, 132, 136, 143, 148, 154, 158, 160, 163, 175, 179, 182, 209, 214, 257
    communication, 115, 116, 134, 209
    compentencies, 92, 94, 104
    context, 29, 92, 96, 99, 257
    distributed, 24
    ethics, 67
    global, 5, 15, 19, 29, 233, 240, 241, 244, 249
    literacy, 93, 272, 274, 283
    training course, 299
    writing, 215, 317
writing process, 40, 121, 209, 211, 289, 292, 295, 299
writing prompts, 303, 316

www.ingramcontent.com/pod-product-compliance
Lightning Source LLC
Chambersburg PA
CBHW031056080526
44587CB00011B/714